Nonlinear Microwave Circuit Design

Nonlinear Microwave Circuit Design

F. Giannini

University of Roma, Tor Vergata, Italy

and

G. Leuzzi

University of L'Aquila, Italy

John Wiley & Sons, Ltd

Other Wiley Editorial Offices

John Wiley & Sons Inc., 111 River Street, Hoboken, NJ 07030, USA

Jossey-Bass, 989 Market Street, San Francisco, CA 94103-1741, USA

Wiley-VCH Verlag GmbH, Boschstr. 12, D-69469 Weinheim, Germany

John Wiley & Sons Australia Ltd, 33 Park Road, Milton, Queensland 4064, Australia

John Wiley & Sons (Asia) Pte Ltd, 2 Clementi Loop #02-01, Jin Xing Distripark, Singapore 129809

John Wiley & Sons Canada Ltd, 22 Worcester Road, Etobicoke, Ontario, Canada M9W 1L1

Wiley also publishes its books in a variety of electronic formats. Some content that appears
in print may not be available in electronic books.

Library of Congress Cataloging-in-Publication Data

Giannini, Franco, 1944-
Nonlinear microwave circuit design / Franco Giannini and Giorgio Leuzzi
    p. cm.
Includes bibliographical references and index.
ISBN 0-470-84701-8 (cloth: alk. paper)
1. Microwave circuits. 2. Electric circuits, Nonlinear. I. Leuzzi,
Giorgio. II. Title.
TK7876.G53 2004
621.381'33 – dc22
                                              2004004941

British Library Cataloguing in Publication Data

A catalogue record for this book is available from the British Library

ISBN 0-470-84701-8

Typeset in 10/12pt Times by Laserwords Private Limited, Chennai, India
Printed and bound in Great Britain by Antony Rowe Ltd, Chippenham, Wiltshire
This book is printed on acid-free paper responsibly manufactured from sustainable forestry
in which at least two trees are planted for each one used for paper production.

Contents

Preface   ix

Chapter 1   Nonlinear Analysis Methods   1

1.1 Introduction   1
1.2 Time-Domain Solution   4
    1.2.1 General Formulation   4
    1.2.2 Steady State Analysis   7
    1.2.3 Convolution Methods   9
1.3 Solution Through Series Expansion   13
    1.3.1 Volterra Series   13
    1.3.2 Fourier Series   22
        1.3.2.1 Basic formulation (single tone)   23
        1.3.2.2 Multi-tone analysis   33
        1.3.2.3 Envelope analysis   43
        1.3.2.4 Additional remarks   45
        1.3.2.5 Describing function   46
        1.3.2.6 Spectral balance   47
1.4 The Conversion Matrix   49
1.5 Bibliography   56

Chapter 2   Nonlinear Measurements   61

2.1 Introduction   61
2.2 Load/Source Pull   62
2.3 The Vector Nonlinear Network Analyser   71
2.4 Pulsed Measurements   74
2.5 Bibliography   80

Chapter 3   Nonlinear Models   83

3.1 Introduction   83
3.2 Physical Models   84

3.2.1 Introduction   84
3.2.2 Basic Equations   86
3.2.3 Numerical Models   88
3.2.4 Analytical Models   92
3.3 Equivalent-Circuit Models   95
  3.3.1 Introduction   95
  3.3.2 Linear Models   96
  3.3.3 From Linear to Nonlinear   102
  3.3.4 Extraction of an Equivalent Circuit from Multi-bias Small-signal
         Measurements   121
  3.3.5 Nonlinear Models   133
  3.3.6 Packages   139
3.4 Black-Box Models   142
  3.4.1 Table-based Models   142
  3.4.2 Quasi-static Model Identified from Time-domain Data   143
  3.4.3 Frequency-domain Models   144
  3.4.4 Behavioural Models   146
3.5 Simplified Models   148
3.6 Bibliography   151

Chapter 4   Power Amplifiers   **159**

4.1 Introduction   159
4.2 Classes of Operation   168
4.3 Simplified Class-A Fundamental-frequency Design for High Efficiency   170
  4.3.1 The Methodology   170
  4.3.2 An Example of Application   180
4.4 Multi-harmonic Design for High Power and Efficiency   182
  4.4.1 Introduction   182
  4.4.2 Basic Assumptions   187
  4.4.3 Harmonic Tuning Approach   194
  4.4.4 Mathematical Statements   197
  4.4.5 Design Statements   205
  4.4.6 Harmonic Generation Mechanisms and Drain Current Waveforms   207
  4.4.7 Sample Realisations and Measured Performances   212
4.5 Bibliography   226

Chapter 5   Oscillators   **229**

5.1 Introduction   229
5.2 Linear Stability and Oscillation Conditions   230
5.3 From Linear to Nonlinear: Quasi-large-signal Oscillation and Stability
     Conditions   243
5.4 Design Methods   252

5.5 Nonlinear Analysis Methods for Oscillators   259
    5.5.1 The Probe Approach   260
    5.5.2 Nonlinear Methods   261
5.6 Noise   269
5.7 Bibliography   276

Chapter 6   Frequency Multipliers and Dividers   **279**

6.1 Introduction   279
6.2 Passive Multipliers   280
6.3 Active Multipliers   282
    6.3.1 Introduction   282
    6.3.2 Piecewise-linear Analysis   283
    6.3.3 Full-nonlinear Analysis   298
        6.3.4 Other Circuit Considerations   306
6.4 Frequency Dividers – the Regenerative (Passive) Approach   308
6.5 Bibliography   311

Chapter 7   Mixers   **315**

7.1 Introduction   315
7.2 Mixer Configurations   318
    7.2.1 Passive and Active Mixers   318
    7.2.2 Symmetry   322
7.3 Mixer Design   329
7.4 Nonlinear Analysis   332
7.5 Noise   337
7.6 Bibliography   339

Chapter 8   Stability and Injection-locked Circuits   **341**

8.1 Introduction   341
8.2 Local Stability of Nonlinear Circuits in Large-signal Regime   341
8.3 Nonlinear Analysis, Stability and Bifurcations   349
    8.3.1 Stability and Bifurcations   349
    8.3.2 Nonlinear Algorithms for Stability Analysis   356
8.4 Injection Locking   359
8.5 Bibliography   368

Appendix   **371**

A.1 Transformation in the Fourier Domain of the Linear Differential Equation 371
A.2 Time-Frequency Transformations   372
A.3 Generalised Fourier Transformation for the Volterra Series Expansion   372

A.4 Discrete Fourier Transform and Inverse Discrete Fourier Transform for
    Periodic Signals   373
A.5 The Harmonic Balance System of Equations for the Example Circuit with
    $N = 3$ 375
A.6 The Jacobian Matrix   378
A.7 Multi-Dimensional Discrete Fourier Transform and Inverse Discrete
    Fourier Transform for Quasi-periodic Signals   379
A.8 Oversampled Discrete Fourier Transform and Inverse Discrete Fourier
    Transform for Quasi-Periodic Signals   380
A.9 Derivation of Simplified Transport Equations   382
A.10 Determination of the Stability of a Linear Network   382
A.11 Determination of the Locking Range of an Injection-Locked Oscillator   384

Index   **387**

Preface

Nonlinear microwave circuits is a field still open to investigation; however, many basic concepts and design guidelines are already well established. Many researchers and design engineers have contributed in the past decades to the development of a solid knowledge that forms the basis of the current powerful capabilities of microwave engineers.

This book is composed of two main parts. In the first part, some fundamental tools are described: nonlinear circuit analysis, nonlinear measurement, and nonlinear modeling techniques. In the second part, basic structure and design guidelines are described for some basic blocks in microwave systems, that is, power amplifiers, oscillators, frequency multipliers and dividers, and mixers. Stability in nonlinear operating conditions is also addressed.

A short description of fundamental techniques is needed because of the inherent differences between linear and nonlinear systems and because of the greater familiarity of the microwave engineer with the linear tools and concepts. Therefore, an introduction to some general methods and rules proves useful for a better understanding of the basic behaviour of nonlinear circuits. The description of design guidelines, on the other hand, covers some well-established approaches, allowing the microwave engineer to understand the basic methodology required to perform an effective design.

The book mainly focuses on general concepts and methods, rather than on practical techniques and specific applications. To this aim, simple examples are given throughout the book and simplified models and methods are used whenever possible. The expected result is a better comprehension of basic concepts and of general approaches rather than a fast track to immediate design capability. The readers will judge for themselves the success of this approach.

Finally, we acknowledge the help of many colleagues. Dr. Franco Di Paolo has provided invaluable help in generating simulation results and graphs. Prof. Tom Brazil, Prof. Aldo Di Carlo, Prof. Angel Mediavilla, and Prof. Andrea Ferrero, Dr. Giuseppe Ocera and Dr. Carlo Del Vecchio have contributed with relevant material. Prof. Giovanni Ghione and Prof. Fabrizio Bonani have provided important comments and remarks, although responsibility for eventual inaccuracies must be ascribed only to the authors. To all these people goes our warm gratitude.

Authors' wives and families are also acknowledged for patiently tolerating the extra work connected with writing a book.

Franco Giannini
Giorgio Leuzzi

1

Nonlinear Analysis Methods

1.1 INTRODUCTION

In this introduction, some well-known basic concepts are recalled, and a simple example is introduced that will be used in the following paragraphs for the illustration of the different nonlinear analysis methods.

Electrical and electronic circuits are described by means of voltages and currents. The equations that fulfil the topological constraints of the network, and that form the basis for the network analysis, are Kirchhoff's equations. The equations describe the constraints on voltages (mesh equations) or currents (nodal equations), expressing the constraint that the sum of all the voltages in each mesh, or, respectively, that all the currents entering each node, must sum up to zero. The number of equations is one half of the total number of the unknown voltages and currents. The system can be solved when the relation between voltage and current in each element of the network is known (constitutive relations of the elements). In this way, for example, in the case of nodal equations, the currents that appear in the equations are expressed as functions of the voltages that are the actual unknowns of the problem. Let us illustrate this by means of a simple example (Figure 1.1).

$$i_s + i_g + i_C = 0 \qquad \text{Nodal Kirchhoff's equation} \qquad (1.1)$$

$$
\begin{aligned}
i_s &= i_s(t) \\
i_g &= g \cdot v \\
i_C &= C \cdot \frac{dv}{dt}
\end{aligned}
\qquad \text{Constitutive relations of the elements} \qquad (1.2)
$$

where $i_s(t)$ is a known, generic function of time. Introducing the constitutive relations into the nodal equation we get

$$C \cdot \frac{dv(t)}{dt} + g \cdot v(t) + i_s(t) = 0 \qquad (1.3)$$

Since in this case all the constitutive relations (eq. (1.2)) of the elements are linear and one of them is differential, the system (eq. (1.3)) turns out to be a linear differential

Nonlinear Microwave Circuit Design   F. Giannini and G. Leuzzi
© 2004 John Wiley & Sons, Ltd   ISBN: 0-470-84701-8

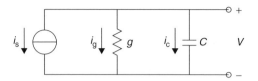

Figure 1.1   A simple example circuit

system in the unknown $v(t)$ (in this case a single equation in one unknown). One of the elements ($i_s(t)$) is a known quantity independent of voltage (known term), and the equation is non-homogeneous. The solution is found by standard solution methods of linear differential equations:

$$v(t) = v(t_0) \cdot e^{-\frac{g}{C} \cdot (t-t_0)} - \int_{t_0}^{t} \frac{e^{-\frac{g}{C} \cdot (t-\tau)}}{C} \cdot i_s(\tau) \cdot d\tau \tag{1.4}$$

More generally, the solution can be written in the time domain as a convolution integral:

$$v(t) = v(t_0) + \int_{t_0}^{t} h(t - \tau) \cdot i_s(\tau) \cdot d\tau \tag{1.5}$$

where $h(t)$ is the impulse response of the system.

The linear differential equation system can be transformed in the Fourier or Laplace domain. The well-known formulae converting between the time domain and the transformed Fourier domain, or frequency domain, and vice versa, are the Fourier transform and inverse Fourier transform respectively:

$$V(\omega) = \frac{1}{\sqrt{2\pi}} \cdot \int_{-\infty}^{+\infty} v(t) \cdot e^{-j\omega t} \cdot dt \tag{1.6}$$

$$v(t) = \frac{1}{\sqrt{2\pi}} \cdot \int_{-\infty}^{+\infty} V(\omega) \cdot e^{j\omega t} \cdot d\omega \tag{1.7}$$

By Fourier transforming eq. (1.3), after simple manipulation (Appendix A.1) we have

$$V(\omega) = H(\omega) \cdot I_s(\omega) \tag{1.8}$$

where $H(\omega)$ and $I_s(\omega)$ are obtained by Fourier transformation of the time-domain functions $h(t)$ and $i_s(t)$; $H(\omega)$ is the transfer function of the circuit.

We can describe this approach from another point of view: if the current $i_s(t)$ is sinusoidal, and we look for the solution in the permanent regime, we can make use of phasors, that is, complex numbers such that

$$v(t) = \text{Im}[V \cdot e^{j\omega t}] \tag{1.9}$$

and similarly for the other electrical quantities; the voltage phasor V corresponds to the $V(\omega)$, defined above. Then, by replacing in eq. (1.3) we get

$$j\omega C \cdot V + g \cdot V + I_s = (g + j\omega C) \cdot V + I_s = Y \cdot V + I_s = 0 \qquad (1.10)$$

and the solution is easily found by standard solution methods for linear equations:

$$V = \frac{I_s}{Y} \qquad \frac{1}{Y(\omega)} = H(\omega) = \frac{1}{g + j\omega C} \qquad (1.11)$$

Let us now introduce nonlinearities. Nonlinear circuits are electrical networks that include elements with a nonlinear relation between voltage and current; as an example, let us consider a nonlinear conductance (Figure 1.2) described by

$$i_g(v) = i_{max} \cdot tgh\left(\frac{g \cdot v}{i_{max}}\right) \qquad (1.12)$$

that is, a conductance saturating to a maximum current value i_{max} (Figure 1.3).

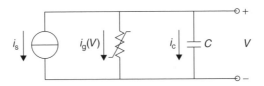

Figure 1.2   The example circuit with a nonlinear conductance

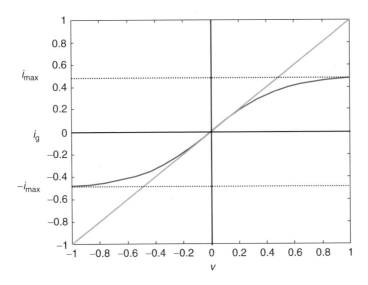

Figure 1.3   The current–voltage characteristic of the nonlinear conductance

When we introduce this relation in Kirchhoff's equation, we have a nonlinear differential equation (in general, a system of nonlinear differential equations)

$$C \cdot \frac{dv(t)}{dt} + i_{max} \cdot tgh \left(\frac{g \cdot v(t)}{i_{max}} \right) + i_s(t) = 0 \qquad (1.13)$$

that has no explicit solution. Moreover, contrary to the linear case, transformation into the Fourier or Laplace domain is not applicable.

Practical solutions to this type of problems fall into two main categories: direct numerical integration in the time domain, and numerical solution through series expansion; they are described in some detail in the following paragraphs.

1.2 TIME-DOMAIN SOLUTION

In this paragraph, the solution of the nonlinear differential Kirchhoff's equations by direct numerical integration in the time domain is described. Advantages and drawbacks are described, together with some improvements to the basic approach.

1.2.1 General Formulation

The time-domain solution of the nonlinear differential equations system that describes the circuit (Kirchhoff's equations) can be performed by means of standard numerical integration methods. These methods require the discretisation of the time variable, and likewise the sampling of the known and unknown time-domain voltages and currents at the discretised time instants.

The time variable, in general a real number in the interval $[t_0, \infty]$, is discretised, that is, considered as a discrete variable:

$$t = t_k \qquad k = 1, 2, \ldots \qquad t \in [t_0, \infty] \qquad (1.14)$$

All functions of time are evaluated only at this set of values of the time variable. The differential equation becomes a finite-difference equation, and the knowledge of the unknown function $v(t)$ is reduced to the knowledge of a discrete set of values:

$$v_k = v(t_k) \qquad k = 1, 2, \ldots \qquad t \in [t_0, \infty] \qquad (1.15)$$

Similarly, the known function $i_s(t)$ is computed only at a discrete set of values:

$$i_{s,k} = i_s(t_k) \qquad k = 1, 2, \ldots \qquad t \in [t_0, \infty] \qquad (1.16)$$

The obvious advantage of this scheme is that the derivative with respect to time becomes a finite-difference incremental ratio:

$$\frac{dv(t)}{dt} = \frac{v_k - v_{k-1}}{t_k - t_{k-1}} \qquad (1.17)$$

Let us apply the discretisation to our example. Equation (1.13) becomes

$$C \cdot \left(\frac{v_k - v_{k-1}}{t_k - t_{k-1}} \right) + i_{max} \cdot tgh \left(\frac{g \cdot v_k}{i_{max}} \right) + i_{s,k} = 0 \qquad k = 1, 2, \ldots \qquad (1.18)$$

where we have replaced the derivative with respect to time, defined in the continuous time, with the incremental ratio, defined in the discrete time. In this formulation, the discrete derivative is computed between the current point k, where also the rest of the equation is evaluated, and the previous point $k - 1$. There is, however, another possibility:

$$C \cdot \left(\frac{v_k - v_{k-1}}{t_k - t_{k-1}} \right) + i_{max} \cdot tgh \left(\frac{g \cdot v_{k-1}}{i_{max}} \right) + i_{s,k-1} = 0 \qquad k = 1, 2, \ldots \qquad (1.19)$$

In the second case, the rest of the equation, including the nonlinear function of the voltage, is evaluated in the previous point $k - 1$. In both cases, if an initial value is known for the problem, that is, if the value $v_0 = v(t_0)$ is known, then the problem can be solved iteratively, time instant after time instant, starting from the initial time instant t_0 at $k = 0$. In the case of our example, the initial value is the voltage at which the capacitance is initially charged.

The two cases of eq. (1.18) and eq. (1.19) differ in complexity and accuracy. In the case of eq. (1.19), the unknown voltage v_k at the current point k appears only in the finite-difference incremental ratio; the equation can be therefore easily inverted, yielding

$$v_k = v_{k-1} - \frac{(t_k - t_{k-1})}{C} \cdot i_{max} \cdot tgh \left(\frac{g \cdot v_{k-1}}{i_{max}} \right) + i_{s,k-1} \qquad k = 1, 2, \ldots \qquad (1.20)$$

This approach allows the explicit calculation of the unknown voltage v_k at the current point k, once the solution at the previous point $k - 1$ is known. The obvious advantage of this approach is that the calculation of the unknown voltage requires only the evaluation of an expression at each of the sampling instants t_k. A major disadvantage of this solution, usually termed as 'explicit', is that the stability of the solution cannot be guaranteed. In general, the solution found by any discretised approach is always an approximation; that is, there will always be a difference between the actual value of the exact (unknown) solution $v(t)$ at each time instant t_k and the values found by this method

$$v(t_k) \neq v_k \qquad v(t_k) - v_k = \Delta v_k \qquad k = 1, 2, \ldots \qquad (1.21)$$

because of the inherently approximated nature of the discretisation with respect to the originally continuous system. The error Δv_k due to an explicit formulation, however, can increase without limits when we proceed in time, even if we reduce the discretisation step $t_k - t_{k-1}$, and the solution values can even diverge to infinity. Even if the values do not diverge, the error can be large and difficult to reduce or control; in fact, it is not guaranteed that the error goes to zero even if the time discretisation becomes arbitrarily dense and the time step arbitrarily small. In fact, for simple circuits the explicit solution is usually adequate, but it is prone to failure for strongly nonlinear circuits. This explicit formulation is also called 'forward Euler' integration algorithm in numerical analysis [1, 2].

In the case of the formulation of eq. (1.18), the unknown voltage v_k appears not only in the finite-difference incremental ratio but also in the rest of the equation, and in particular within the nonlinear function. At each time instant, the unknown voltage v_k must be found as a solution of the nonlinear implicit equation:

$$C \cdot \left(\frac{v_k - v_{k-1}}{t_k - t_{k-1}} \right) + i_{max} \cdot tgh \left(\frac{g \cdot v_k}{i_{max}} \right) + i_{s,k} = F(v_k) = 0 \qquad k = 1, 2, \ldots \qquad (1.22)$$

This equation in general must be solved numerically, at each time instant t_k. Any zero-searching numerical algorithm can be applied, as for instance the fixed-point or Newton–Raphson algorithms. A numerical search requires an initial guess for the unknown voltage at the time instant t_k and hopefully converges toward the exact solution in a short number of steps; the better the initial guess, the shorter the number of steps required for a given accuracy. As an example, the explicit solution can be a suitable initial guess. The iterative algorithm is stopped when the current guess is estimated to be reasonably close to the exact solution. This approach is also called 'backward Euler' integration scheme in numerical analysis [1, 2].

An obvious disadvantage of this approach w.r.t. the explicit one is the much higher computational burden, and the risk of non-convergence of the iterative zero-searching algorithm. However, in this case the error Δv_k can be made arbitrarily small by reducing the time discretisation step $t_k - t_{k-1}$, at least in principle. Numerical round-off errors due to finite number representation in the computer is however always present.

The discretisation of the t_k can be uniform, that is, with a constant step Δt, so that

$$t_{k+1} = t_k + \Delta t \qquad t_k = t_0 + k \cdot \Delta t \qquad k = 1, 2, \ldots \qquad (1.23)$$

This approach is not the most efficient. A variable time step is usually adopted with smaller time steps where the solution varies rapidly in time and larger time steps where the solution is smoother. The time step is usually adjusted dynamically as the solution proceeds; in particular, a short time step makes the solution of the nonlinear eq. (1.22) easier. A simple procedure when the solution of eq. (1.22) becomes too slow or does not converge at all consists of stopping the zero-searching algorithm, reducing the time step and restarting the algorithm.

There is an intuitive relation between time step and accuracy of the solution. For a band-limited signal in permanent regime, an obvious criterion for time discretisation is given by Nykvist's sampling theorem. If the time step is larger than the sampling time required by Nykvist's theorem, the bandwidth of the solution will be smaller than that of the actual solution and some information will be lost. The picture is not so simple for complex signals, but the principle still holds: the finer the time step, the more accurate the solution. Since higher frequency components are sometimes negligible for practical applications, a compromise between accuracy and computational burden is usually chosen. In practical algorithms, more elaborate schemes are implemented, including modified nodal analysis, advanced integration schemes, sophisticated adaptive time-step schemes and robust zero-searching algorithms [3–7].

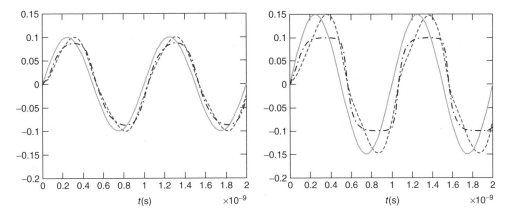

Figure 1.4 Currents and voltages in the example circuit for two different amplitudes of a sinusoidal input current

With the view to illustrate, the time-domain solution of our example circuit is given for a sinusoidal input current, for the following values of the circuit elements (Figure 1.4):

$$g = 10 \text{ mS} \qquad C = 500 \text{ fF} \qquad f = 1 \text{ GHz} \tag{1.24}$$

A simple implicit integration scheme is used, with a uniform time step of $\Delta t = 33.3$ ps (30 discretisation points per period). The plots show the input current i_s (———), the voltage v (- - - -) and the current in the nonlinear resistor i_g (-·-·-·), for an input current of $i_{s,max} = 100$ mA (a) and for a larger input of $i_{s,max} = 150$ mA (b).

As an additional example, the response of the same circuit to a 1 mA input current step is shown in Figure 1.5, where a uniform time step of $\Delta t = 10$ ps is used.

Time-domain direct numerical integration is very general. No limitation on the type or stiffness of the nonlinearity is imposed. Transient as well as steady state behaviour are computed, making it very suitable, for instance, for oscillator analysis, where the determination of the onset of the oscillations is required. Instabilities are also well predicted, provided that the time step is sufficiently fine. Also, digital circuits are easily analysed.

1.2.2 Steady State Analysis

Direct numerical integration is not very efficient when the steady state regime is sought, especially when large time constant are present in a circuit, like those introduced by the bias circuitry. In this case, a large number of microwave periods must be analysed before the reactances in the bias circuitry are charged, starting from an arbitrary initial state. Since the time step must be chosen small enough in order that the microwave voltages and currents are sufficiently well sampled, a large number of time steps must be computed before the steady state is reached. The same is true when the spectrum of the signal includes components both at very low and at very high frequencies, as in the case

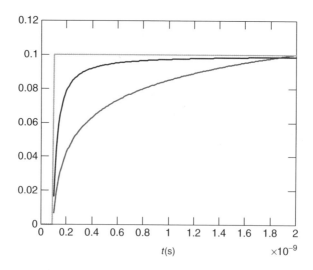

Figure 1.5   Currents and voltages in the example circuit for a step input current

of two sinusoids with very close frequencies, or of a narrowband modulated carrier. The time step must be small enough to accurately sample the high-frequency carrier, but the overall repetition time, that is, the period of the envelope, is comparatively very long.

The case when a long time must be waited for the steady state to be reached can be coped with by a special arrangement of the time-domain integration, called 'shooting method' [8–11]. It is interesting especially for non-autonomous circuits, when an external periodic input signal forces the circuit to a periodic behaviour; in fact, in autonomous circuits like oscillators, the analysis of the transient is also interesting, for the check of the correct onset of the oscillation and for the detection of spurious oscillations and instabilities. In the shooting method, the period of the steady state solution must be known in advance: this is usually not a problem, since it is the period of the input signal. The time-domain integration is carried over only for one period starting from a first guess of the initial state, that is, the state at the beginning of a period in steady state conditions, and then the state at the end of the period is checked. In the case of our example, the voltage at the initial time t_0 is guessed as

$$v_0 = v(t_0) \tag{1.25}$$

and the voltage at the end of the period T is computed after integration over one period:

$$v(T) = v(t_K) = v_K \qquad (k = 1, 2, \ldots, K) \tag{1.26}$$

This final voltage is a numerical function of the initial voltage:

$$v(T) = f(v_0) \tag{1.27}$$

If the initial voltage is the actual voltage at which the capacitor is charged at the beginning of a period $t = t_0$ in the periodic steady state regime, that is, if it is the solution to our periodic problem, the final voltage after a period must be identical to it:

$$v(T) = f(v_0) = v_0 \qquad (1.28)$$

In case this is not true, the correct value of the initial voltage is searched by adjustment of the initial guess v_0 until the final value $v(T)$ comes out to be equal to it. This can be done automatically by a zero-searching algorithm, where the unknown is the initial voltage v_0, and the function to be made equal to zero is

$$F(v_0) = v(T) - v_0 = f(v_0) - v_0 \qquad (1.29)$$

Each computation of the function $F(v_0)$ consists of the time-domain numerical integration over one period T from the initial value of the voltage v_0 to the final value $v(T) = v_K$. The zero-searching algorithm will require several iterations, that is, several integrations over a period; if the number of iterations required by the zero-searching algorithm to converge to the solution is smaller than the number of periods before the attainment of the steady state by standard integration from an initial voltage, then the shooting algorithm is a convenient alternative.

1.2.3 Convolution Methods

The time-domain numerical integration method has in fact two major drawbacks: on the one hand the number of equations grows with the dimension of the circuit, even when the largest part of it is linear. On the other hand, all the circuit elements must have a time-domain constitutive relation in order for the equations to be written in time domain. It is well known that in many practical cases the linear part of nonlinear microwave circuits is large and that it is best described in the frequency domain; as an example, consider the matching and bias networks of a microwave amplifier. In particular, distributed elements are very difficult to represent in the time domain. A solution to these problems is represented by the 'convolution method' [12–17]. By this approach, a linear subcircuit is modelled by means of frequency-domain data, either measured or simulated; then, the frequency-domain representation is transformed into time-domain impulse response, to be used for convolution in the time domain with the rest of the circuit. In fact, this mixed time-frequency domain approach is somehow dual to the harmonic balance method, to be described in a later paragraph. In order to better understand the approach, a general scheme of time-frequency domain transformations for periodic and aperiodic functions is shown in Appendix A.2.

The basic scheme of the convolution approach is based on the application of eq. (1.5), with the relevant impulse response, to the linear subcircuit. Let us illustrate this principle with our test circuit, where a shunt admittance has been added (Figure 1.6).

Equation (1.13) becomes

$$C \cdot \frac{dv(t)}{dt} + i_{\max} \cdot tgh\left(\frac{g \cdot v(t)}{i_{\max}}\right) + i_y(t) + i_s(t) = 0 \qquad (1.30)$$

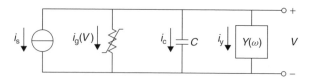

Figure 1.6   The example nonlinear circuit with an added shunt network

where the current through the shunt admittance is defined in the frequency domain:

$$I_y(\omega) = Y(\omega) \cdot V(\omega) \tag{1.31}$$

The time-domain current through the shunt admittance is expressed by means of the convolution integral (1.5) as

$$i_y(t) = i_y(t_0) + \int_{t_0}^{t} y(t - \tau) \cdot v(\tau) \cdot d\tau \tag{1.32}$$

where

$$y(t) = \frac{1}{\sqrt{2\pi}} \cdot \int_{-\infty}^{+\infty} Y(\omega) \cdot e^{j\omega t} \cdot d\omega \tag{1.33}$$

The integral in eq. (1.32) is computed numerically; if the impulse response $y(t)$ is limited in time, this becomes

$$i_y(t_k) = i_{y,k} = \sum_{m=0}^{M} y_m \cdot v_{k-m} \tag{1.34}$$

Discretisation of eq. (1.30) then yields

$$C \cdot \left(\frac{v_k - v_{k-1}}{t_k - t_{k-1}} \right) + i_{max} \cdot tgh \left(\frac{g \cdot v_k}{i_{max}} \right) + \sum_{m=0}^{M} y_m \cdot v_{k-m} + i_{s,k} = F(v_k) = 0 \tag{1.35}$$

where the unknown v_k appears also in the convolution summation with a linear term. This is a modified form of eq. (1.18) and must be solved numerically with the same procedure.

A first remark on this approach is that the algorithm becomes heavier: on the one hand, the convolution with past values of the electrical variables must be recomputed at each time step k, increasing computing time; on the other hand, the values of the electrical variables must be stored for as many time instants as corresponding to the duration of the impulse response, increasing data storage requirements.

An additional difficulty is related to the time step. A time-domain solution may use an adaptive time step for better efficiency of the algorithm; however, the time step of the discrete convolution in eq. (1.34) is a fixed number. This means that at the time instants where the convolution must be computed, the quantities to be used in the convolution

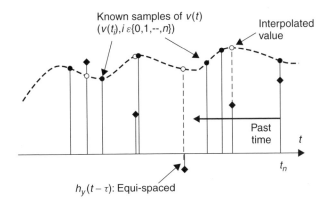

Figure 1.7 Sampling time instants and convolution time instants

are not available. An interpolating algorithm must be used to allow for the convolution to be computed, introducing an additional computational overhead and additional error (Figure 1.7).

The assumption of an impulse response limited in time requires some comments. An impulse response of infinite duration corresponds to an infinite bandwidth of the frequency-domain admittance. The latter however is usually known only within a limited frequency band, both in the case of experimental data and in the case of numerical modelling. A truncated frequency-domain admittance produces a non-causal impulse response when the inverse Fourier transform (eq. (1.33)) is applied (Figure 1.8).

As an alternative, the frequency-domain admittance can be 'windowed' by means, for example, of a low-pass filter, forcing the admittance to (almost) zero just before the limiting frequency f_m; however, this usually produces a severe distortion in phase, so that the accuracy will be unacceptably affected.

An alternative approach is to consider the impulse response as a discrete function of time, with finite duration in time. From the scheme in Appendix A.2, the corresponding spectrum is periodic in the frequency domain. Therefore, the admittance must be extended periodically in the frequency domain (Figure 1.9).

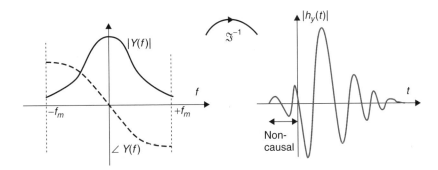

Figure 1.8 Non-causal impulse response generated by artificially band-limited frequency data

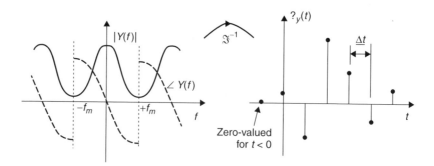

Figure 1.9   Periodical extension of frequency-domain data

In order to satisfy causality, however, the periodic extension must satisfy the Hilbert transform:

$$\hat{Y}(\omega) = \hat{G}(\omega) + j\hat{B}(\omega) \tag{1.36a}$$

$$\hat{G}(\omega) = \hat{G}(0) + \frac{1}{2\pi} \cdot \int_{-\pi}^{+\pi} \hat{B}(\alpha) \cdot \cot\left(\frac{\omega - \alpha}{2}\right) \cdot d\alpha$$

$$\hat{B}(\omega) = -\frac{1}{2\pi} \cdot \int_{-\pi}^{+\pi} \hat{G}(\alpha) \cdot \cot\left(\frac{\omega - \alpha}{2}\right) \cdot d\alpha \tag{1.36b}$$

This can be done by suitable procedures [18]. Care must be taken that the frequency-domain data be available in a band wide enough to make the extension error negligible. This is true if the spectrum of the voltages and currents in the circuit are narrower than the frequency 'window'. In practice, the frequency data must extend to frequencies where the signal spectrum has a negligible amplitude (Figure 1.10).

Several microwave or general CAD programmes are now commercially available implementing this scheme, allowing easy inclusion of passive networks described in the frequency domain; as an example, ultra-wide-band systems using short pulses often require the evaluation of pulse propagation through the transmit antenna/channel/receive antenna path, typically described in the frequency domain.

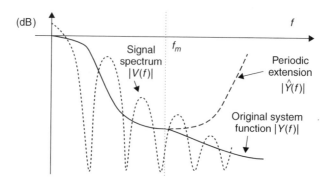

Figure 1.10   Approximation in the periodical extension of frequency-domain data

1.3 SOLUTION THROUGH SERIES EXPANSION

An alternative to direct discretisation of a difficult equation is the assumption of some hypotheses on the solution, in this case, on the unknown function $v(t)$. A typical hypothesis is that the solution can be expressed as an infinite sum of simple terms, and that the terms are chosen in such a suitable way that the first ones already include most of the information on the function. The series is therefore truncated after the first few terms. When replaced in the original equation, the solution in the form of a series allows the splitting of the original equation into infinite simpler equations (one per term of the series). Only a few of the simpler equations are solved however, corresponding to the first terms of the series.

In the following sections, two types of series expansions will be described: the Volterra and the Fourier series expansions, which are the only ones currently used.

1.3.1 Volterra Series

In this paragraph, the solution of the nonlinear differential Kirchhoff's equations by means of the Volterra series is described. Advantages and drawbacks are illustrated, together with some examples.

It has been shown above that the solution of our example circuit in the linear case is (eq. (1.4))

$$v(t) = v(t_0) \cdot e^{-\frac{g}{C} \cdot (t-t_0)} - \int_{t_0}^{t} \frac{e^{-\frac{g}{C} \cdot (t-\tau)}}{C} \cdot i_s(\tau) \cdot d\tau \tag{1.37}$$

that can be put in the general form known as the convolution integral (eq. (1.5)):

$$y(t) = y(t_0) + \int_{t_0}^{t} h(t-\tau) \cdot x(\tau) \cdot d\tau \tag{1.38}$$

Equation (1.38) can be interpreted in the following way: the output signal of a linear system is the infinite sum (integral) of all contributions due to the input signal at all the time instants in the past, weighted by a function called impulse response, representing the effect of the transfer through the system. In fact, the transfer function represents the 'memory' of the system, and normally becomes smaller as the time elapsed from the time instant of the input contribution to the current time instant becomes larger. If the system is instantaneous (e.g. a resistance), the impulse response is a delta function $k \cdot \delta(t)$, and the integral becomes a simple product:

$$y(t) = k \cdot x(t) \tag{1.39}$$

In this case, the output signal at any given time responds only to the input at that time and has no memory of past values of the input itself.

As we have seen above, given the linearity of the system, its response can be transformed in the Laplace or Fourier domain:

$$y(t) = \frac{1}{\sqrt{2\pi}} \cdot \int_{-\infty}^{+\infty} Y(\omega) \cdot e^{j\omega t} \cdot d\omega = \frac{1}{\sqrt{2\pi}} \cdot \int_{-\infty}^{+\infty} H(\omega) \cdot X(\omega) \cdot e^{j\omega t} \cdot d\omega \quad (1.40)$$

where

$$X(\omega) = \frac{1}{\sqrt{2\pi}} \cdot \int_{-\infty}^{+\infty} x(t) \cdot e^{-j\omega t} \cdot dt \qquad H(\omega) = \frac{1}{\sqrt{2\pi}} \cdot \int_{-\infty}^{+\infty} h(t) \cdot e^{-j\omega t} \cdot dt \quad (1.41)$$

Equation (1.40) can be interpreted in the following way: the output signal of a linear system is the infinite sum (integral) of all spectral contributions of the input signal, weighted by a function of frequency called transfer function that represents the effect of the transfer through the system. We note explicitly that the spectrum occupancy of the output signal is the same of the spectrum of the input signal, or smaller if the transfer function suppresses a part of it, as for example in a filter. If the system is instantaneous, the transfer function is a constant k and does not alter the harmonic content of the input signal:

$$Y(\omega) = k \cdot X(\omega) \quad (1.42)$$

An extension of this type of formulation to nonlinear circuits has been proposed by the mathematician Vito Volterra early in last century [19–29], in the form

$$y(t) = \int_{-\infty}^{t} h_1(t - \tau_1) \cdot x(\tau_1) \cdot d\tau_1$$

$$+ \int_{-\infty}^{t} \int_{-\infty}^{t} h_2(t - \tau_1, t - \tau_2) \cdot x(\tau_1) \cdot x(\tau_2) \cdot d\tau_2 \cdot d\tau_1 + \cdots \quad (1.43)$$

where the first term is the linear one (first-order term) and the following ones are higher-order terms that take into account the effect on nonlinearities. The hypothesis in this case of series expansion is that the nonlinearities are weak and that only a few higher-order terms will be sufficient to describe their effect. The generalised transfer functions of nth order $h_n(t_1, \ldots, t_n)$ are called nuclei of nth order. In order to compute the nuclei analytically, it is also required that the nonlinearity be expressed as a power series:

$$i_g(v) = g_0 + g_1 \cdot v + g_2 \cdot v^2 + g_3 \cdot v^3 + \cdots \quad (1.44)$$

a requirement that will be justified below. It is clear that any nonlinearity can be expanded in power series, but only within a limited voltage and current range.

The Volterra series can be interpreted in the following way: the output signal of a nonlinear system is composed by an infinite number of terms of increasing order; each term is the infinite sum (integral) of all contributions due to the input signal multiplied by itself n times, where n is the order of the term, in any possible combination of time instants in the past, weighted by a function called nucleus of nth order, representing the effect of the transfer through the system for that order. The nuclei represent also, in this case, the 'memory' of the system, and they represent the way in which the system responds to the presence of an input signal at different time instants in the past; since the system is nonlinear, its response to the input signal applied at a certain time instant is not independent of the value of the input signal at a different time instant. All the combinations must therefore be taken into account through multiple integration. The nuclei become normally smaller as the time elapsed from the time instants of the input contributions and the current time instant becomes larger. If the system is instantaneous, the nuclei are delta functions, and the nth order integral becomes the nth power of the input:

$$y(t) = k_1 \cdot x(t) + k_2 \cdot x^2(t) + \cdots \tag{1.45}$$

A generalisation of the Fourier transform can be defined for the nonlinear case: if we define

$$H_n(\omega_1, \ldots, \omega_n) = \frac{1}{\sqrt{2\pi}} \cdot \int_{-\infty}^{\infty} \cdots \int_{-\infty}^{\infty} h_n(\tau_1, \ldots, \tau_n)$$

$$\times e^{-j(\omega_1 \tau_1 + \cdots \omega_n \tau_n)} \cdot d\tau_n, \cdots, d\tau_1 \tag{1.46a}$$

$$h_n(\tau_1, \ldots, \tau_n) = \frac{1}{\sqrt{2\pi}} \cdot \int_{-\infty}^{\infty} \cdots \int_{-\infty}^{\infty} H_n(\omega_1, \ldots, \omega_n)$$

$$\times e^{j(\omega_1 \tau_1 + \cdots \omega_n \tau_n)} \cdot d\omega_n, \ldots, d\omega_1 \tag{1.46b}$$

the Volterra series becomes (Appendix A.3)

$$Y(\omega) = \cdots + \int_{-\infty}^{\infty} \cdots \int_{-\infty}^{\infty} H_n(\omega_1, \ldots, \omega_n) \cdot X(\omega_1) \cdot \ldots \cdot X(\omega_n)$$

$$\times \delta(\omega - \omega_1 - \cdots - \omega_n) \cdot d\omega_n, \ldots, d\omega_1 + \cdots \tag{1.47}$$

Equation (1.47) can be interpreted in the following way: the output signal of a nonlinear system is the sum of an (infinite) number of terms of given orders; each term is the infinite sum (integral) of all spectral contributions of the input signal multiplied by itself n times, where n is the order of the term, in any possible combination of frequencies, weighted by a function of frequency called frequency-domain nucleus of nth order, which represents the effect of the transfer through the system for that order. The frequency of each spectral contribution to the output signal is the algebraic sum of the frequencies of the contributing terms of the input signal; in other words, the spectrum of the output signal will not be zero at a given frequency if there is a combination of the input frequency n times that equals this frequency. We note explicitly that the spectrum occupancy of the output signal is now broader than that of the spectrum of the input signal.

Let us clarify these concepts by illustrating the special case of periodic signals. If the input signal is a periodic function, its spectrum is discrete and the integrals become summations; in the case of an ideal, complex single tone

$$x(t) = A \cdot e^{j\omega_0 t} \qquad X(\omega) = A \cdot \delta(\omega - \omega_0) \tag{1.48}$$

the output signal is given by

$$y(t) = A \cdot H_1(\omega_0) \cdot e^{j\omega_0 t} + A^2 \cdot H_2(\omega_0, \omega_0) \cdot e^{j2\omega_0 t} + \cdots \tag{1.49a}$$

$$Y(\omega) = A \cdot H_1(\omega_0) \cdot \delta(\omega - \omega_0) + A^2 \cdot H_2(\omega_0, \omega_0) \cdot \delta(\omega - 2\omega_0) + \cdots \tag{1.49b}$$

The second-order term generates a signal component at second-harmonic frequency, and so on for higher-order terms. In the case of a real single tone, that is, a couple of ideal single tones at opposite frequencies,

$$x(t) = A \cdot \frac{(e^{j\omega_0 t} + e^{-j\omega_0 t})}{2} = A \cdot \cos(\omega_0 t) \tag{1.50a}$$

$$X(\omega) = \frac{A}{2} \cdot \delta(\omega - \omega_0) + \frac{A}{2} \cdot \delta(\omega + \omega_0) \tag{1.50b}$$

The output signal is given by

$$y(t) = y_1(t) + y_2(t) + y_3(t) + \cdots \tag{1.51}$$

$$y_1(t) = A \cdot \overline{H_1(\omega_0)} \cdot \cos(\omega_0 t)$$

$$y_2(t) = \frac{A^2}{2} \cdot \overline{H_2(\omega_0, -\omega_0)} + \frac{A^2}{2} \cdot \overline{H_2(\omega_0, \omega_0)} \cdot \cos(2\omega_0 t)$$

$$y_3(t) = \frac{3A^3}{4} \cdot \overline{H_3(\omega_0, \omega_0, -\omega_0)} \cdot \cos(\omega_0 t) + \frac{A^3}{4} \cdot \overline{H_3(\omega_0, \omega_0, \omega_0)} \cdot \cos(3\omega_0 t)$$

$$Y(\omega) = Y_1(\omega) + Y_2(\omega) + Y_3(\omega) + \cdots$$

$$Y_1(\omega) = \frac{A}{2} \cdot H_1(\omega_0) \cdot \delta(\omega - \omega_0) + \frac{A}{2} \cdot H_1(-\omega_0) \cdot \delta(\omega + \omega_0)$$

$$Y_2(\omega) = \frac{A^2}{4} \cdot H_2(-\omega_0, -\omega_0) \cdot \delta(\omega + 2\omega_0) + \frac{A^2}{4} \cdot H_2(\omega_0, \omega_0) \cdot \delta(\omega - 2\omega_0)$$

$$+ \frac{A^2}{4} \cdot H_2(-\omega_0, \omega_0) \cdot \delta(\omega) + \frac{A^2}{4} \cdot H_2(\omega_0, -\omega_0) \cdot \delta(\omega)$$

$$Y_3(\omega) = \frac{A^3}{8} \cdot H_3(-\omega_0, -\omega_0, -\omega_0) \cdot \delta(\omega + 3\omega_0) + \frac{A^3}{8} \cdot H_2(\omega_0, \omega_0, \omega_0) \cdot \delta(\omega - 3\omega_0)$$

$$+ \frac{A^3}{8} \cdot H_2(-\omega_0, -\omega_0, \omega_0) \cdot \delta(\omega + \omega_0) + \frac{A^3}{8} \cdot H_2(-\omega_0, \omega_0, -\omega_0) \cdot \delta(\omega + \omega_0)$$

$$+ \frac{A^3}{8} \cdot H_2(\omega_0, -\omega_0, -\omega_0) \cdot \delta(\omega + \omega_0) + \frac{A^3}{8} \cdot H_2(-\omega_0, \omega_0, \omega_0) \cdot \delta(\omega - \omega_0)$$

$$+ \frac{A^3}{8} \cdot H_2(\omega_0, -\omega_0, \omega_0) \cdot \delta(\omega - \omega_0) + \frac{A^3}{8} \cdot H_2(\omega_0, \omega_0, -\omega_0) \cdot \delta(\omega - \omega_0)$$

The first-order terms generate the linear output signal at input frequency; the second-order terms generate a zero-frequency signal (rectified signal) and a double-frequency signal (second harmonic); the third-order terms generate a signal at input frequency (compression or expansion) and at triple frequency (third harmonic); and so on. The higher-order terms are the nonlinear contribution to the distortion of the signal and are proportional to the nth power of the input where n is the order of the term. A graphical representation of the spectra is depicted in Figure 1.11.

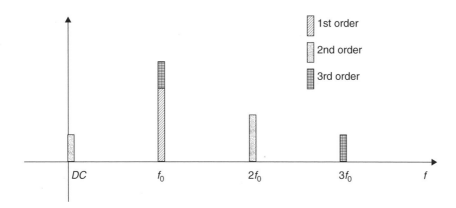

Figure 1.11 Contributions of the terms of the Volterra series to the spectrum of a single-tone signal

Let us now consider a two-tone input signal, in the form

$$x(t) = A_1 \cdot \cos(\omega_1 t) + A_2 \cdot \cos(\omega_2 t) \tag{1.52}$$

The output signal is given by

$$y(t) = y_1(t) + y_2(t) + y_3(t) + \cdots \tag{1.53}$$

$$y_1(t) = A_1 \cdot \overline{H_1(\omega_1)} \cdot \cos(\omega_1 t) + A_2 \cdot \overline{H_1(\omega_2)} \cdot \cos(\omega_2 t)$$

$$y_2(t) = \frac{A_1^2}{2} \cdot \overline{H_2(\omega_1, -\omega_1)} + \frac{A_2^2}{2} \cdot \overline{H_2(\omega_2, -\omega_2)}$$

$$+ \frac{A_1^2}{2} \cdot \overline{H_2(\omega_1, \omega_1)} \cdot \cos(2\omega_1 t) + \frac{A_1 A_2}{2} \cdot \overline{H_2(\omega_1, \omega_2)} \cdot \cos((\omega_1 + \omega_2)t)$$

$$+ \frac{A_2^2}{2} \cdot \overline{H_2(\omega_2, \omega_2)} \cdot \cos(2\omega_2 t)$$

$$y_3(t) = \frac{3A_1^3}{4} \cdot \overline{H_3(\omega_1, \omega_1, -\omega_1)} \cdot \cos(\omega_1 t) + \frac{3A_2^3}{4} \cdot \overline{H_3(\omega_2, \omega_2, -\omega_2)} \cdot \cos(\omega_2 t)$$

$$+ \frac{3A_1 A_2^2}{4} \cdot \overline{H_3(\omega_1, \omega_2, -\omega_2)} \cdot \cos(\omega_1 t) + \frac{3A_1^2 A_2}{4} \cdot \overline{H_3(\omega_2, \omega_1, -\omega_1)} \cdot \cos(\omega_2 t)$$

$$+ \frac{3A_1^2 A_2}{4} \cdot \overline{H_3(\omega_1, \omega_1, -\omega_2)} \cdot \cos((2\omega_1 - \omega_2)t)$$

$$+ \frac{3A_1 A_2^2}{4} \cdot \overline{H_3(\omega_2, \omega_2, -\omega_1)} \cdot \cos((2\omega_2 - \omega_1)t)$$

$$+ \frac{3A_1^3}{4} \cdot \overline{H_3(\omega_1, \omega_1, \omega_1)} \cdot \cos(3\omega_1 t) + \frac{3A_1^2 A_2}{4} \cdot \overline{H_3(\omega_1, \omega_1, \omega_2)} \cdot \cos((2\omega_1 + \omega_2)t)$$

$$+ \frac{3A_1 A_2^2}{4} \cdot \overline{H_3(\omega_1, \omega_2, \omega_2)} \cdot \cos((2\omega_2 - \omega_1)t) + \frac{A_2^3}{4} \cdot \overline{H_3(\omega_2, \omega_2, \omega_2)} \cdot \cos(3\omega_2 t)$$

The first-order terms generate the linear output signals at input frequencies. The second-order terms generate three components: a zero-frequency signal that is the rectification of both input signals; a difference-frequency signal and three second-harmonic or mixed-harmonic signals. The third-order terms generate four components: two compression components at input frequencies; two desensitivisation components again at input frequencies, due to the interaction of the two input signals, that add to compression; two intermodulation signals at $2\omega_1 - \omega_2$ and at $2\omega_2 - \omega_1$ and four third-harmonic or mixed-harmonic signals. The higher-order terms are proportional to suitable combinations of powers of the input signals. A graphical representation of the spectrum is depicted in Figure 1.12.

From the formulae above, it is clear that the output signal is easily computed when the nuclei are known. In fact, the nuclei are computed by a recursive method if the nonlinearity is expressed as a power series [23, 29]; in the case of our example (eq. (1.44))

$$i_g(v) = g_0 + g_1 \cdot v + g_2 \cdot v^2 + g_3 \cdot v^3 + \cdots \tag{1.54}$$

An input 'probing' signal in the form of an ideal tone of unit amplitude (eq. (1.48)) is first used:

$$i_s(t) = e^{j\omega_1 t} \tag{1.55}$$

The output can formally be written as (see eq. (1.49))

$$v(t) = H_1(\omega_1) \cdot e^{j\omega_1 t} + H_2(\omega_1, \omega_1) \cdot e^{j2\omega_1 t} + \cdots \tag{1.56}$$

where the nuclei are still unknown. Kirchhoff's equation (eq. (1.3)) with the nonlinearity in power-series form (eq. (1.54), limited to second order for brevity) is

$$C \cdot \frac{dv(t)}{dt} + g_1 \cdot v(t) + g_2 \cdot v^2(t) + \cdots + i_s(t) = 0 \tag{1.57}$$

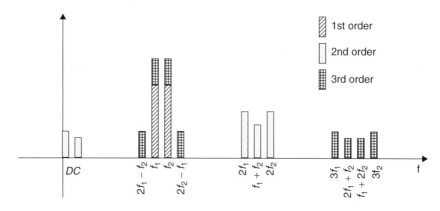

Figure 1.12   Contributions of the terms of the Volterra series to the spectrum of a two-tone signal

When the voltage as in eq. (1.56) is replaced into Kirchhoff's equation (eq. (1.57)), we get

$$C \cdot (j\omega_1 \cdot H_1(\omega_1) \cdot e^{j\omega_1 t} + j2\omega_1 \cdot H_2(\omega_1, \omega_1) \cdot e^{j2\omega_1 t} + \cdots)$$
$$+ g_1 \cdot (H_1(\omega_1) \cdot e^{j\omega_1 t} + H_2(\omega_1, \omega_1) \cdot e^{j2\omega_1 t} + \cdots)$$
$$+ g_2 \cdot (H_1(\omega_1) \cdot e^{j\omega_1 t} + H_2(\omega_1, \omega_1) \cdot e^{j2\omega_1 t} + \cdots)^2 + i_s \cdot e^{j\omega_1 t} = 0 \quad (1.58)$$

Expanding the expressions in the parentheses, we get

$$C \cdot j\omega_1 \cdot H_1(\omega_1) \cdot e^{j\omega_1 t} + C \cdot j2\omega_1 \cdot H_2(\omega_1, \omega_2) \cdot e^{j2\omega_1 t} + g_1 \cdot H_1(\omega_1) \cdot e^{j\omega_1 t}$$
$$+ g_1 \cdot H_2(\omega_1, \omega_1) \cdot e^{j2\omega_1 t} + g_2 \cdot H_1^2(\omega_1) \cdot e^{j2\omega_1 t} + g_2 \cdot 2H_1(\omega_1) \cdot H_2(\omega_1, \omega_1) \cdot e^{j3\omega_1 t}$$
$$+ g_2 \cdot H_2^2(\omega_1, \omega_1) \cdot e^{j4\omega_1 t} + i_s \cdot e^{j\omega_1 t} = 0 \quad (1.59)$$

where the first and third rows are the linear terms, and the second row is the nonlinear term truncated to the second power.

We now try to split eq. (1.59) into several simpler equations. The terms in this equation depend on time through exponential terms at different frequencies. In order that their sum be zero for all time instants t, all terms depending on time with the same frequency must sum up to zero because of the orthogonality of the sinusoidal functions with respect to time. We can therefore equate the sums of all the terms at the same frequency to zero:

$$C \cdot j\omega_1 \cdot H_1(\omega_1) \cdot e^{j\omega_1 t} + g_1 \cdot H_1(\omega_1) \cdot e^{j\omega_1 t} + i_s \cdot e^{j\omega_1 t} = 0 \quad (1.60a)$$
$$C \cdot j2\omega_1 \cdot H_2(\omega_1, \omega_1) \cdot e^{j3\omega_1 t} + g_1 \cdot H_2(\omega, \omega_{1_1}) \cdot e^{j2\omega_1 t}$$
$$+ g_2 \cdot H_1^2(\omega_1) \cdot e^{j2\omega_1 t} = 0 \ldots \quad (1.60b)$$

where we have limited ourselves to the second-harmonic frequency. From the first equation (eq. (1.60a)), we immediately have

$$H_1(\omega_1) = -\frac{1}{g_1 + j\omega_1 C} \quad (1.61)$$

which is nothing but the solution of the linear part of the circuit.

The second-order nucleus does appear in eq. (1.60b), but not in its general form $H_2(\omega_1, \omega_2)$; to get it, we use a two-tone unit-amplitude ideal input of the form

$$i_s(t) = e^{j\omega_1 t} + e^{j\omega_2 t} \quad (1.62)$$

For this type of input, the output is written as

$$v(t) = H_1(\omega_1) \cdot e^{j\omega_1 t} + H_1(\omega_2) \cdot e^{j\omega_2 t} + H_2(\omega_1, \omega_1) \cdot e^{j2\omega_1 t}$$
$$+ H_2(\omega_1, \omega_2) \cdot e^{j(\omega_1 + \omega_2)t} + H_2(\omega_2, \omega_2) \cdot e^{j2\omega_2 t} + \cdots \quad (1.63)$$

When the voltage in the form of eq. (1.63) is replaced into Kirchhoff's eq. (1.57) we get

$$
\begin{aligned}
&C \cdot j\omega_1 \cdot H_1(\omega_1) \cdot e^{j\omega_1 t} + C \cdot j\omega_2 \cdot H_1(\omega_2) \cdot e^{j\omega_2 t} \\
&\quad + C \cdot j2\omega_1 \cdot H_2(\omega_1, \omega_1) \cdot e^{j2\omega_1 t} + C \cdot j(\omega_1 + \omega_2) \cdot H_2(\omega_1, \omega_2) \cdot e^{j(\omega_1+\omega_2)t} \\
&\quad + C \cdot j2\omega_2 \cdot H_2(\omega_2, \omega_2) \cdot e^{j2\omega_2 t} + g_1 \cdot H_1(\omega_1) \cdot e^{j\omega_1 t} + g_1 \cdot H_1(\omega_2) \cdot e^{j\omega_2 t} \\
&\quad + g_1 \cdot H_2(\omega_1, \omega_1) \cdot e^{j2\omega_1 t} + g_1 \cdot H_2(\omega_1, \omega_2) \cdot e^{j(\omega_1+\omega_2)t} + g_1 \cdot H_2(\omega_2, \omega_2) \cdot e^{j2\omega_2 t} \\
&\quad + g_2 \cdot H_1^2(\omega_1) \cdot e^{j2\omega_1 t} + g_2 \cdot H_1^2(\omega_2) \cdot e^{j2\omega_2 t} + g_2 \cdot 2 \cdot H_1(\omega_1) \cdot H_1(\omega_2) \cdot e^{j(\omega_1+\omega_2)t} \\
&\quad + g_2 \cdot H_2^2(\omega_1, \omega_1) \cdot e^{j4\omega_1 t} + g_2 \cdot H_2^2(\omega_1, \omega_2) \cdot e^{j2(\omega_1+\omega_2)t} + g_2 \cdot H_2^2(\omega_2, \omega_2) \cdot e^{j4\omega_2 t} \\
&\quad + g_2 \cdot 2 \cdot H_2(\omega_1, \omega_1) \cdot H_2(\omega_1, \omega_2) \cdot e^{j(3\omega_1+\omega_2)t} \\
&\quad + g_2 \cdot 2 \cdot H_2(\omega_1, \omega_1) \cdot H_2(\omega_2, \omega_2) \cdot e^{j(2\omega_1+2\omega_2)t} \\
&\quad + g_2 \cdot 2 \cdot H_2(\omega_1, \omega_2) \cdot H_2(\omega_2, \omega_2) \cdot e^{j(\omega_1+3\omega_2)t} \\
&\quad + g_2 \cdot 2 \cdot H_1(\omega_1) \cdot H_2(\omega_1, \omega_1) \cdot e^{j3\omega_1 t} + g_2 \cdot 2 \cdot H_1(\omega_1) \cdot H_2(\omega_1, \omega_2) \cdot e^{j(2\omega_1+\omega_2)t} \\
&\quad + g_2 \cdot 2 \cdot H_1(\omega_1) \cdot H_2(\omega_2, \omega_2) \cdot e^{j(\omega_1+2\omega_2)t} + g_2 \cdot 2 \cdot H_1(\omega_2) \cdot H_2(\omega_1, \omega_1) \cdot e^{j(2\omega_1+\omega_2)t} \\
&\quad + g_2 \cdot 2 \cdot H_1(\omega_2) \cdot H_2(\omega_1, \omega_2) \cdot e^{j(\omega_1+2\omega_2)t} + g_2 \cdot 2 \cdot H_1(\omega_2) \cdot H_2(\omega_2, \omega_2) \cdot e^{j3\omega_2 t} \\
&\quad + e^{j\omega_1 t} + e^{j\omega_2 t} = 0
\end{aligned}
\tag{1.64}
$$

We see that eq. (1.64) can be split into several equations, relative to the dependence on time. The equations relative to the terms in $e^{j\omega_1 t}$ and in $e^{j\omega_2 t}$ yield the same result as in the case of a single tone probing signal (eq. (1.60a)):

$$
H_1(\omega_1) = -\frac{1}{g_1 + j\omega_1 C} \qquad H_1(\omega_2) = -\frac{1}{g_1 + j\omega_2 C}
\tag{1.65}
$$

The two expressions given above yield equivalent expressions for the first-order nucleus (eq. (1.61)). The second-order nucleus is explicitly computed from the equation in $e^{j(\omega_1+\omega_2)t}$:

$$
\begin{aligned}
&C \cdot j(\omega_1 + \omega_2) \cdot H_2(\omega_1, \omega_2) \cdot e^{j(\omega_1+\omega_2)t} + g_1 \cdot H_2(\omega_1, \omega_2) \cdot e^{j(\omega_1+\omega_2)t} \\
&\quad + g_2 \cdot 2 \cdot H_1(\omega_1) \cdot H_1(\omega_2) \cdot e^{j(\omega_1+\omega_2)t} = 0
\end{aligned}
\tag{1.66}
$$

from which we get

$$
H_2(\omega_1, \omega_2) = \frac{2g_2 \cdot H_1(\omega_1) \cdot H_1(\omega_2)}{g_1 + j(\omega_1 + \omega_2)C}
\tag{1.67}
$$

The second-order nucleus is an explicit expression that includes only the first-order nucleus, already found from eq. (1.65) (or equivalently from eq. (1.61)). By substitution we get

$$
H_2(\omega_1, \omega_2) = \frac{2g_2}{(g_1 + j\omega_1 C) \cdot (g_1 + j\omega_2 C) \cdot (g_1 + j(\omega_1 + \omega_2)C)}
\tag{1.68}
$$

It is easy to see that higher-order nuclei are found recursively by the same procedure. When the nuclei up to the nth order must be computed, a probing signal of n independent ideal tones must be used, as shown above.

When all the nuclei are known (up to a certain order), the output of the considered circuit is available as a Volterra series in a general form, that is, the output can be written in an analytical form for any input signal. The time-domain version of the nuclei is easily computed by means of eq. (1.46), from which the general time-domain formulation (eq. 1.43) is expressed.

It is clear from the described procedure that the nonlinearity must be expressed as a power series in order to compute the nuclei explicitly. In other words, the powers of exponential terms are immediately and explicitly expressed as exponential terms and can therefore be grouped by frequency. Other functions of exponential terms, for example, the hyperbolic tangent as in the example above, cannot be explicitly expressed as sum of exponentials, and do not allow for the explicit solution of the problem with the probing method.

As an example, a third-order Volterra-series response of our example circuit to a sinusoidal input with amplitude $i_{\mathrm{s,max}} = 80$ mA is shown in Figure 1.13; the left plot shows the input current i_{s} (———), the voltage v (- - - -) and the current in the nonlinear resistor i_{g} (·····). The right plot shows the current–voltage characteristic of the nonlinear resistor as a hyperbolic tangent (———) and as a third-order polynomial approximation (- - - -) for the current and voltage swing in the example.

In Figure 1.14, the same quantities are shown, but for an input current amplitude of $i_{\mathrm{s,max}} = 100$ mA.

Comparison with time-domain analysis (see Figure 1.4) reveals that already for moderate nonlinearities both Volterra series expansion of the output signal and the power-series expansion of the current–voltage nonlinear element limit the accuracy of the method. For increased accuracy, a high number of nuclei is necessary; however, their

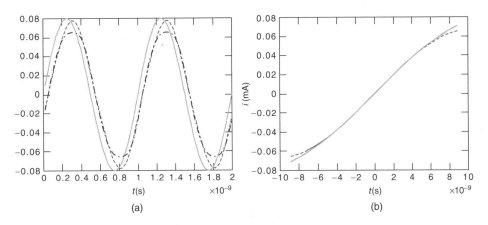

Figure 1.13 Voltages and currents in the example circuit from Volterra analysis (a) and the voltage–current characteristic of the nonlinear conductance in the correct and approximated form (b)

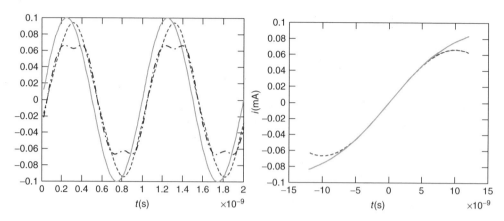

Figure 1.14   Same as in Figure 1.13, but for a larger input signal

calculation becomes impractical for two reasons. First, the 'probing signal' calculation method becomes cumbersome. Second and more important, the high-order nuclei depend on high-order terms in the power-series expansion of the nonlinear element. These terms come from the experimental characterisation of the nonlinearity (see Chapters 2 and 3) and are affected by an increasing degree of inaccuracy. Therefore, practical applications of the Volterra series must be limited to a small order and consequently to mild nonlinear problems.

From what has been said, we can conclude that the Volterra series is an easy and elegant method, suitable for the analysis of mildly nonlinear problems only. Its natural application is the analysis of intermodulation in linear power amplifiers with two-tone or multiple-tone input signal; a special development of this technique has been applied to mixer analysis [30].

1.3.2 Fourier Series

In this paragraph, the solution of the nonlinear differential Kirchhoff's equations by use of the Fourier series expansion is described. The harmonic balance technique, belonging to this category, is especially considered.

It has been shown above (Section 1.3.1) that when the input signal of a nonlinear system is a sinusoid, the output signal is again a sinusoid together with all harmonics, plus a rectified component at zero frequency (DC). In fact, this is not always true, because subharmonic generation or chaotic or quasi-chaotic behaviour is in general possible in nonlinear circuits [31]. We will not consider this case for the moment (see Chapter 5) and we will assume that a periodic behaviour is established in the circuit, with the same period of the input signal. In this case, we are interested in the steady state response of circuits driven by a sinusoidal (or periodic) input and can therefore assume that the output signal ($v(t)$ in our example circuit) be expressed as a Fourier series. We now look at the special form that our nonlinear Kirchhoff's equation assumes when this assumption is made.

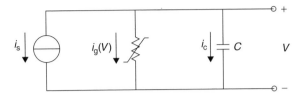

Figure 1.15   The example circuit, repeated here for convenience

1.3.2.1 Basic formulation (single tone)

Let us describe first the harmonic balance formulation with single-tone input signal [33–40]. The output signal is first written as a Fourier series, and so is the input periodic signal; for our example circuit (Figure 1.15):

$$v(t) = \sum_{n=-\infty}^{\infty} V_n \cdot e^{jn\omega_0 t} \qquad i_s(t) = I_s \cdot \left(\frac{e^{jn\omega_0 t} + e^{-jn\omega_0 t}}{2} \right) = I_s \cdot \cos(\omega_0 t) \qquad (1.69)$$

In this particular case, we assume that the input current is a simple cosinusoid, for simplicity. We also assume here that the input signal has zero phase; this is usually the case, since the input signal acts as a reference for the time scale. The output signal is completely known once the infinite complex phasors V_n are known: these are the unknowns of our analysis problem. The Fourier series representing the output signal (eq. (1.69)) is replaced into the nonlinear Kirchhoff's equation; then, we try to split the single 'difficult' equation into several 'simpler' equations.

By replacing eq. (1.69) into the nonlinear Kirchhoff's eq. (1.13), we get,

$$C \cdot \frac{d \left(\sum_{n=-\infty}^{\infty} V_n \cdot e^{jn\omega_0 t} \right)}{dt} + i_{max} \cdot tgh \left(\frac{g \cdot \sum_{n=-\infty}^{\infty} V_n \cdot e^{jn\omega_0 t}}{i_{max}} \right)$$

$$+ I_s \cdot \left(\frac{e^{j\omega_0 t} + e^{-j\omega_0 t}}{2} \right) = 0 \qquad (1.70)$$

The unknowns of the equation are the phasors that appear in the Fourier series expansion of the voltage (eq. (1.69)). We now rewrite eq. (1.70) with all the terms expressed in the form of Fourier series:

$$\sum_{n=-\infty}^{\infty} jn\omega_0 C \cdot V_n \cdot e^{jn\omega_0 t} + \sum_{n=-\infty}^{\infty} I_{g,n} \cdot e^{jn\omega_0 t} + I_s \cdot \left(\frac{e^{j\omega_0 t} + e^{-j\omega_0 t}}{2} \right) = 0 \qquad (1.71)$$

Some terms in eq. 1.71 are immediately and explicitly available: the first term representing the currents in the linear capacitor i_C and the third term representing the current source i_s, that is, the linear elements of the circuit. The Fourier series expansion of the current in the nonlinear conductance (i_g) on the other hand cannot be computed explicitly because of the nonlinearity of the element, and is only indicated symbolically so far:

$$i_g(t) = i_{max} \cdot tgh \left(\frac{g \cdot \sum_{n=-\infty}^{\infty} V_n \cdot e^{jn\omega_0 t}}{i_{max}} \right) = \sum_{n=-\infty}^{\infty} I_{g,n} \cdot e^{jn\omega_0 t} \qquad (1.72)$$

The (periodic) current in the nonlinear conductance however can be expressed in the time domain, and its phasors computed from it by means of a Fourier transform:

$$i_g(t) = i_{max} \cdot tgh \left(\frac{g \cdot \sum_{n=-\infty}^{\infty} V_n \cdot e^{jn\omega_0 t}}{i_{max}} \right) \qquad i_g(t) \Rightarrow \mathfrak{I} \Rightarrow I_{g,n} \qquad (1.73)$$

The phasors can now be written into the eq. (1.71) for the solution of the equation itself. It is clear that in general each phasor $I_{g,n}$ is a function of all the (still unknown) voltage phasors V_n.

The terms in eq. (1.71) above have sinusoidal or cosinusoidal time dependence at different frequencies. In order that their sum be zero for all time instants t, all terms depending on time with the same frequency must sum up to zero, because of the orthogonality of the sinusoidal functions with respect to time. We can therefore equate to zero the sum of all the terms at the same frequency, obtaining an infinite set of complex equations:

$$jn\omega_0 C \cdot V_n + I_{g,n} + \frac{I_s}{2} = 0 \qquad n = -\infty, \ldots, 0, \ldots, \infty \qquad (1.74)$$

where I_s is present only for $n = -1$ and $n = 1$ (input frequency); however, in the case that the input signal in periodic but not purely sinusoidal, the phasors of the Fourier series expansion of the input current will be present also in the equations relative to other frequencies. The system eq. (1.74) of infinite equations is equivalent to the original problem under the hypothesis of periodic response of the nonlinear circuit.

As stated above, we assume that the first few harmonics are sufficient to describe the behaviour of the electrical quantities; in other words, we assume that the output signal has a limited bandwidth (or a limited number of harmonics). We therefore truncate the Fourier series expansion after N terms, obtaining a finite number of equations (and harmonics). In this case, the Fourier transform is computed by evaluating the nonlinear current at a suitable number of points in time, according to Nykvist's sampling theorem:

$$i_{g,k} = i_g(t_k) = i_{max} \cdot tgh \left(\frac{g \cdot \sum_{n=-N}^{N} V_n \cdot e^{jn\omega_0 t_k}}{i_{max}} \right) \qquad t_k = t_0 + \frac{T}{K_{max} + 1} \cdot k$$

$$k = 0, \ldots, K_{max} \qquad (1.75)$$

The $4N + 1$ real numbers that make up the $2N$ complex plus 1 real current phasors are then found by a simple discrete Fourier transform (DFT) method (Appendix A.4).

The system is now solved for the $4N + 1$ real unknowns, that is, the $2N$ complex and 1 real coefficients of the Fourier series expansion (phasors) of the unknown function $v(t)$:

$$v(t) = \sum_{n=-N}^{N} V_n \cdot e^{jn\omega_0 t} \tag{1.76}$$

System eq. (1.74) becomes

$$jn\omega_0 C \cdot V_n + I_{g,n}(\vec{V}) + \frac{I_s}{2} = 0 \qquad n = -N, \ldots, N \tag{1.77}$$

where again I_s is present only for $n = -1$ and $n = 1$ (input frequency), and

$$\vec{V} = [V_{-N}, V_0, \ldots, V_N]^T \tag{1.78}$$

Equation (1.77) is the nonlinear complex equation system in the unknown voltage phasors V_n that yields the solution to our nonlinear circuit. It states that the currents flowing into the linear and nonlinear part of the circuit must be balanced at each harmonic, as required by Kirchhoff's current law in this special formulation; this gives the name 'harmonic balance' to this method. The system is solved numerically by an iterative method, where the values of the voltage phasors are first estimated and then iteratively corrected until the error is considered to be negligible. The error vector is the left-hand-side of the equation system (1.77), that is, the sum of the currents at the node at each harmonic frequency; it should be zero, after Kirchhoff's current law, but a value below, for example 1 μA, can often be considered adequate for accurate results.

In fact, there is no general guarantee that a nonlinear system has a unique solution or that it has a solution at all; however, it is usually true that a harmonic problem in this form, especially if not too stiff, has at least a solution corresponding to the linear solution of a linearised problem.

The above formulation is in fact redundant. Since the Fourier coefficients of a real function are Hermitean, the phasors and equations at negative frequencies are the complex conjugate of the phasors and equations at positive frequencies: only the phasors and equations relative to positive or negative frequencies (plus zero frequency, DC) must therefore be retained.

By exploiting this property and retaining positive frequencies only (plus DC), voltage and currents are rewritten as

$$v(t) = \text{Re}\left[\sum_{n=0}^{N} V_n \cdot e^{jn\omega_0 t}\right] = V_0 + \text{Re}\left[\sum_{n=1}^{N} V_n \cdot e^{jn\omega_0 t}\right]$$

$$= V_0 + \sum_{n=1}^{N} \{V_n^r \cos(n\omega_0 t) - V_n^i \sin(n\omega_0 t)\} \tag{1.79a}$$

$$i_g(t) = i_{max} \cdot tgh \left(\frac{g \cdot \displaystyle\sum_{n=-\infty}^{\infty} V_n \cdot e^{jn\omega_0 t}}{i_{max}} \right) = \text{Re} \left[\sum_{n=0}^{N} V_{g,n} \cdot e^{jn\omega_0 t} \right] = I_{g,0}$$

$$+ \text{Re} \left[\sum_{n=1}^{N} I_n \cdot e^{jn\omega_0 t} \right] = I_{g,0} + \sum_{n=1}^{N} \{ I_{g,n}^r \cos(n\omega_0 t) - I_{g,n}^i \sin(n\omega_0 t) \} \quad (1.79\text{b})$$

$$i_s(t) = I_s \cdot \cos(\omega_0 t) \tag{1.79c}$$

By replacing eq. (1.79) into eq. (1.13), the equation system (1.77) is rewritten as

$$I_{g,0}(\overline{V}) = 0 \qquad\qquad n = 0 \tag{1.80a}$$

$$-\omega_0 C \cdot V_1^i + I_{g,1}^r(\overline{V}) + I_s = 0 \qquad\qquad n = 1 \tag{1.80b}$$

$$-\omega_0 C \cdot V_1^r + I_{g,1}^i(\overline{V}) = 0 \tag{1.80c}$$

$$-n\omega_0 C \cdot V_n^i + I_{g,n}^r(\overline{V}) = 0 \qquad\qquad 2 \leqslant n \leqslant N \tag{1.80d}$$

$$-n\omega_0 C \cdot V_n^r + I_{g,n}^i(\overline{V}) = 0 \tag{1.80e}$$

We have now only $2N + 1$ real equations in the $2N + 1$ unknowns (the voltage phasors): two equations for the input frequency and for each of the harmonic frequencies, and a real equation for $n = 0$, since no phase information is needed for a DC signal. In Appendix A.5, the system for $N = 3$ is described in detail.

In the case of a real formulation, the Fourier transform assumes the form described in Appendix A.4. In particular, the inverse transformation is not analytical, with consequences on the numerical treatment of the equation system.

The number of sampling time instants where the nonlinear current is evaluated, indicated with $K_{max} + 1$ above, requires a comment. A discrete Fourier transform is exact if the signal to be transformed has a limited band, and if the number of sampling points is chosen according to Nykvist's sampling theorem. A basic hypothesis in the harmonic balance algorithm is also that the unknown signal (a voltage, in our case) can be represented with a limited number of harmonics, previously indicated by N. In this ideal case, $K_{max} = 2N$. However, it may be beneficial to use a larger number of points, and therefore a higher number of harmonics, during the discrete Fourier transform in order to avoid aliasing; the current harmonics higher than N are then discarded from successive operations. This procedure can be useful when a stiff current nonlinearity is present in the circuit; it is called 'oversampling', and some commercial CAD packages allow the user to introduce it.

In Figure 1.16, the solution of our example circuit is given for the following values of the circuit elements:

$$g = 10 \text{ mS} \quad C = 500 \text{ fF} \quad f = 1 \text{ GHz}$$

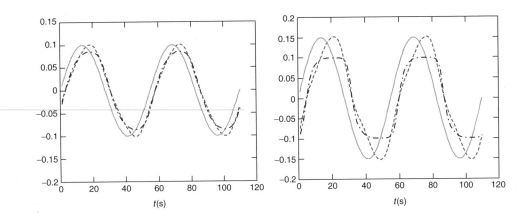

Figure 1.16   Currents and voltages in the example circuit for two different amplitudes of a sinusoidal input current

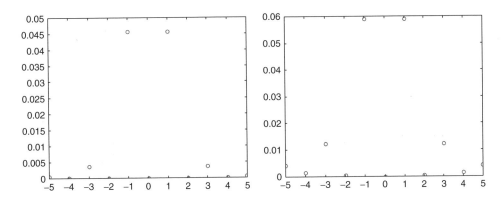

Figure 1.17   Current and voltage spectra in the example circuit for two different amplitudes of a sinusoidal input current

and with the Fourier series truncated after the fifth harmonic ($N = 5$). The plots show input current (———), output voltage (- - - -) and current in the nonlinear resistor (·-·-·-·), for an input current of $i_s = 100$ mA (a) and for a larger input of $i_s = 150$ mA (b).

   The results are very similar to those of the direct integration method (see Figure 1.4). In Figure 1.17, the amplitudes of the voltage spectra are shown for the same two cases: the larger relative amplitude of higher harmonics in the case of larger input current amplitude is clearly shown. The presence of an odd current nonlinearity with respect to voltage results in the absence of even harmonics in the spectrum. The amplitudes of spectral lines at negative frequencies are obviously identical to those at positive frequencies.

   In Figure 1.18, voltage and currents are shown only at the $2N + 1 = 11$ sampling points in time domain for the case of the larger input current.

   The number of points corresponds to the number of real values of the spectrum: five complex phasors at $\omega = n\omega_0, n = 1, \ldots, 5$ (fundamental frequency and the first three

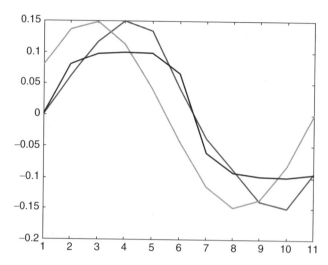

Figure 1.18   Currents and voltages in the example circuit at the sampling times only, as computed from Fourier transform

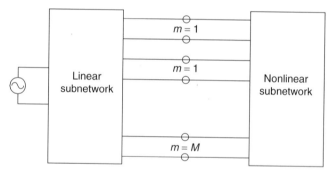

Figure 1.19   A general nonlinear network as partitioned for the harmonic balance analysis

harmonics) and a real phasor at DC ($n = 0$). The continuous curves in the previous figures have been plotted by means of eq. (1.69) and eq. (1.73) once the values of the phasors are known.

In general terms, a nonlinear circuit is divided into two parts connected by M ports (Figure 1.19): a part including only linear elements and a part including only nonlinear ones; the voltages at the connecting ports are expressed by Fourier series expansions:

$$v_m(t) = \mathrm{Re}\left[\sum_{n=0}^{N} V_{m,n} \cdot e^{jn\omega_0 t}\right] = V_{m,0} + \mathrm{Re}\left[\sum_{n=1}^{N} V_{m,n} \cdot e^{jn\omega_0 t}\right] = V_{m,0}$$

$$+ \sum_{n=1}^{N} \{V_{m,n}^r \cos(n\omega_0 t) - V_{m,n}^i \sin(n\omega_0 t)\} \quad m = 1, \ldots, M \tag{1.81}$$

The voltages at the connecting ports are the unknowns of Kirchhoff's node equations. In our formulation, the unknowns are actually the phasors that appear in their Fourier series expansion; since the series is truncated, they are $M \cdot (2N + 1)$. In vector form,

$$\vec{V} = \big[V_{1,0} \quad . \quad . \quad V_{M,0} \quad , \quad V_{1,1}^r \quad V_{1,1}^i \quad .. \quad V_{M,1}^r \quad V_{M,1}^i \quad , \quad V_{1,N}^r$$
$$V_{1,N}^i \quad .. \quad V_{M,N}^r \quad V_{M,N}^i \big]^T \tag{1.82}$$

The linear part of the circuit is replaced by its Norton equivalent; the currents flowing into it are computed by simple multiplication of the (still unknown) vector of the voltage phasors by the Norton equivalent admittance matrix, plus the (known) Norton equivalent current sources due to the input signal (Figure 1.20).

$$\vec{I}_L = \overset{\leftrightarrow}{Y} \cdot \vec{V} + \vec{I}_{L,0} \tag{1.83}$$

where

$$\vec{I} = \big[I_{1,0,L} \quad . \quad . \quad I_{M,0,L} \quad , \quad I_{1,1,L}^r \quad I_{1,1,L}^i \quad .. \quad I_{M,1,L}^r \quad I_{M,1,L}^i \quad , \quad I_{1,N,L}^r$$
$$I_{1,N,L}^i \quad .. \quad I_{M,N,L}^r \quad I_{M,N,L}^i \big]^T \tag{1.84}$$

When the voltages and currents are ordered as in eqs. (1.82) and (1.84), the admittance matrix, relative to the linear subcircuit, is block-diagonal

$$\overset{\leftrightarrow}{Y} = \begin{bmatrix} \overset{\leftrightarrow}{Y}_L(0) & 0 & 0 & 0 \\ 0 & \overset{\leftrightarrow}{Y}_L(\omega_0) & 0 & 0 \\ 0 & 0 & .. & 0 \\ 0 & 0 & 0 & \overset{\leftrightarrow}{Y}_L(N\omega_0) \end{bmatrix} \tag{1.85}$$

where

$$\overset{\leftrightarrow}{Y}_L(\omega) = \begin{bmatrix} y_{11}(\omega) & . & y_{1M}(\omega) \\ . & . & . \\ y_{M1}(\omega) & . & y_{MM}(\omega) \end{bmatrix} \tag{1.86}$$

is the $m \times m$ standard linear admittance matrix of the linear subnetwork at frequency ω.

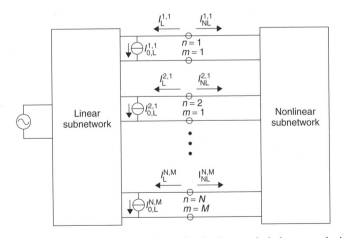

Figure 1.20   Currents and voltages for the harmonic balance analysis

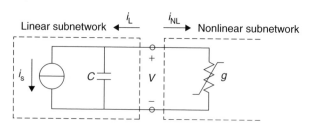

Figure 1.21   The example circuit partitioned for the harmonic balance analysis

In the case of our example circuit, the linear part is already a one-port Norton equivalent network (Figure 1.21).

The currents flowing into the nonlinear part of the circuit are computed as stated above. The time-domain currents are first computed at each connecting port

$$i_{m,\mathrm{NL}}(t) = G_m(\vec{v}(t)) \qquad m = 1, \ldots, M \tag{1.87}$$

where $G_m(v)$ is the nonlinear current–voltage characteristic of the nonlinear subnetwork at port m, and the voltage vector is

$$\vec{v}(t) = \begin{bmatrix} v_1(t) \\ .. \\ v_M(t) \end{bmatrix} \tag{1.88}$$

that is, the vector of the time-domain voltages at all ports. The latter is computed from the voltage phasors by means of an inverse Fourier transform:

$$\vec{v}(t) = \Im^{-1}(\vec{V}) \tag{1.89}$$

The phasors of the currents flowing into the nonlinear subnetwork are then computed by means of a Fourier transform:

$$I_{m,n,\mathrm{NL}} = \Im(i_{m,\mathrm{NL}}(t)) \tag{1.90}$$

In an actual harmonic balance algorithm, the time-domain voltages and currents are sampled at a set of time instant satisfying Nykvist's theorem, for calculation of the phasors by means of a DFT. A detailed description is not given here, but it can easily be deduced by generalisation to an M-port problem from the formulae reported in the Appendix A.4 for $M = 1$.

The solving system is now written at each connecting port and for each harmonic:

$$I^r_{m,n,\mathrm{L}} + I^r_{m,n,\mathrm{NL}} = 0 \tag{1.90a}$$

$$I^i_{m,n,\mathrm{L}} + I^i_{m,n,\mathrm{NL}} = 0 \qquad m = 1, \ldots, M \qquad n = 0, \ldots, N \tag{1.90b}$$

The unknowns of the system are the voltages, or more exactly the phasors of the truncated Fourier series expansions of the voltages at the ports connecting the linear and the nonlinear subnetworks. The values of the phasors are found by an iterative numerical algorithm, given the nonlinearity of the equations. The real and imaginary parts must be equated separately because of the non-analyticity of the dependence of currents on voltages, as stated above.

For the numerical analysis, the nonlinear equation system (1.90) is written as

$$\overline{I}_{\mathrm{L}}(\overline{V}) + \overline{I}_{\mathrm{NL}}(\overline{V}) = \overline{F}(\overline{V}) = \overline{0} \qquad (1.91)$$

This system is usually solved by means of the zero-searching iterative algorithm known as Newton–Raphson's method [1, 2, 41]. A first guess for the value of the voltage phasors must be given; let us call it

$$\vec{V}_{\mathrm{first\_guess}} = \vec{V}^{(0)} \qquad (1.92)$$

Obviously, this will not be the exact solution of eq. (1.91). An improved value will be found by the recursive formula

$$\overline{V}^{(k+1)} = \overline{V}^{(k)} - \left(\overline{J}^{(k)}\right)^{-1} \cdot F\left(\overline{V}^{(k)}\right) \qquad (1.93)$$

which is the vector form of the well-known Newton–Raphson's tangent method. The J matrix is the Jacobian matrix of eq. (1.91), corresponding to the derivative of the scalar function in a scalar Newton–Raphson's method:

$$\overset{\leftrightarrow}{J}(\vec{V}) = \frac{\partial \vec{F}(\vec{V})}{\partial \vec{V}} \qquad (1.94)$$

The Jacobian matrix can be computed analytically, if the nonlinear function is known in analytical form, or numerically by incremental ratio, if the nonlinearity is available as a look-up table or if analytical derivation is unpractical. The analytical derivation, however, has better numerical properties, and it is advisable when available. A more detailed description of the Jacobian matrix is given in the Appendix A.6. The inversion of the Jacobian matrix is a computationally heavy step of the algorithm; several approaches have been developed to improve its efficiency [42–44]. The algorithm will hopefully converge towards the correct solution, and will be stopped when the error decreases below a limit value. The error is actually the vector of the error currents, real and imaginary parts, at each node and for each harmonic frequency; convergence will be assumed to be reached when its norm will be lower than a desired accuracy level in the currents:

$$|\overline{F}^{(k)}| < \varepsilon \qquad (1.95)$$

The actual value of ε will normally vary with the current levels in the circuit: a value below 100 μA will probably be satisfactory in most cases. Alternatively, the algorithm is stopped when the solution does not vary any more:

$$|\overline{V}^{(k+1)} - \overline{V}^{(k)}| < \delta \qquad (1.96)$$

Once more, a reasonable value for δ depends on the voltage levels in the circuit, but a value below 1 mV will probably be adequate in most cases.

Another critical point in the algorithm is the choice of the first guess. A well-chosen first guess will considerably ease the convergence of the algorithm to the correct solution. If the circuit is mildly nonlinear, the linear solution, obtained for a low-level input, will probably be a good first guess. If the circuit is driven into strong nonlinearity, a continuation method will probably be the best approach. The level of the input signal is first reduced to a quasi-linear excitation and a mildly nonlinear analysis is performed; then, the input level is increased stepwise, using the result of the previous step as a first guess. In most cases the intermediate results will also be of practical interest, as in the case of a power amplifier driven from small-signal level into compression. Most commercially available CAD programmes automatically enforce this method when convergence becomes difficult or when it is not reached at all.

The described nodal formulation is based on Kirchhoff's voltage law: the unknown is the voltage, and the circuit elements are described as admittances. Alternatively, Kirchhoff's current law can be used, with the current being the unknown, and the circuit elements described as impedances. While no problem usually arises for the linear elements, the nonlinear elements are usually voltage-controlled nonlinear conductances (e.g. a junction, or the output characteristics of a transistor) or capacitances (e.g. junction capacitances in a diode or in a transistor). This is why the nodal formulation (KVL) is the standard form. However, any alternative form of Kirchhoff's equations is allowed as a basis for the harmonic balance algorithm in the cases in which the nonlinear elements have a different representation.

Another alternative formulation is obtained when the nonlinear equation (1.13) is rewritten as

$$C \cdot \left.\frac{dv(t)}{dt}\right|_{t=t_k} + i_{max} \cdot tgh\left(\frac{g \cdot v(t_k)}{i_{max}}\right) + i_s(t_k) = 0 \qquad k = 0, 1, \ldots, 2N \qquad (1.97)$$

where the unknowns are the time-domain voltage samples at the $2N + 1$ sampling instants:

$$v_k = v(t_k) \qquad k = 0, 1, \ldots, 2N \qquad (1.98)$$

In this formulation, eq. (1.13) must be satisfied only at $2N + 1$ time instants. The formulation is similar to that of the time-domain solution (Section 1.2), but in this case the derivative with respect to time is expressed as

$$\left.\frac{dv(t)}{dt}\right|_{t=t_k} = \left.\frac{d \sum_{n=-N}^{N} V_n \cdot e^{jn\omega_0 t}}{dt}\right|_{t=t_k} = \sum_{n=-N}^{N} jn\omega_0 V_n \cdot e^{jn\omega_0 t_k} \qquad (1.99)$$

according to the assumption of a periodic solution with limited bandwidth. The voltage phasors are computed from the time-domain voltage samples by means of a DFT:

$$v_k = v(t_k) \Rightarrow \Im \Rightarrow V_n \qquad (1.100)$$

The nonlinear equation system (1.97) is again solved by an iterative numerical method. This formulation of the nonlinear problem is known as *waveform balance*, since in eq. (1.97) the current waveforms of the linear and nonlinear subcircuits must be balanced at a finite set of time instants. In fact, it can easily be seen that the standard harmonic balance formulation also satisfies eq. (1.13) only at the sampling instants where the DFT is computed.

There are two other formulations of the kind: in the first, Kirchhoff's equations are written in the time domain (eq. (1.97) above), but the unknowns are the voltage phasors; in the second, Kirchhoff's equations are written in the frequency domain (eq. (1.77) or eq. (1.90) above), and the unknowns are the time-domain voltage samples. The four formulations are actually completely equivalent, at least in principle; one or the other may be more convenient in some cases, when special problems must be dealt with.

1.3.2.2 Multi-tone analysis

So far, only strictly periodic excitation and steady-state have been considered. In the real world, however, many important phenomena occur when two or more periodic signals with different periods excite a nonlinear circuit, as shown in Section 1.3.1 on the Volterra series. In some cases a single-tone analysis gives enough information to the designer, but in many other cases a more realistic picture is needed, especially when distortion or intermodulation is a critical issue. Moreover, the behaviour of circuits like mixers can by no means be reduced to a simply periodic one. A first step towards a more realistic picture is the introduction of a more complex Fourier series for the signal, composed of two tones:

$$v(t) = \sum_{n_1=-\infty}^{\infty} \sum_{n_2=-\infty}^{\infty} V_{n_1,n_2} \cdot e^{j(n_1\omega_1+n_2\omega_2)t} = \sum_{n_1=-\infty}^{\infty} \sum_{n_2=-\infty}^{\infty} V_{n_1,n_2} \cdot e^{j\omega_{n_1,n_2}t} \qquad (1.101)$$

where the two frequencies ω_1 and ω_2 are the frequencies of the input signal or signals: for instance, two equal tones at closely spaced frequencies in the case of intermodulation analysis in a power amplifier; or the local oscillator and the RF signal in the case of a mixer. The unknowns of the problem are still the phasors of the voltage, but now they are not relative to the harmonics of a periodic signal: they rather represent a complex spectrum, as shown in the case of Volterra series analysis. The series in eq. (1.101) must be truncated so that only important terms are retained: a proper choice increases the accuracy of the analysis while limiting the numerical effort.

If the two basic frequencies ω_1 and ω_2 are incommensurate, the signal is said to be quasi-periodic. On the other hand, when the two basic frequencies ω_1 and ω_2 are commensurate, they can be considered as the harmonics of a dummy fundamental frequency ω_0, and the problem can formally be taken back to the single-tone case [45]. However, if the two basic frequencies are closely spaced or are very different from one another, a very large number of harmonics must be included in the analysis. For instance, when two input tones at 1 GHz and 1.01 GHz are applied for intermodulation analysis of an amplifier, at least *300* harmonics of the signal at the dummy 10 MHz fundamental frequency must be included for third-order analysis of the signal. This redundance can be reduced by retaining only the meaningful terms in the Fourier series expansion; in this case, however, the DFT described above experiences the same severe numerical problems as in the quasi-periodic case, as explained in the following. A special two-tones form of the harmonic balance algorithm is therefore usually adopted also in these cases.

The formalism for two-tone analysis is easily extended to multi-tone analysis, when more than two periodic signals at different frequencies are present in the circuit; however, the computational burden increases quickly, usually limiting the effective analysis capabilities to no more than three tones. For more complex signals, different techniques are used to extend the algorithm, which are shortly described in the following paragraphs.

The harmonic balance method requires some adjustments for two-tone analysis. First of all, a suitable truncation of the Fourier series must be defined [11]. The expansion of a single-tone signal is truncated so that the neglected harmonics have negligible amplitude. The same principle holds for a multi-tone analysis. The frequency spectrum includes all the frequencies that are combinations of the two basic frequencies, or more than two for multi-tone analysis:

$$\omega_{n_1,n_2} = n_1\omega_1 + n_2\omega_2 \qquad (1.102)$$

The sum of the absolute values of the two indices $n = |n_1| + |n_2|$ is the order of the harmonic component.

Not all the lines of the spectrum, however, have significant amplitude. It is a reasonable assumption that the amplitude of a spectral component decreases as its order increases. However, the picture can vary for different cases. A typical example is given in Figure 1.12, where two equal-amplitude signals at closely spaced frequencies are fed to a power amplifier, generating distortion. A partially different situation occurs when a mixer is considered. Typically, the local oscillator has a much higher amplitude than the input signal (e.g. at RF) or the output signal (e.g. at IF), and the situation is rather as in Figure 1.22.

A first example of truncation of the expansion is the so-called box truncation scheme: all the combinations of the two indices n_1 and n_2 are retained for values of the indices less than N_1 and N_2 respectively. This truncation can be illustrated graphically as shown in Figure 1.23.

Since real signals have Hermitean spectral coefficients, only half of them are actually needed. This is obtained, for instance, by retaining only the terms whose indices satisfy the following conditions:

$$n_1 \geq 0; \quad n_2 \geq 0 \text{ if } n_1 = 0 \qquad (1.103)$$

The resulting reduced spectrum is shown in Figure 1.24.

In this case the terms with maximum order are those with $n_1 = N_1$, $n_2 = \pm N_2$ and $n_{max} = N_1 + N_2$. The number of terms retained in the Fourier series expansion for a real signal is $2 \cdot N_1 \cdot N_2 + N_1 + N_2 + 1$.

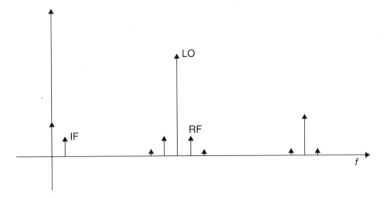

Figure 1.22   Typical spectrum of the electrical quantities in a mixer

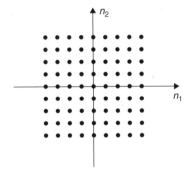

Figure 1.23   Box truncation scheme for two-tone Fourier series expansion

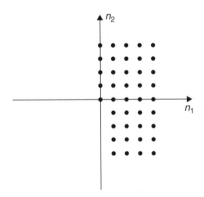

Figure 1.24   Reduced box truncation scheme for real signals

An example of the spectrum resulting from a box truncation scheme with $N_1 = 3$ and $N_2 = 2$ is shown in Figure 1.25, where frequency f_1 is much larger than frequency f_2; both the complete (light) and reduced (dark) spectra are depicted.

The choice of a box truncation is very simple, but not necessarily the most effective one. As said above, a reasonable assumption is that the amplitude of a spectral component

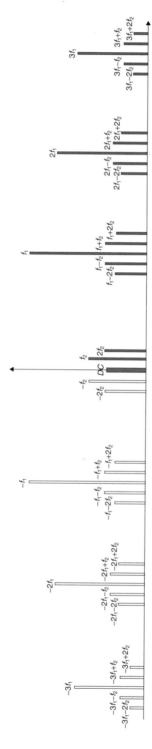

Figure 1.25 Complete (light) and reduced (dark) spectra of a two-tone signal after box truncation with $N_1 = 3$ and $N_2 = 2$

decays with its order. A reasonable truncation scheme therefore drops all terms with $n > n_{max}$, retaining all those with $n \leq n_{max}$. This is called diamond truncation scheme, as apparent from Figure 1.26; the scheme for real signals is also indicated.

The number of terms retained in the Fourier series expansion is approximately one half that of the box truncation scheme. An example of the spectrum resulting from a diamond truncation scheme with $N_1 = N_2 = 3$ is shown in Figure 1.27.

A further variation of the truncation scheme is illustrated in Figure 1.28.

This scheme allows an independent choice of the number of harmonics of the two input tones, and of the maximum number of intermodulation products as in the diamond truncation scheme. An example of the spectrum resulting from a mixed truncation scheme is shown in Figure 1.29.

The general structure of the harmonic balance algorithm, as described above, still holds. The main modification is related to the Fourier transform that becomes severely inaccurate unless special schemes are used. The main difficulty is related to the choice of the sampling time instants. In principle, a number of time instants equal to the number of variables to be determined (the coefficients in the Fourier series expansion in eq. (1.101)) always allows for a Fourier transformation from time to frequency domain, by inversion of a suitable matrix similar to that described in the Appendix A.4. However, the matrix becomes very ill-conditioned unless the sampling time instants are

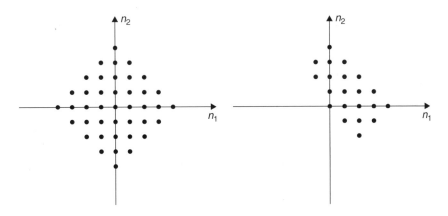

Figure 1.26 Diamond truncation scheme for general (a) and real signals (b), and $N_1 = N_2$

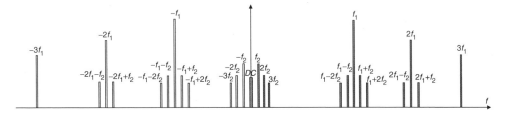

Figure 1.27 Spectrum relative to the diamond truncation scheme with $N_1 = N_2 = 3$

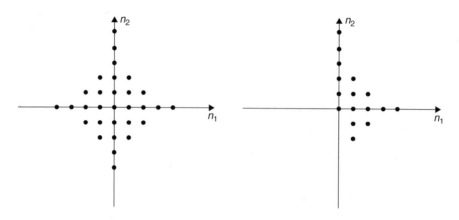

Figure 1.28   Mixed truncation scheme general (a) and real signals (b)

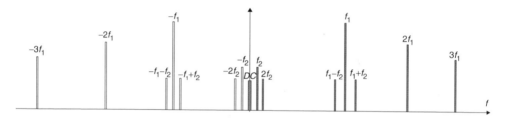

Figure 1.29   Spectrum relative to the mixed truncation scheme

properly chosen. Several schemes have been proposed for overcoming the problem: over-sampling and orthonormalisation [46–50], multi-dimensional Fourier transform [51–52], frequency remapping [37, 53, 54], and others [55, 56]. They are described in some detail in the following text.

The multi-dimensional Fourier transform is actually defined for a function $\bar{v}(t_1, t_2, \ldots)$ of several variables, each with its own periodicity; we limit the number of variables to only two in our case for simplicity of notation.

$$\bar{v}(t_1, t_2) = \sum_{n_1=-\infty}^{\infty} \sum_{n_2=-\infty}^{\infty} \bar{V}_{n_1,n_2} \cdot e^{j(n_1\omega_1 t_1 + n_2\omega_2 t_2)} \tag{1.104}$$

Each variable is sampled over its own periodicity, in analogy with what has been described above: in this way, a two-dimensional grid of samples is obtained. If the signal has a limited frequency spectrum, and if the number of samples satisfies Nykvist's theorem, we can compute the two-dimensional grid of coefficients in the two-dimensional Fourier series expansion (eq. (1.101)); the details are given in the Appendix A.7. The samples are taken at the sampling time instants,

$$t_{k_1} = \frac{T_1}{2N_1 + 1} \cdot k_1, \quad k_1 = -N_1, \ldots, N_1, \quad t_{k_2} = \frac{T_2}{2N_2 + 1} \cdot k_2, \quad k_2 = -N_2, \ldots, N_2 \tag{1.105}$$

summing up to a number of samples:

$$N_{tot} = (2N_1 + 1)(2N_2 + 1) \tag{1.106}$$

Once the phasors are computed, the original two-tone voltage is readily obtained as

$$v(t) = \overline{v}(t, t) \tag{1.107}$$

as can be seen from eq. (1.104). This transform is widely used in commercial simulators.

As stated above, the main problem in a multi-tone harmonic balance analysis lies in the difficult choice of the sampling time instants for Fourier transformation. An improper choice will lead to a severely ill-conditioned DFT matrix. An effective and simple strategy consists of the random selection of a number of sampling time instants two or three times in excess of the minimum required by Nykvist's theorem. The system of equations relating the sampled values and the coefficients of the Fourier series expansion (see Appendix A.8) therefore becomes rectangular, having more equations than unknowns, and the unknown coefficients are overdetermined. Among all equations, only the 'best' ones are retained to form a square system suitable for inversion; the other equations, in excess of the minimum number and the corresponding time samples, are discarded. The 'best' equations are selected on the basis of their orthogonality, in order to have a well-conditioned system of equations. A standard orthonormalisation scheme is described in the Appendix A.8.

In Figure 1.30, the solution of our example circuit is given for the following values of the circuit elements:

$$g = 10 \text{ mS} \quad C = 500 \text{ fF} \quad f_1 = 1 \text{ GHz} \quad f_2 = 1.05 \text{ GHz}$$

with a box truncation scheme with $n_{max} = 5$; the waveforms are oversampled by a factor 6. The plots show input current (———), output voltage (------) and current in the nonlinear resistor (- - - -), for an input current of $i_{s,1} = i_{s,2} = 100$ mA.

In Figure 1.31, the spectra of voltages and currents in the example circuit are given.

In order to clarify the oversampling principle, the current waveform is shown in Figure 1.32; the samples taken at uniform times are shown as black circles, the randomly taken samples are shown as black crosses, while the selected samples after orthonormalisation are shown as grey circles dotted.

For comparison, voltages and currents in the same circuit are shown in Figure 1.33 as computed with a time-domain analysis with uniform step; two pseudoperiods have been computed, and are shown in the figure.

It has been stated above that the analysis of a system driven by two (or more) tones with commensurate frequencies can be approached by a single-tone analysis by choosing the minimum common divider of the frequencies as the fundamental frequency of the analysis. As said, this may lead to an impractically high numbers of harmonics to be included in the analysis. An alternative approach to reduce the number of harmonics within an equivalent single-tone analysis, not limited to incommensurate frequencies,

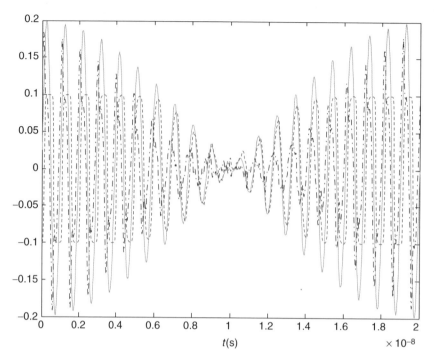

Figure 1.30 Voltages and currents in the example circuit for a two-tone input signal

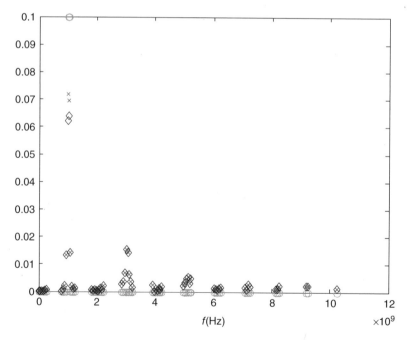

Figure 1.31 Spectra of voltages (crosses) and currents (diamonds) in the example circuit for a two-tone input signal

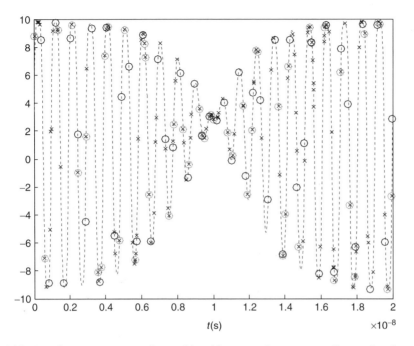

Figure 1.32 Nonlinear current waveform with uniform samples, oversampling, and optimum samples after orthonormalisation

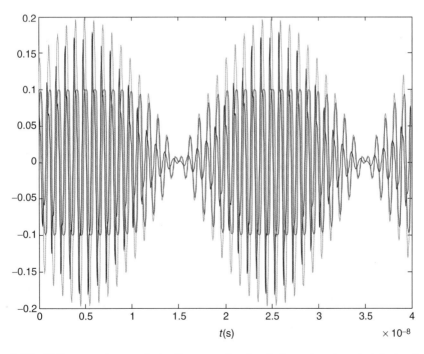

Figure 1.33 Voltages and currents in the example circuit for a two-tone input signal and a time-domain analysis

requires the remapping of the multi-tone frequencies. It can be seen (Section 1.3.1) that a resistive nonlinear element generates a spectrum that includes the sum and difference frequencies of the input ones; this is true independently of the actual values of the frequencies. We can therefore replace the two input frequencies by another couple of (commensurate) input frequencies such that their harmonics and intermodulation products occupy the harmonics of a single fundamental frequency, provided that the correspondence with the original ones is univocal, and that the resulting spectrum is dense. Since the new remapped fundamental frequencies are arbitrary, they can be integer numbers for convenience.

Let us illustrate this with an example. Suppose that two input tones at 100 MHz and 2 GHz are fed to a nonlinear circuit, and that we want to adopt a box truncation scheme with $N_1 = 3$ and $N_2 = 5$; a typical application could be an up-converting mixer from 100 MHz to, for example, 2.1 GHz. The maximum order of the intermodulation products is $n_{max} = N_1 + N_2 = 8$. Two suitable remapped basis frequencies can be chosen as

$$f_1' = 1 \text{ Hz} \quad \text{and} \quad f_2' = n_{max} - 1 = 7 \text{ Hz}$$

The remapped spectrum is obtained in the same way as the original from the two remapped fundamental frequencies:

$$f' = n_1 \cdot f_1' + n_2 \cdot f_2' \tag{1.108}$$

This relation establishes a univocal correspondence and produces a dense spectrum, as shown in Table 1.1; the correspondence is also depicted in Figure 1.34. It is also apparent that all the spectral lines are the harmonics of the remapped fundamental frequency f_2'. The analysis can now be performed by means of a standard single-tone algorithm, as described earlier, with a fundamental frequency $f_0 = f_1' = 1$ Hz and a maximum number of harmonics $N_{max} = 38$ to be included in the analysis.

Similar schemes can be found for different truncation methods [53], even though not always a dense remapped spectrum is obtained.

Table 1.1   The remapped frequencies

f' (Hz)	n_1	n_2	f (MHz)	f'	n_1	n_2	f	f'	n_1	n_2	f	f'	n_1	n_2	f
0	0	0	0	11	−3	2	3700	22	1	3	6100	33	−2	5	9800
1	1	0	100	12	−2	2	3800	23	2	3	6200	34	−1	5	9900
2	2	0	200	13	−1	2	3900	24	3	3	6300	35	0	5	10000
3	3	0	300	14	0	2	4000	25	−3	4	7700	36	1	5	10100
4	−3	1	1700	15	1	2	4100	26	−2	4	7800	37	2	5	10200
5	−2	1	1800	16	2	2	4200	27	−1	4	7900	38	3	5	10300
6	−1	1	1900	17	3	2	4300	28	0	4	8000				
7	0	1	2000	18	−3	3	5700	29	1	4	8100				
8	1	1	2100	19	−2	3	5800	30	2	4	8200				
9	2	1	2200	20	−1	3	5900	31	3	4	8300				
10	3	1	2300	21	0	3	6000	32	−3	5	9700				

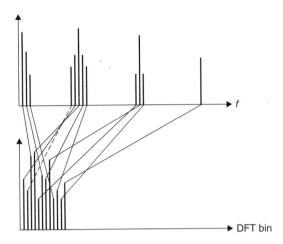

Figure 1.34   The original and remapped frequency spectrum

1.3.2.3 Envelope analysis

Harmonic balance algorithms so far described are quite general and flexible. However, there are a few types of signals that are very difficult or nearly impossible to treat with them: in particular, periodic or quasi-periodic signals with slowly varying amplitude, as in phase-lock loops during locking, or in variable-gain amplifiers, or in narrowband multi-carrier communications systems; or signals that cannot be easily represented by sine-wave-based representations as digitally modulated signals; all these cannot be easily handled by what has been seen so far. A harmonic balance–based approach has been developed for these cases, which treats the slowly varying amplitude (or envelope) of the fast carrier signals separately from the carrier themselves [57–61].

We shortly outline in the following the algorithm for a single carrier modulated by a slowly varying 'envelope' signal for our test circuit; extension to a multi-carrier signal in a general nonlinear circuit is straightforward. For this signal, the expressions in eq. (1.69) are replaced by

$$v(t) = \sum_{n=-\infty}^{\infty} V_n(t) \cdot e^{jn\omega_0 t} \qquad i_s(t) = I_s(t) \cdot \cos(\omega_0 t) \tag{1.109}$$

where the phasors $V_n(t)$ and $I_s(t)$ are assumed to vary slowly with respect to the period of the carrier $T_0 = \dfrac{2\pi}{\omega_0}$. Kirchhoff's nodal equation (1.70) becomes

$$C \cdot \frac{d\left(\displaystyle\sum_{n=-\infty}^{\infty} V_n(t) \cdot e^{jn\omega_0 t}\right)}{dt} + i_{max} \cdot tgh\left(\frac{g \cdot \displaystyle\sum_{n=-\infty}^{\infty} V_n(t) \cdot e^{jn\omega_0 t}}{i_{max}}\right)$$

$$+ I_s(t) \cdot \left(\frac{e^{j\omega_0 t} + e^{-j\omega_0 t}}{2}\right) = 0 \tag{1.110}$$

In analogy with eq. (1.109), we can rewrite eq. (1.72) as

$$i_g(t) = i_{max} \cdot tgh \left(\frac{g \cdot \sum_{n=-\infty}^{\infty} V_n(t) \cdot e^{jn\omega_0 t}}{i_{max}} \right) = \sum_{n=-\infty}^{\infty} I_{g,n}(t) \cdot e^{jn\omega_0 t} \qquad (1.111)$$

Consequently, eq. (1.110) becomes

$$C \cdot \frac{d\left(\sum_{n=-\infty}^{\infty} V_n(t) \cdot e^{jn\omega_0 t} \right)}{dt} + \sum_{n=-\infty}^{\infty} I_{g,n}(t) \cdot e^{jn\omega_0 t} + I_s(t) \cdot \left(\frac{e^{j\omega_0 t} + e^{-j\omega_0 t}}{2} \right) = 0$$

$$(1.112)$$

By part differentiation of the first term, eq. (1.112) becomes

$$\sum_{n=-\infty}^{\infty} C \frac{dV_n(t)}{dt} \cdot e^{jn\omega_0 t} + \sum_{n=-\infty}^{\infty} jn\omega_0 C \cdot V_n(t) \cdot e^{jn\omega_0 t} + \sum_{n=-\infty}^{\infty} I_{g,n}(t) \cdot e^{jn\omega_0 t}$$

$$+ I_s(t) \cdot \left(\frac{e^{j\omega_0 t} + e^{-j\omega_0 t}}{2} \right) = 0 \qquad (1.113)$$

By separating the harmonics in a way similar to what has been done for eq. (1.74), eq. (1.113) becomes a system of equations:

$$C \cdot \frac{dV_n(t)}{dt} + jn\omega_0 C \cdot V_n(t) + I_{g,n}(t) + \frac{I_s}{2}(t) = 0 \qquad n = -\infty, \ldots, 0, \ldots, \infty \quad (1.114)$$

In vector notation,

$$C \cdot \frac{d\vec{V}(t)}{dt} + j\Omega C \cdot \vec{V}(t) + \vec{I}_{g,n}(t) + \vec{I}_s(t) = 0 \qquad (1.115)$$

This is a nonlinear differential equation system in the unknown vector of the voltage phasors $\vec{V}(t)$, which is a function of time. The system can be solved by direct time integration, for example, by the backward Euler implicit formulation:

$$C \cdot \frac{\vec{V}(t_k) - \vec{V}(t_{k-1})}{t_k - t_{k-1}} + j\Omega C \cdot \vec{V}(t_k) + \vec{I}_{g,n}(t_k) + \vec{I}_s(t_k) = 0 \qquad k = 0, 1, \ldots \quad (1.116)$$

As has been said in Section 1.2, at each time step t_k a system of nonlinear equations must be solved, which actually is a harmonic balance system of equations. The analysis is therefore transformed in a succession of harmonic balance problems.

We can now attempt an intuitive explanation of this formalism. Since the envelope of the signal, and therefore the phasors of the carrier frequency, vary slowly with respect to the carrier, we can assume that value of the phasors is almost constant over several periods of the carrier. We can therefore sample them at some time t_1, and keep these values constant for a time interval up to some other time instant t_2, such that

$$t_2 - t_1 \gg T_0 \qquad (1.117)$$

Equation (1.110) can therefore be rewritten as

$$C \cdot \frac{d\left(\sum_{n=-\infty}^{\infty} V_n(t_k) \cdot e^{jn\omega_0 t}\right)}{dt} + i_{max} \cdot tgh\left(\frac{g \cdot \sum_{n=-\infty}^{\infty} V_n(t_k) \cdot e^{jn\omega_0 t}}{i_{max}}\right)$$

$$+ I_s(t_k) \cdot \left(\frac{e^{j\omega_0 t} + e^{-j\omega_0 t}}{2}\right) = 0 \qquad k = 1, 2, \ldots \qquad (1.118)$$

This equation is equivalent to a harmonic balance (Kirchhoff's) equation for a single-tone excitation, for each time interval during which the envelope is assumed to be constant. However, even if the envelope varies slowly with respect to the carrier, this is not true with respect to the reactances of the circuit. They keep memory of its past behaviour, and affect the envelope behaviour as described by the differential equation (1.115). A pictorial representation of the method is shown in Figure 1.35.

1.3.2.4 Additional remarks

The harmonic balance method has several advantageous features that have determined its success: the linear parts of the circuit are reduced to an equivalent network that is

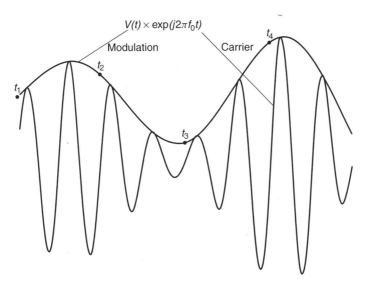

Figure 1.35 Waveforms in an envelope harmonic balance analysis

evaluated in the frequency domain. This allows the reduction of usually large passive subnetworks to a minimum number of connecting nodes, or ports, and the reduction of numerical complexity. Moreover, the evaluation in frequency domain is very practical for most linear microwave components, both lumped and distributed. On the other hand, the voltage–current characteristics of the nonlinear elements can be represented by any function, even numerically by means of look-up tables with interpolation, provided that it is continuous; however, the numerical solution of the system has better properties if the first derivative is also continuous.

There are, however, also drawbacks. First of all, it is not possible to detect instabilities of the circuit at frequencies not correlated to the excitation frequency, at least with this simple formulation. This is a natural consequence of the assumption of periodic voltages with the same period of excitation. On the other hand, if the circuit is unstable at any harmonic frequency of the excitation, the iterative numerical algorithm does not converge. However, the opposite is not true: the algorithm can fail to converge for other reasons. With harmonic balance analysis, the study of the stability of the circuit must be performed with special methods, which will be dealt with in Chapter 5.

Another drawback is the difficulty to represent a frequency dispersive behaviour of the nonlinear device; this is a natural consequence of the time-domain analysis of the nonlinear subnetwork. Special representations of the active device can however help with this problem.

An obvious limitation of the method is that only periodic, steady state circuits can be analysed. In fact, transients such as, for example, the step response can be analysed by periodic repetition of the step [62] (Figure 1.36).

The repetition time must be longer than the transient phenomena, and the duty cycle must be such that the DC component is close to the actual one. However, the number of harmonics required for an accurate analysis of a step makes this method unpractical in many cases. Moreover, care must be taken in order to define a correct DC value for the excitation.

1.3.2.5 Describing function

A simplified version of the harmonic balance algorithm has been developed and used in the past, and it has been long neglected for CAD applications [63]. Basically, referring to eq. (1.69), the higher harmonics of the voltage are neglected, and the latter is assumed

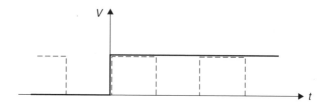

Figure 1.36 Periodic repetition of a step excitation for analysis with the harmonic balance method

to be a single-frequency signal:

$$v(t) \cong V \cdot \cos(\omega_0 t + \varphi_v) \qquad i_s(t) = I_s \cdot \cos(\omega_0 t) \qquad (1.119)$$

Obviously, this is an approximation. Accordingly, eq. (1.71) becomes

$$\omega_0 C \cdot V \cdot \sin(\omega_0 t + \varphi_v) + I_g \cdot \cos(\omega_0 t + \varphi_g) + I_s \cdot \cos(\omega_0 t) = 0 \qquad (1.120)$$

where the current in the nonlinear element is computed through the time domain

$$i_g(t) = i_{max} \cdot tgh \left(\frac{g \cdot V \cdot \cos(\omega_0 t + \varphi_v)}{i_{max}} \right) \cong I_g \cdot \cos(\omega_0 t + \varphi_g) \qquad (1.121)$$

and the phasor of the current is still computed by means of a Fourier transform, as in a standard harmonic balance algorithm:

$$i_g(t) \Rightarrow \Im \Rightarrow I_g \qquad (1.122)$$

but only the first term is retained now. In fact, the current in the nonlinear element is computed with all harmonics as a nonlinear response to the applied sinusoidal voltage; only its fundamental-frequency sinusoidal component is retained for solving (balancing) Kirchhoff's equation (eq. (1.120)). In electrical terms, the nonlinear conductance is considered as a linear equivalent large-signal conductance for a fundamental-frequency sinusoidal signal. The unknown voltage is found by solving eq. (1.120). By repeating the analysis for increasing amplitudes of the input current, the relation between output voltage and input current is numerically found; when voltage and current are expressed as phasors, the relation relates complex numbers and can be written as

$$V = DF(I_s) \qquad (1.123)$$

The complex function DF is the describing function. In practice, it is of practical importance when the linear part of the circuit behaves as a narrowband filtering structure that filters out the harmonics generated inside the nonlinear element. It is the simulation equivalent of the popular AM/AM, AM/PM experimental characterisation of narrowband amplifiers or nonlinear systems in general. The practical application of this approach extends to the case of slowly modulated sinusoidal signal in a narrowband circuit: a formulation very similar to the envelope analysis can be set up for the describing function also, with big savings in terms of computation time.

1.3.2.6 Spectral balance

Yet another different approach is obtained if the nonlinear element has a polynomial current–voltage characteristic (eq. (1.44)) [64, 65]:

$$i_g(v) = g_0 + g_1 \cdot v + g_2 \cdot v^2 + g_3 \cdot v^3 + \cdots \qquad (1.124)$$

In this case, the nonlinear Kirchhoff's equation (eq. (1.13)) reads as in eq. (1.57):

$$C \cdot \frac{dv(t)}{dt} + g_1 \cdot v(t) + g_2 \cdot v^2(t) + \cdots + i_s(t) = 0 \qquad (1.125)$$

The voltage is once more expanded in Fourier series, then replaced into eq. (1.125):

$$C \cdot \frac{d\left(\sum_{n=-\infty}^{\infty} V_n \cdot e^{jn\omega_0 t}\right)}{dt} + g_1 \cdot \sum_{n=-\infty}^{\infty} V_n \cdot e^{jn\omega_0 t} + g_2 \cdot \left(\sum_{n=-\infty}^{\infty} V_n \cdot e^{jn\omega_0 t}\right)^2$$

$$+ \cdots + I_s \cdot \left(\frac{e^{j\omega_0 t} + e^{-j\omega_0 t}}{2}\right) = 0 \qquad (1.126)$$

The system must be brought to the form of eq. (1.71):

$$\sum_{n=-\infty}^{\infty} jn\omega_0 C \cdot V_n \cdot e^{jn\omega_0 t} + \sum_{n=-\infty}^{\infty} I_{g,n} \cdot e^{jn\omega_0 t} + I_s \cdot \left(\frac{e^{j\omega_0 t} + e^{-j\omega_0 t}}{2}\right) = 0 \qquad (1.127)$$

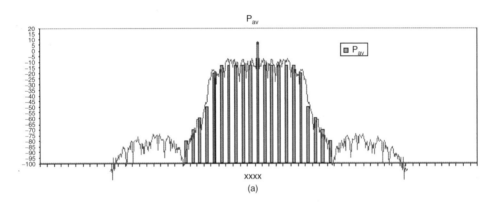

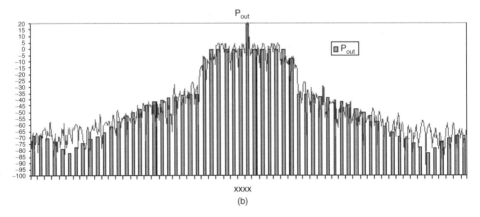

Figure 1.37 Spectral regrowth of a modulated signal computed by a multi-tone spectral balance algorithm

In this case, however, the phasors of the nonlinear current $I_{g,n}$ are analytically computed from the powers of the voltage Fourier series (see the 'probing method' for the Volterra series), without any Fourier transform. In fact, the current phasor computation requires a good deal of formalism, but it can be easily handled by suitable numerical arrangements. The rest of the procedure is similar to the standard harmonic balance method, and the nonlinear equation system is formally identical to eq. (1.77) or eq. (1.90). The method is called spectral balance because only manipulations of spectra are involved, and no time-domain waveforms are computed [66–70].

The method is very efficient especially for multi-tone input signal, up to several tens of input tones. By way of illustration, the spectral regrowth of a pseudorandom modulated signal is shown in Figure 1.37. Both the input and the output to the nonlinear system are shown, as computed by a multi-tone spectral balance algorithm.

1.4 THE CONVERSION MATRIX

In this paragraph, a linearised representation of a nonlinear circuit in large-signal operations is described for small-signal applications at non-harmonic frequencies. This method is also called the large-signal/small- signal analysis.

So far, one or more large signals have been applied to a nonlinear circuit. The case when one signal is large and another is small has important applications, and will be described in the following. Typically, the large signal drives the nonlinear element(s) into nonlinear operations and must be treated numerically as seen above; the small signal applies a small perturbation to the nonlinear operating regime, which can be linearised if its amplitude is small enough. In fact, this is also the case of standard small-signal S-parameters or any other equivalent small-signal parameters: the large signal is the bias voltages at zero frequency (DC), and the small signal is any signal with a generic spectrum. The large-signal operating point, that is, the DC quiescent point, is found by means of nonlinear analysis, typically the graphical load-line method, or a numerical iterative method as in direct time-domain analysis; the device is linearised around it by means of small perturbations, which in a standard experimental set-up is sinusoidal (vector network analyser with bias Ts). In the case of the conversion matrix, the large signal is periodic and the operating regime is not a DC quiescent point but a time-varying periodic state, and the effect of the perturbation must be computed with some care [71–74].

Let us illustrate this case with our example circuit. The circuit is driven into nonlinear, periodic regime by a large sinusoidal signal, in our case a large-amplitude current generator. Voltage and currents have already been found with several methods; they can be expressed as Fourier series expansions of the form

$$v_{LS}(t) = \sum_{n=-\infty}^{\infty} V_{LS,n} \cdot e^{jn\omega_{LS}t} \tag{1.128}$$

$$i_{C,LS}(t) = C \cdot \frac{dv_{LS}(t)}{dt} = \sum_{n=-\infty}^{\infty} jn\omega_{LS}C \cdot V_{LS,n} \cdot e^{jn\omega_{LS}t} \tag{1.129}$$

$$i_{g,LS}(t) = i_{max} \cdot tgh \left(\frac{g \cdot v_{LS}(t)}{i_{max}} \right) = i_{max} \cdot tgh \left(\frac{g \cdot \displaystyle\sum_{n=-\infty}^{\infty} V_{LS,n} \cdot e^{jn\omega_{LS}t}}{i_{max}} \right)$$

$$= \sum_{n=-\infty}^{\infty} I_{g,LS,n} \cdot e^{jn\omega_{LS}t} \tag{1.130}$$

where the subscript LS has been added to identify the large-signal quantities.

Let us now add a small input current $i_{ss}(t)$ (Figure 1.38).

The voltage will be perturbed by a small component $v_{ss}(t)$:

$$v(t) = v_{LS}(t) + v_{ss}(t) \tag{1.131}$$

The currents in the capacitance and in the nonlinear resistor will also be perturbed by small components; in the case of the linear elements (in this case the capacitance), it is immediately found by the superposition principle:

$$i_C(t) = i_{C,LS}(t) + i_{C,ss}(t) = C \cdot \frac{dv_{LS}(t)}{dt} + C \cdot \frac{dv_{ss}(t)}{dt} \tag{1.132}$$

The small perturbation component of the current in the nonlinear elements (in this case the nonlinear resistor) is computed by linearisation around the steady state:

$$i_g(t) = i_{g,LS}(t) + i_{g,ss}(t) \cong i_{g,LS}(t) + \left. \frac{di_g(v)}{dv} \right|_{v=v_{LS}(t)} \cdot v_{ss}(t) + \cdots \tag{1.133}$$

From an electrical point of view, the linearised perturbation can be seen as a small deviation from the 'bias' point (i_{LS}, v_{LS}) along the nonlinear I/V characteristic of the resistor (Figure 1.39).

If the perturbation is small enough, the I/V curve can be replaced by its tangent in the 'bias' point $v = v_{LS}$, whose slope is the dynamic conductance. The small perturbation current is therefore expressed as the perturbation voltage times the dynamic conductance:

$$i_{g,ss}(t) \cong \left. \frac{di_g(v)}{dv} \right|_{v=v_{LS}(t)} \cdot v_{ss}(t) = g_{ss}(v)|_{v=v_{LS}(t)} \cdot v_{ss}(t) = g_{ss}(t) \cdot v_{ss}(t) \tag{1.134}$$

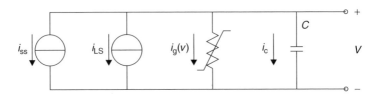

Figure 1.38   The example circuit for the calculation of the conversion matrix

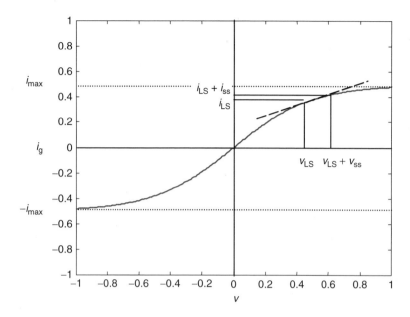

Figure 1.39   Definition of the dynamic conductance

Since the 'bias' point is time varying, the small-signal dynamic conductance also varies with time. For our example circuit (Figure 1.40),

$$g_{ss}(v) = \frac{di_g(v)}{dt} = g \cdot \left(1 - tgh^2\left(\frac{g \cdot v}{i_{max}}\right)\right) \tag{1.135}$$

Since the large-signal voltage is known and periodic, the behaviour of the dynamic conductance in the time domain is known and periodic:

$$g_{ss}(t) = g_{ss}(v_{LS}(t)) = g \cdot \left(1 - tgh^2\left(\frac{g \cdot \sum\limits_{n=-\infty}^{\infty} V_{LS,n} \cdot e^{jn\omega_{LS}t}}{i_{max}}\right)\right) = \sum\limits_{m=-\infty}^{\infty} G_{ss,m} \cdot e^{jm\omega_{LS}t}$$

$$\tag{1.136}$$

In the above formula, the dynamic conductance has been expanded in Fourier series with respect to time, where the coefficients $G_{ss,m}$ are computed by means of a Fourier transform.

The small perturbation currents must fulfil Kirchhoff's current law,

$$i_{ss}(t) + i_{C,ss}(t) + i_{g,ss}(t) = 0 \tag{1.137}$$

for all time instants; in our case,

$$i_{ss}(t) + C \cdot \frac{dv_{ss}(t)}{dt} + g(t) \cdot v_{ss}(t) = 0 \tag{1.138}$$

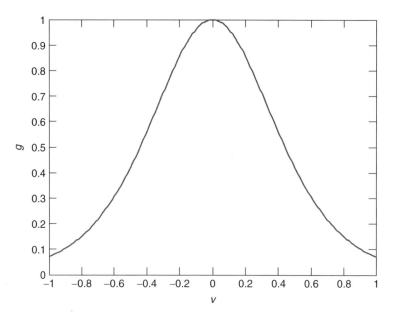

Figure 1.40   Dynamic conductance as a function of large-signal voltage

where $i_{ss}(t)$ is the known excitation and $v_{ss}(t)$ is the unknown voltage. The equation is linear with time-varying coefficients (the dynamic conductance of the nonlinear resistance); it corresponds to a small-signal equivalent circuit as in Figure 1.41.

So far nothing has been said on the time dependence or spectrum of the small input current. Since the circuit equation is linear, the solution can be obtained in the spectral domain by Fourier transform as the superposition of sinusoidal spectral components, as for any linear system (see Section 1.1 and Appendix A.1). However, the time-varying coefficients give a different turn to the circuit analysis. Let us show this with an example.

Let us assume a sinusoidal input current:

$$i_{ss}(t) = I_{ss} \cdot \cos(\omega_{ss}t) \tag{1.139}$$

The small-signal voltage will have a component with the same frequency, which we will identify by the subscript (0):

$$v_{ss,0}(t) = V_{ss,0} \cdot e^{j\omega_{ss}t} + V_{ss,-0} \cdot e^{-j\omega_{ss}t} \tag{1.140}$$

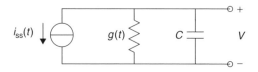

Figure 1.41   Small-signal time-variant equivalent circuit of the example circuit

The current in the capacitor will also have a component with the same frequency:

$$i_{C,ss,0}(t) = C \cdot \frac{dv_{ss,0}(t)}{dt} = j\omega_{ss}C \cdot V_{ss,0} \cdot e^{j\omega_{ss}t} - j\omega_{ss}C \cdot V_{ss,-0} \cdot e^{-j\omega_{ss}t} \qquad (1.141)$$

The current in the nonlinear resistor will be

$$i_{g,ss}(t) = g_{ss}(t) \cdot v_{ss,0}(t) = \sum_{m=-\infty}^{\infty} (G_{ss,m} \cdot V_{ss,0} \cdot e^{j(m\omega_{LS}+\omega_{ss})t} + G_{ss,m} \cdot V_{ss,-0} \cdot e^{j(m\omega_{LS}-\omega_{ss})t})$$

$$(1.142)$$

Therefore, a small sinusoidal voltage at the input frequency ω_{ss} generates a whole set of new frequency components by interaction with a periodically varying conductance. If the spectrum of the (time-varying) conductance is composed by the fundamental frequency ω_{LS} with all its harmonics, the spectrum of the small current will be composed by the same spectral components shifted upwards and downwards by the small voltage frequency ω_{ss} (Figure 1.42).

The expression for the current is complicated and redundant: as is well known, for a real signal, the phasors at a negative frequency are the complex conjugate of the phasors at the corresponding positive frequency. In our case, we can rewrite the previous expression as

$$i_{g,ss}(t) = \sum_{m=-\infty}^{0} G_{ss,m} \cdot V_{ss,0} \cdot e^{j(m\omega_{LS}+\omega_{ss})t} + \sum_{m=1}^{\infty} G_{ss,m} \cdot V_{ss,0} \cdot e^{j(m\omega_{LS}+\omega_{ss})t}$$

$$+ \sum_{m=-\infty}^{-1} G_{ss,m} \cdot V_{ss,-0} \cdot e^{j(m\omega_{LS}-\omega_{ss})t} + \sum_{m=0}^{\infty} G_{ss,m} \cdot V_{ss,-0} \cdot e^{j(m\omega_{LS}-\omega_{ss})t} \qquad (1.143)$$

from which it is clear that the first and fourth terms are complex conjugate, and so are the second and third. To make this point clearer, the previous figure is repeated with the four terms marked in different shadings (Figure 1.43).

The first and third terms represent the lower sideband of the signal, while the second and fourth represent the upper sideband. In order to have a compact representation,

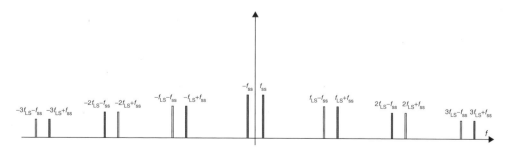

Figure 1.42   Spectrum of the current in the time-variant dynamic conductance

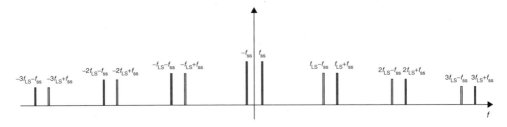

Figure 1.43   Spectrum of the current in the time-variant dynamic conductance with the four terms in eq. (1.143) marked in different shadings

we will omit the third and fourth terms, retaining only the first and second ones. The signal written in this simplified way is not physical, but it contains all the information necessary for the reconstruction of the real signal. From now on, we will use this representation for voltages and currents within this section without any specification. We will therefore rewrite

$$i_{ss}(t) = I_{ss} \cdot e^{j\omega_{ss}t} \tag{1.144}$$

$$v_{ss,0}(t) = V_{ss,0} \cdot e^{j\omega_{ss}t} \tag{1.145}$$

and so on.

The complete spectrum of the small perturbation voltage has to include the same spectral components as the current generated in the nonlinear elements, since the current will flow through the linear part of the circuit generating a voltage with the same spectrum. Its Fourier series expansion with the convention above is

$$v_{ss}(t) = \sum_{m=-\infty}^{\infty} V_{ss,m} \cdot e^{j(m\omega_{LS}+\omega_{ss})t} \tag{1.146}$$

With this new formalism, Kirchhoff's equation for the small perturbation reads as follows:

$$\sum_{n=-\infty}^{\infty} I_{ss,n} \cdot e^{j(n\omega_{LS}+\omega_{ss})t} + \sum_{n=-\infty}^{\infty} j(\omega_{LS} + \omega_{ss}) \cdot C \cdot V_{ss,n} \cdot e^{j(n\omega_{LS}+\omega_{ss})t}$$

$$+ \sum_{m=-\infty}^{\infty} G_{ss,m} \cdot e^{jm\omega_{LS}t} \cdot \sum_{n=-\infty}^{\infty} V_{ss,n} \cdot e^{j(n\omega_{LS}+\omega_{ss})t} \tag{1.147}$$

This time-domain equation can be split into infinite equations, each one balancing the current at a different frequency $n\omega_{LS} + \omega_{ss}$, as seen in the previous paragraph. The generic equation becomes

$$I_{ss,n} \cdot e^{j(n\omega_{LS}+\omega_{ss})t} + j(\omega_{LS} + \omega_{ss}) \cdot C \cdot V_{ss,n} \cdot e^{j(n\omega_{LS}+\omega_{ss})t}$$

$$+ \sum_{m=-\infty}^{\infty} G_{ss,n-m} \cdot V_{ss,m} \cdot e^{j(n\omega_{LS}+\omega_{ss})t} \tag{1.148}$$

This expression can be put in matrix form by truncating the series expansions. Voltages and currents become vectors

$$\vec{I}_{ss} = \begin{bmatrix} I_{ss,N} \\ \cdot \\ I_{ss,0} \\ \cdot \\ I_{ss,-N} \end{bmatrix} \qquad \vec{V}_{ss} = \begin{bmatrix} V_{ss,N} \\ \cdot \\ V_{ss,0} \\ \cdot \\ V_{ss,-N} \end{bmatrix} \qquad (1.149)$$

the linear capacitance becomes a diagonal matrix

$$\ddot{C} = \begin{bmatrix} j(N\omega_{LS} + \omega_{ss})C & \cdot & 0 & \cdot & 0 \\ \cdot & \cdot & \cdot & \cdot & \cdot \\ 0 & \cdot & j\omega_{ss}C & \cdot & 0 \\ \cdot & \cdot & \cdot & \cdot & \cdot \\ 0 & \cdot & 0 & \cdot & j(-N\omega_{LS} + \omega_{ss})C \end{bmatrix} \qquad (1.150)$$

and the nonlinear resistance becomes a matrix in the form

$$\ddot{G} = \begin{bmatrix} G_{ss,0} & \cdot & G_{ss,N} & \cdot & G_{ss,2N} \\ \cdot & \cdot & \cdot & \cdot & \cdot \\ G_{ss,-N} & \cdot & G_{ss,0} & \cdot & G_{ss,N} \\ \cdot & \cdot & \cdot & \cdot & \cdot \\ G_{ss,-2N} & \cdot & G_{ss,-N} & \cdot & G_{ss,0} \end{bmatrix} \qquad (1.151)$$

The system of KCL equations for the small perturbation becomes

$$\vec{I}_{ss} + \ddot{C} \cdot \vec{V}_{ss} + \ddot{G} \cdot \vec{V}_{ss} = 0 \qquad (1.152)$$

The matrix $\ddot{G}$, or its generalised expression in case of a complex nonlinear network, is the conversion matrix. As seen before, a nonlinear element driven by a large signal at a frequency ω_{LS} acts as a time-periodic linear conductance that converts the frequency of a signal to many different frequencies, shifted upwards and downwards by the harmonics of the large signal. In other words, the output frequencies are the sum and difference of the input frequency with the harmonic frequencies of the large signal. The complete spectrum of voltages and currents is as in Figure 1.44.

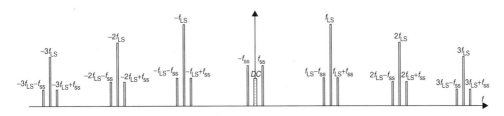

Figure 1.44 Complete spectrum of currents and voltages in the circuit

In the case that no large periodic signal is driving the nonlinear element into nonlinear regime, the conversion matrix becomes diagonal, and the diagonal elements are the small-signal conductance at the bias point $G_{ss,0}$. The circuit becomes a standard small-signal circuit.

The conversion matrix is very convenient in the case of the small amplitude of the input signal to be frequency converted. When the amplitude of the input signal increases, a Volterra series analysis can be implemented by extending the series expansion of the current in the nonlinear elements (eq. (1.133)) to second-, third- and possibly higher-order terms [30]. In this way, the dynamic range of the conversion is found and the intermodulation distortion is also predicted. This approach extends the Volterra series formalism from time-independent to time-dependent nonlinear circuits.

1.5 BIBLIOGRAPHY

[1] L.O. Chua, P.-M. Lin, *Computer-aided Analysis of Electronic Circuits: Algorithms and Computational Techniques*, Prentice Hall, Englewood Cliffs (NJ), 1975.
[2] M. Abramowitz, I.A. Stegun, *Handbook of Mathematical Functions with Formulas, Graphs and Mathematical Tables*, Dover Publications, New York (NY), 1972.
[3] C.-W. Ho, A.E. Ruehli, P.A. Brennan, 'The modified nodal approach to network analysis', *IEEE Trans. Circuits Syst.*, **CAS-22**(6), 504–509, 1975.
[4] C.W. Gear, 'Simultaneous numerical solution of differential algebraic equations', *IEEE Trans. Circuit Theory*, **18**(1), 89–95, 1971.
[5] M.I. Sobhy, A.K. Jastrzebski, 'Direct integration methods of nonlinear microwave circuits', *Proc. European Microwave Conf.*, Paris (France), Sep. 1975, pp. 1110–1118.
[6] M.I. Sobhy, A.K. Jastrzebski, 'Computer-aided design of microwave integrated circuits', *Proc. 14th European Microwave Conf.*, Liège (Belgium), Sep. 1984, pp. 705–710.
[7] D.O. Pederson, 'A historical review of circuit simulation', *IEEE Trans. Circuits Syst.*, **CAS-31**, 103–111, 1984.
[8] H.B. Keller, *Numerical Solution of Two Point Boundary-Value Problems*, SIAM, Blaisdell, Waltham (MA), 1976.
[9] T.J. Aprille, T.N. Trick, 'Steady-state analysis for nonlinear circuits with periodic inputs', *Proc. of the IEEE*, Jan. 1972, pp. 108–114.
[10] R. Telichevesky, K. Kundert, J. White, 'Efficient steady-state analysis based on matrix-free Krylov subspace methods', Proc. of the 32nd Design Automation Conference, June 1995, pp. 480–484.
[11] L.O. Chua, A. Ushida, 'Algorithms for computing almost-periodic steady-state response of nonlinear systems to multiple input frequencies', *IEEE Trans. Circuits Syst.*, **CAS-28**, 953–971, 1981.
[12] M. Silverberg, O. Wing, 'Time-domain computer solutions for networks containing limped nonlinear elements', *IEEE Trans. Circuit Theory*, **CT-15**, 292–294, 1968.
[13] A.R. Djordjevic, T.K. Sarkar, 'Analysis of time response of lossy multiconductor transmission line networks', *IEEE Trans. Microwave Theory Tech.*, **MTT-35**, 898–908, 1987.
[14] J.F. Whitaker, T.B. Norris, G. Mourou, T.Y. Hsiang, 'Pulse dispersion and shaping in microstrip lines', *IEEE Trans. Microwave Theory Tech.*, **MTT-35**, n.1, pp. 41–46, Jan. 1987.
[15] T.J. Brazil, 'A new method for the transient simulation of causal linear systems described in the frequency domain', *IEEE Microwave Theory Tech. Symp. Dig.*, 1992, pp. 1485–1488.
[16] T.J. Brazil, 'Causal convolution - a new method for the transient analysis of linear systems at microwave frequencies', *IEEE Trans. Microwave Theory Tech.*, **MTT-43**, 315–323, 1995.

[17] C.F. Christoffersen, M. Ozkar, M.B. Steer, M.G. Case, M. Rodwell, 'State-variable-based transient analysis using convolution', *IEEE Trans. Microwave Theory Tech.*, **MTT-47**(6), 882–889, 1999.

[18] P. Perry, T.J. Brazil, 'Hilbert-transform-derived relative group delay', *IEEE Trans. Microwave Theory Tech.*, **45**, 1214–1225, 1997.

[19] V. Volterra, *Theory of Functionals and of Integro-differential Equations*, Dover, New York, 1959.

[20] N. Wiener, Nonlinear Problems in Random Theory, MIT Press, Cambridge (MA), 1959.

[21] S. Narayanan, 'Application of Volterra series to intermodulation distortion analysis of transistor feedback amplifiers', *IEEE Trans. Circuit Theory*, **CT-17**, 518–527, 1970.

[22] E. Bedrosian, S.O. Rice, 'The output properties of Volterra systems (nonlinear systems with memory) driven by harmonic and Gaussian inputs', *Proc. IEEE*, **59**, 1688–1707, 1971.

[23] J. Bussgang, L. Ehrman, J. Graham, 'Analysis of nonlinear systems with multiple inputs', *Proc. IEEE*, **62**, 1088–1119, 1974.

[24] D.D Weiner, J.F. Spina, 'A scattering variable approach to the Volterra analysis of nonlinear systems', *IEEE Trans. Microwave Theory Tech.*, **MTT-24**(7), 422–433, 1976.

[25] R.A. Minasian, 'Intermodulation distortion analysis of MESFET amplifiers using Volterra series representation', *IEEE Trans. Microwave Theory Tech.*, **MTT-28**(1), 1–8, 1980.

[26] D.D. Weiner, J.F. Spina, *Sinusoidal Analysis and Modelling of Weakly Nonlinear Circuit*, Van Nostrand Reinhold, New York (NY), 1980.

[27] M. Schetzen, 'Multilinear theory of nonlinear networks', *J. Franklin Inst.*, **320**(5), 221–228, 1985.

[28] C.L. law, C.S. Aitchison, 'Prediction of wideband power performance of MESFET distributed amplifiers using the Volterra series representation', *IEEE Trans. Microwave Theory Tech.*, **MTT-34**, 1038–1317, 1986.

[29] S.A. Maas, *Nonlinear Microwave Circuits*, Artech House, New York (NY), 1988.

[30] S.A. Maas, 'Two-tone intermodulation in diode mixers', *IEEE Trans. Microwave Theory Tech.*, **MTT-35**(3), 307–314, 1987.

[31] J.M.T. Thomson, H.B. Stewart, *Nonlinear Dynamics and Chaos*, John Wiley & Sons, New York (NY), 1997.

[32] M.S. Nakhla, J. Vlach, 'A piecewise harmonic balance technique for the determination of the periodic response of nonlinear systems', *IEEE Trans. Circuits Syst.*, **CAS-26**, 85–91, 1976.

[33] F. Filicori, V.A. Monaco, C. Naldi, 'Simulation and design of microwave Class-C amplifiers through harmonic analysis', *IEEE Trans. Microwave Theory Tech.*, **MTT-27**, 1043–1051, 1979.

[34] V. Rizzoli, A. Lipparini, E. Marazzi, 'A general-purpose program for nonlinear microwave circuit design', *IEEE Trans. Microwave Theory Tech.*, **MTT-31**, 762–769, 1983.

[35] K.S. Kundert, A. Sangiovanni-Vincentelli, 'Simulation of nonlinear circuits in the frequency domain', *IEEE Trans. Comput.-Aided Des.*, **CAD-5**(4), 521–535, 1986.

[36] R.J. Gilmore, M.B. Steer, 'Nonlinear circuit analysis using the method of harmonic balance – a review of the art: part I and part II', *Int. J. Microwave Millimetre-Wave Comput.-Aided Eng.*, **1**(2), 159–180, 1991.

[37] K. Kundert, J. White, A. Sangiovanni-Vincentelli, *Steady-state Methods for Simulating Analog and Microwave Circuits*, Kluwer, Norwell (MA), 1990.

[38] S. El-Rabaie, V.F. Fusco, C. Stewart, 'Harmonic balance evaluation of nonlinear microwave circuits – a tutorial approach', *IEEE Trans. Educ.*, **31**(3), 181–192, 1988.

[39] F. Filicori, V.A. Monaco, 'Computer-aided design of nonlinear microwave circuits', *Alta Frequenza*, **57**(7), 355–378, 1988.

[40] V. Rizzoli, A. Lipparini, A. Costanzo, F. Mastri, C. Cecchetti, A. Neri, D. Masotti, 'State-of-the-art harmonic-balance simulation of forced nonlinear microwave circuits by the piecewise technique', *IEEE Trans. Microwave Theory Tech.*, **MTT-40**(1), 12–28, 1992.

[41] C.P. Silva, 'Efficient and reliable numerical algorithms for nonlinear microwave computer-aided design', *IEEE Microwave Theory Tech. Symp. Dig.*, 1992, pp. 422–428.

[42] H.G. Brachtendorf, G. Welsch, R. Laur, 'Fast simulation of the steady-state of circuits by the harmonic balance technique', *Proc. IEEE Int. Symp. on Circuits Syst.*, 1995, pp. 1388–1391

[43] V. Rizzoli, F. Mastri, C. Cecchetti, F. Sgallari, 'Fast and robust inexact Newton approach to the harmonic-balance analysis of nonlinear microwave circuits', *IEEE Microwave Guided Wave Lett.*, **7**(10), 359–361, 1997.

[44] B. Troyanovsky, Z.-P. Yu, R.W. Dutton, 'Physics-based simulation of nonlinear distortion in semiconductor devices using the harmonic balance method', *Comput. Methods Appl. Mech. Eng.*, **181**, 467–482, 2000.

[45] W.R. Curtice, 'Nonlinear analysis of GaAs MESFET amplifiers, mixers, and distributed amplifiers using the harmonic balance technique', *IEEE Trans. Microwave Theory Tech.*, **MTT-35**(4), 441–447, 1987.

[46] A. Ushida, L.O. Chua, 'Frequency-domain analysis of nonlinear circuits driven by multi-tone signals', *IEEE Trans. Circuits Syst.*, **CAS-31**(9), 766–779, 1984.

[47] K. Kundert, G.B. Sorkin, A. Sangiovanni Vincentelli, 'Applying harmonic balance to almost-periodic circuits', *IEEE Trans. Microwave Theory Tech.*, **MTT-36**, 366–378, 1988.

[48] X.-D. Zhang, X.-N. Hong, B.-X. Gao, 'Accurate Fourier transform method for almost periodic response simulation of microwave nonlinear circuits', *Electron. Lett.*, **25**, 404–406, 1989.

[49] P.J.C. Rodrigues, 'An orthogonal almost-periodic Fourier transform for use in nonlinear circuit simulation', *IEEE Microwave Guided Wave Lett.*, **4**, 74–76, 1994.

[50] E. Ngoya, J. Rousset, M. Gayral, R. Quere, J. Obregon, 'Efficient algorithms for spectra calculations in nonlinear microwave circuits simulators', *IEEE Trans. Circuits Syst.*, **CAS-37**(11), 1339–1355, 1990.

[51] V. Rizzoli, C. Cecchetti, A. Lipparini, F. Mastri, 'General-purpose harmonic balance analysis of nonlinear microwave circuits under multitone excitation', *IEEE Trans. Microwave Theory Tech.*, **36**, 1650–1659, 1988.

[52] P.L. Heron, M.B. Steer, 'Jacobian calculation using the multidimensional fast Fourier transform in the harmonic balance analysis of nonlinear circuits', *IEEE Trans. Microwave Theory Tech.*, **38**, 429–431, 1990.

[53] D. Hente, R.H. Jansen, 'Frequency domain continuation method for the analysis and stability investigation of nonlinear microwave circuits', *IEE Proc. Part H*, **133**(5), 351–362, 1986.

[54] P.J.C. Rodrigues, 'A general mapping technique for Fourier transform computation in nonlinear circuit analysis', *IEEE Microwave Guided Wave Lett.*, **7**(11), 374–376, 1997.

[55] R.J. Gilmore, 'Nonlinear circuit design using the modified harmonic-balance algorithm', *IEEE Trans. Microwave Theory Tech.*, **MTT-34**, 1294–1307, 1986.

[56] F. Filicori, V.A. Monaco, G. Vannini, 'Computationally efficient multitone analysis of nonlinear microwave circuits', *Proc. 21st European Microwave Conf.*, 1991, pp. 1550–1555.

[57] P. Feldmann, J. Roychowdhury, 'Computation of circuit waveform envelopes using an efficient, matrix-decomposed harmonic balance algorithm', *Proc. ICCAD*, 1996, pp. 269–274.

[58] V. Borich, J. East, G. Haddad, 'The method of envelope currents for rapid simulation of weakly nonlinear communication circuits', *IEEE Microwave Theory Tech. Int. Microwave Symp. Dig.*, Anaheim (CA), June 1999, pp. 981–984.

[59] D. Sharrit, 'New method of analysis of communication systems', *IEEE Microwave Theory and Techniques Symposium WMFA: Nonlinear CAD Workshop*, June 1996.

[60] V. Rizzoli, A. Neri, F. Mastri, A. Lipparini, 'A Krylov-subspace technique for the simulation of RF/microwave subsystems driven by digitally modulated carriers', *Int. J. RF Microwave Comput.-Aided Eng.*, **9**, 490–505, 1999.

[61] V. Rizzoli, A. Costanzo, F. Mastri, 'Efficient Krylov-subspace simulation of autonomous RF/microwave circuits driven by digitally modulated carriers', *IEEE Microwave Wireless Comp. Lett.*, **11**(7), 308–310, 2001.

[62] V. Rizzoli et al., 'Pulsed-RF and transient analysis of nonlinear microwave circuits by harmonic balance techniques', *IEEE MTT-S Int. Microwave Symp. Dig.*, Boston (MA), June 1991, pp. 607–610.

[63] L. Gustafsson, G.H.B. Hansson, K.I. Lundstrom, 'On the use of describing functions in the study of nonlinear active microwave circuits', *IEEE Trans. Microwave Theory Tech.*, **MTT-20**, 402–409, 1972.

[64] G.L. Heiter, 'Characterisation of nonlinearities in microwave devices and systems', *IEEE Trans. Microwave Theory Tech.*, **MTT-21**, 797–805, 1973.

[65] R.S. Tucker, 'Third-order intermodulation distortion and gain compression of GaAs FET's', *IEEE Trans. Microwave Theory Tech.*, **MTT-27**, 400–408, 1979.

[66] G.W. Rhyne, M.B. Steer, B.D. Bates, 'Frequency domain nonlinear circuit analysis using generalised power series', *IEEE Trans. Microwave Theory Tech.*, **36**, 379–387, 1988.

[67] C.-R. Chang, M.B. Steer, G.W. Rhyne, 'Frequency-domain spectral balance using the arithmetic operator method', *IEEE Trans. Microwave Theory Tech.*, **MTT-37**, 1681–1688, 1989.

[68] M.B. Steer, C.-R. Chang, G.W. Rhyne, 'Computer-aided analysis of nonlinear microwave circuits using frequency-domain nonlinear analysis techniques: the state of the art', *Int. J. Microwave Millimetre-Wave Comput.-Aided Eng.*, **1**(2), 181–200, 1991.

[69] T. Närhi, 'frequency-domain analysis of strongly nonlinear circuits using a consistent large-signal model', *IEEE Trans. Microwave Theory Tech.*, **MTT-44**, 182–192, 1996.

[70] N. Borges deCarvalho, J.C. Pedro, 'Multitone frequency-domain simulation of nonlinear circuits in large- and small-signal regimes', *IEEE Trans. Microwave Theory Tech.*, **MTT-46**(12), 2016–2024, 1998.

[71] S. Egami, 'Nonlinear-linear analysis and computer-aided design of resistive mixers', *IEEE Trans. Microwave Theory Tech.*, **MTT-22**, 270–275, 1974.

[72] D.N. Held, A.R. Kerr, 'Conversion loss and noise of microwave abdìınd millimetre-wave mixers: part 1 - theory', *IEEE Trans. Microwave Theory Tech.*, **MTT-26**, 49–55, 1978.

[73] A.R. Kerr, 'Noise and loss in balanced and subharmonically pumped mixers: part 1 - theory', *IEEE Trans. Microwave Theory Tech.*, **MTT-27**, 938–943, 1979.

[74] S.A. Maas, 'Theory and analysis of GaAs mixers', *IEEE Trans. Microwave Theory Tech.*, **MTT-32**, 1402–1406, 1984.

2

Nonlinear Measurements

2.1 INTRODUCTION

This introduction describes the main applications of nonlinear measurement methods, to be described more in detail in the following paragraphs.

Nonlinear components and circuits are measured for three main purposes: for the characterisation of the nonlinear device and inclusion of its nonlinear model in an analysis algorithm as described in Chapter 1; for a direct, experimental evaluation of the performances of the device or circuit on the measurement bench, in order to identify the optimum operating conditions; or for the verification of the actual performances of the fabricated circuit to be compared with the designed or expected ones.

The same measurement equipment or set-up can serve more than one of the three different purposes; this chapter is therefore divided by measurement type. For each of them, the different applications will be described. The particular application of nonlinear device modelling will be described in greater detail in the next chapter; it is not only the most critical bottleneck in nonlinear CAD, but also a somewhat tricky exercise.

Three main nonlinear measurement types are described hereafter: the load/source pull, the vector nonlinear network analysis and the pulsed measurements. The first two types involve large-signal RF and DC quantities, while the third one involves large-signal quasi-DC quantities and optionally also small-signal quantities. In fact, the boundary between load/source pull and nonlinear vector network analysis can vanish, and the same set-up can be made to perform both functions. In general, a great deal of work is going on in the field, and things are moving fast, as the awareness of the importance of detailed and reliable nonlinear data gains ground. In this chapter, we will therefore try to describe the basic principles more than the most advanced results or set-up.

As a general description, it will be enough to say here that load/source pull is mainly devoted to the experimental identification of the optimum performances of a device or circuit under nonlinear regime and of the corresponding operating conditions. Nonlinear network vector analysis mainly aims at the direct measurement or optimisation of a nonlinear circuit, or at the identification of a nonlinear model; pulsed measurements

Nonlinear Microwave Circuit Design   F. Giannini and G. Leuzzi
© 2004 John Wiley & Sons, Ltd   ISBN: 0-470-84701-8

usually serve the purpose of gathering data for the extraction of a correct and accurate nonlinear device model.

A general remark can be made here: nonlinear measurements, as opposed to linear ones, cannot exploit the superposition principle. This implies that any representation in the form of an equivalent linear matrix is meaningless: no Thévenin (or Norton) equivalent representation as the impedance (or admittance) matrix is allowed. As a consequence, data measured in a 50 Ω environment cannot be extrapolated to different loading conditions. Another consequence is that nonlinear measurements are dependent on the amplitude of voltages and currents. The amount of measurements and measurement data is consequently much higher than that in the linear case. The amount and complexity of the measuring hardware is also correspondingly greater than that in the linear case, making the total cost of both equipment and human resources quite high.

2.2 LOAD/SOURCE PULL

In this paragraph, the main schemes for load/source pulling are reviewed. The required equipment and the performances of each approach are described and compared. Examples of measured data are also given.

The object of load/source-pull measurements is the experimental determination of the performances of a device in large-signal operations and the identification of the conditions that yield the optimum or desired results. Typical performances of interest for power amplifiers are not only large-signal input and output match, output power, power gain, efficiency and distortion (see Section 4.1) but also conversion gain and conversion efficiency for mixers and frequency multipliers or frequency pulling for oscillators. The performances are evaluated as functions of input and output loads, bias point and input power for power amplifiers and of local oscillator power for mixers. Sometimes the temperature also is a controlled parameter. In fact, the most important parameters, and the ones that give the name to the technique, are the input and output loads: load pulling means varying the output load, that is, the load at the output side of the device under test (DUT), and source pulling means varying the input load, that is, the load at the source side. The loads must be controlled, so that their value is known as accurately as possible while making a measurement. By no means is this an easy job, and several solutions are available.

A first classification of the load/source-pulling techniques indicates whether the measured quantities are measured with scalar or vectorial techniques. Scalar load/source pulling requires only power meters and/or a spectrum analyser, while vectorial load/source pulling requires also a network vector analyser (VNA). In fact, a VNA is also required in a scalar source/load pull for the preliminary (linear) characterisation of the loads, but not during the measurement itself, as described in the following. Calibration techniques are also obviously different in the two cases. The scalar solution is usually cheaper and easier to implement, while the vectorial is more accurate and complete.

Another classification of the techniques indicates whether the loads at the input and output sides are implemented by means of passive or active arrangements; in any

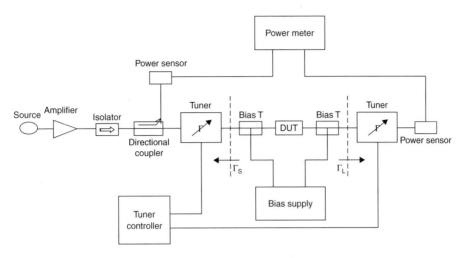

Figure 2.1 A typical set-up for scalar, passive load pull

of the two cases, the loads can be controlled either at fundamental frequency only or at harmonic frequencies also. All combinations are briefly described in the following text.

A typical set-up for scalar, passive, fundamental-frequency load/source pull is shown in Figure 2.1 [1, 2].

The signal from the source is amplified, then sampled by means of a directional coupler and fed to the input of the DUT through a tuner; the sampled signal is measured by the power sensor of a power meter. The input tuner acts as an adjustable matching network, whose output reflection coefficient Γ_S can be set to any desired value within a certain region of the Smith Chart, provided that the other port of the tuner is matched. This Γ_S is actually the source impedance for the DUT, impedance that is 'pulled' by adjusting the tuner. Similarly, at the output, the load-reflection coefficient for the DUT Γ_L is set by another tuner, after which the signal is fed to another power sensor. In fact, the source/load-pull set-up replicates a power amplifier, where the signal source with the amplifier and the isolator act as a matched input large-signal source, the tuners act as adjustable matching networks, the bias Ts act as bias networks and the output power sensor acts as a matched output load; the directional coupler is inserted only for measurement purposes. Provided that the attenuation and the impedances of the tuners are accurately known, input and output power at the DUT for the given loading conditions are evaluated from the power sensors; power gain and efficiency are also immediately computed if the DC power from the power supply is evaluated. The tuners are therefore the centrepieces of the set-up; they must be reliable, repeatable and flexible enough to ensure accurate data.

Passive tuners can be mechanical or electronic. Mechanical tuners are made of a piece of rectangular waveguide with a longitudinal slot in which one or two slugs are inserted (see Figure 2.2).

The position along the slot and the depth of insertion of the slug(s) in the slot control the phase and the amplitude of the reflection coefficient respectively. In manual

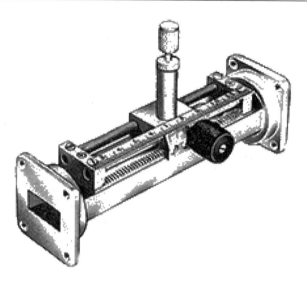

Figure 2.2   A mechanical tuner

tuners, the insertion of each slug is controlled by a micrometric screw, mounted on a slide for position control. In automatic tuners, the same is done by means of step motors, controlled by a computer. The tuners are preliminarily characterised in terms of two-port S-parameter by means of a VNA, for a great number of positions of the slugs, in order that the impedance covers the region of the Smith Chart to be investigated; the data is stored for further processing. An example of the loads presented by a tuner is shown in Figure 2.3.

During the measurement, the slugs are set to the same positions as they are during the pre-characterisation. The repeatability of the tuner is therefore of primary importance for the accuracy of the measurements.

Electronic tuners include PIN diodes that can be switched on or off by an external control, providing a similar set of loads as in Figure 2.3.

The advantage of mechanical tuners consists essentially in their superior power handling. Repeatability is a matter of accurate design for any of the two solutions. Robust and accurate mechanical structure and fabrication ensure high repeatability and reliability of a mechanical tuner; on the other hand, a suitable design of the circuitry ensures stability with respect to power level and temperature to an electronic tuner. Costs, both in terms of initial investment and in terms of measurement time, are usually comparable.

The calibration of the set-up requires a vectorial linear two-port characterisation of the tuners in a large number of slug positions, large enough to cover the region of the Smith Chart to be investigated. In fact, all parts in the set-up (except the DUT, of course) must be characterised beforehand in terms of two-port S-parameters (directional coupler, bias Ts) or reflection coefficient (input source block, output power sensor), in order that the impedances and power levels at the DUT be properly evaluated. The power level is calibrated by an absolute power measurement with the DUT replaced by

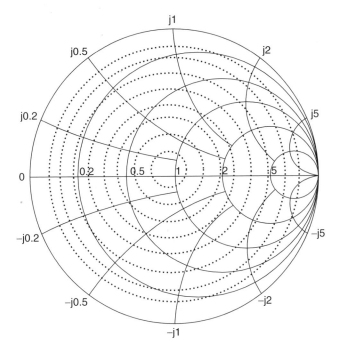

Figure 2.3 Loads synthesised by a mechanical tuner

a through connection. A standard cascaded-networks de-embedding technique is used for the impedance and power level calculation [3–6].

The output power is evaluated for each output load at a specified input power condition: for instance, all data points can be measured for the same input power. Alternatively, the input power can be individually adjusted for each load in such a way that for each data point the transistor works at the same gain compression level (e.g. 1-dB gain compression). The data points on the Smith Chart that yield the same output power level are connected by a constant output power contour; usually, the load yielding the maximum output power is shown individually, and the power contours corresponding to decreasing levels in 1-dB steps are also shown. The contours are usually traced automatically by means of interpolation routines. The contours are almost circles for low compression levels, when nonlinearities are weak; they become more and more distorted as the nonlinear effects increase with the compression level.

Intermodulation is evaluated in very much the same way by replacing the sinusoidal, single-tone input source with a two-tone input source [7, 8]. The output power sensor is also replaced by a spectrum analyser for the evaluation of the third-order spectral line power with respect to the fundamental-frequency spectral line power. The input-signal power amplifier must be very linear in order to prevent its distortion from being amplified by the transistor and getting added to that of the transistor itself. In fact, the cost of the power amplifier becomes a significant portion of the total cost because of the linearity requirements.

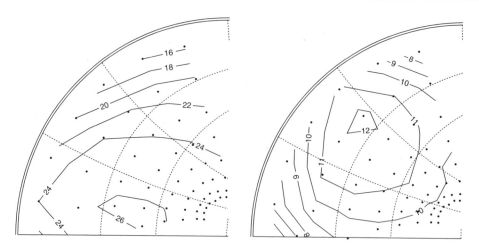

Figure 2.4 Load-pull data with interpolated contours: output power (a) and power gain at 1-dB gain compression (b)

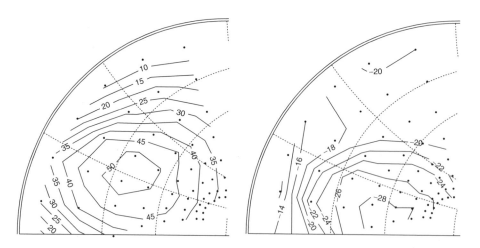

Figure 2.5 Load-pull data with interpolated contours: power-added efficiency at 1-dB gain compression (a) and C/I_3 at 1-dB gain compression (b)

As an example of the capabilities of the set-up, output power and power gain data at 1-dB gain compression are shown in Figure 2.4, while power-added efficiency and C/I_3 data at 1-dB gain compression are shown in Figure 2.5.

A few variations on this general scheme are possible. First of all, another directional coupler can be added before the input tuner, but reversed in direction, in order to monitor the reflected wave at input also: in this way, the large-signal input match can be adjusted by suitably setting the input tuner. An alternative to tuner pre-characterisation is given by the insertion of a switch between each tuner and the corresponding bias T, in order to measure the tuner impedance directly on the site; however, the switch

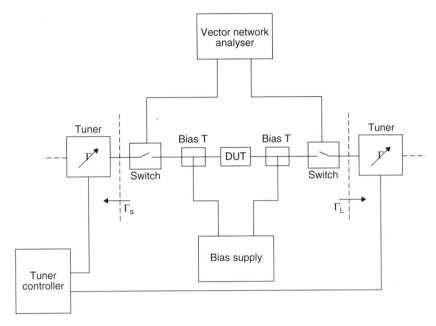

Figure 2.6 A typical set-up for scalar, passive source/load pull with online impedance measurement

repeatability affects the accuracy of the measurement. Moreover, a VNA must be available during the measurements, and the input signal must be switched off when the input reflection coefficient is measured (see Figure 2.6); the measurement time also increases. Alternatively, directional couplers are inserted between the tuners and the DUT for online impedance evaluation [9].

The approach just described does not account for harmonic frequency loading, both at the input and at the output of the DUT. In fact, the impedance value at harmonic frequencies affects the performances of the DUT especially for strongly nonlinear operations, as for instance in high compression [10]. Of course, the harmonic impedance presented by a passive tuner can be accurately measured, but cannot be controlled, unless a special arrangement is made. For instance, two tuners can be series-connected, in order that enough degrees of freedom are available for simultaneous control of first- and second-harmonic impedance. However, losses and characterisation time increase substantially. Harmonic tuners including a resonant structure are available, that partially limit the drawbacks of the solution. A more radical solution consists of two multiplexers at either end of the DUT block and of as many tuners as the harmonics to be controlled (Figure 2.7)

Also, in this case, the additional losses introduced by the multiplexer limit the performance of the system.

The main consequence of the losses associated with passive tuners is that the reflection coefficient cannot reach the edge of the Smith Chart. Typical maximum values for the reflection coefficient amplitude are 0.9 at low microwave frequencies and down to 0.8

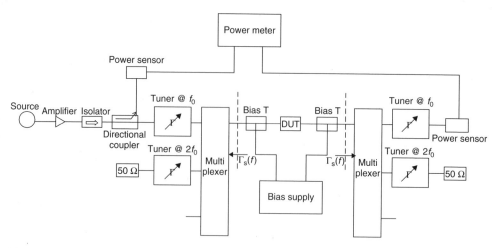

Figure 2.7 A typical set-up for scalar, passive source/load pull with individual control of the harmonic impedances

or even 0.75 in the higher microwave frequency range. This is a real problem, especially when characterising high-power transistors, whose optimum input and output impedances are very low and lie close to the edge of the Smith Chart. A simple solution includes pre-matching circuits, that is, impedance transformers, between the DUT and the tuners; this solution allows for a better accuracy within the transformed region. The pre-matching can also be included in the tuners. Obviously, the pre-matching circuits must be characterised in terms of S-parameters and replaced by different ones whenever the region of interest changes.

A radical solution to the losses problem is the active-load approach. A possible scheme is the two-path technique [11]: the signal source is split and fed both to the input of the DUT and to the output of the DUT after amplification and phase shift (Figure 2.8).

The output reflection coefficient seen by the DUT is the ratio of incident to reflected wave at its output port:

$$\Gamma_L = \frac{a_L}{b_L} \tag{2.1}$$

Now, b_L is the wave coming out of the DUT, while a_L is the wave injected from the output path; the amplitude and phase of the latter are easily set by means of the variable attenuator and phase shifter in the output path. In this way, any ratio can be synthesised, even greater than one in amplitude, since the amplifier in the output path overcomes all the losses. The value of the output reflection coefficient is checked on-site by means of the output directional couplers and a VNA.

The two-path technique is an easy and stable technique for active-load synthesis. However, when simulating an actual power amplifier, the value of the output load must be kept constant for increasing input power levels and also for the associated increasing DUT temperature. This implies that, while b_L changes because of the above, a_L must

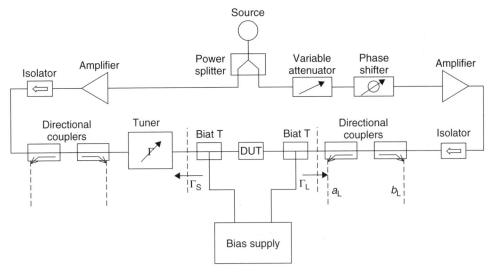

Figure 2.8 A typical set-up for two-path active load pull with passive source pull

also be adjusted, in order to keep their ratio constant. The procedure becomes very time consuming and is not very stable.

In order to overcome the problem, a different active-load configuration is available, the so-called active-loop technique [12, 13] (Figure 2.9).

The wave b_L coming out of the DUT is now sampled, amplified, phase shifted and re-injected at the output of the DUT. The ratio of this wave a_L to the wave coming out of the DUT b_L is fixed once the variable attenuator and the phase shifter in the output loop are fixed. The main problem associated with this approach is the possible onset of oscillations within the output loop, especially at frequencies outside the coupling (and decoupling) band of the directional coupler. A passband loop filter is inserted to prevent the presence of spurious signal propagation around the loop outside the signal-frequency band.

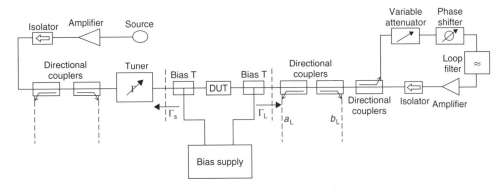

Figure 2.9 A typical set-up for active-loop load pull with passive source pull

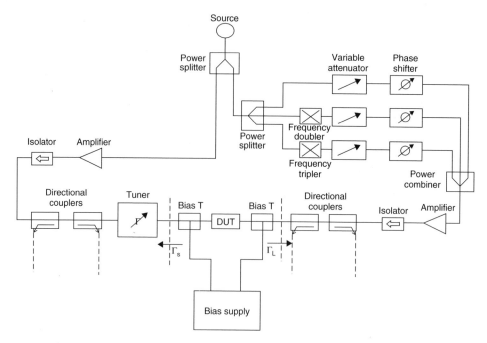

Figure 2.10   A typical set-up for two-path active load pull with individual control of harmonic impedances and passive source pull

Both the active-loop techniques can be extended to cover harmonic frequencies. The two-path approach requires frequency multipliers to generate harmonic signals [14–17] (Figure 2.10).

The active-loop techniques simply add other loops because the harmonics are generated by the DUT itself (Figure 2.11) [18, 19].

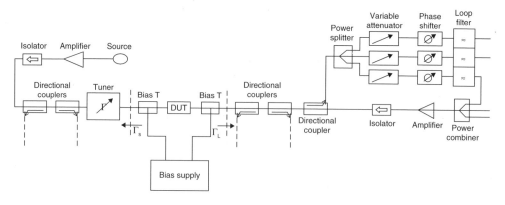

Figure 2.11   A typical set-up for active-loop load pull with individual control of harmonic impedances and passive source pull

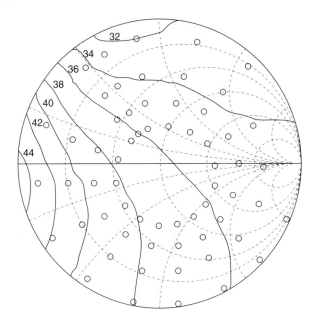

Figure 2.12   Second-harmonic load-pull results for power-added efficiency

As an example, measurements of power-added efficiency as a function of second-harmonic load with interpolated constant-efficiency contours are shown in Figure 2.12.

2.3 THE VECTOR NONLINEAR NETWORK ANALYSER

In this paragraph, the main techniques for nonlinear vector network analysis are reviewed. The required equipment and the performances of each approach are described and compared. Examples of measured data are also given.

It has been shown above (see Section 1.3.1) that when the input signal of a nonlinear system is a sinusoid, the output signal is a sinusoid with all its harmonics, plus a rectified component at zero frequency (DC). In fact, this is not always true, because sub-harmonic components can arise as a result of nonlinear instability; the former case, however, is the common one. A vector nonlinear network analyser (VNNA) is a measurement set-up that is able to measure periodic large-signal waveforms with all their harmonics. In order that the measurements be of any interest, loads different from 50 Ω must be supplied at all harmonic frequencies; this makes the nonlinear VNA actually very close to a vectorial source/load pull.

Let us consider a load/source-pull scheme as seen in the previous paragraph; so far, only scalar measurements have been assumed. Actually, since all incident and reflected waves are available, vectorial measurements are feasible, and in fact very useful. For vectorial measurements, a scheme similar to that of a linear VNA is in place [20, 21]. The input signal is switched between input and output, and a four-port harmonic converter is used as a receiver (Figure 2.13).

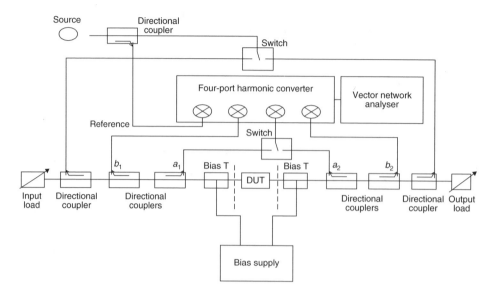

Figure 2.13 Vector nonlinear network analyser with added source/load-pulling facility

The scheme allows for vector error correction in a similar way as for linear measurements; an additional power measurement is required for absolute power evaluation. Typically, a combination of short, open, through, line and matched terminations are used together with a correction algorithm. The combination of vector measurements with harmonic source/load-pulling schemes allows for nonlinear waveforms to be reconstructed with high accuracy under strong nonlinear operations. The higher the number of harmonics, the higher the accuracy of the reconstructed waveforms. As an example, the collector voltage and current waveforms of a power transistor loaded for high-efficiency power amplification are shown in Figure 2.14.

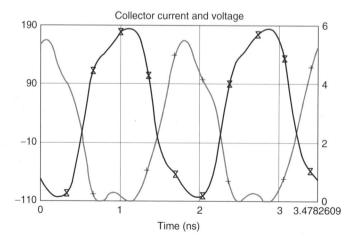

Figure 2.14 Current and voltage waveforms in a power transistor with power loads

An alternative to the linear VNA within a similar type of measurement environment is the waveform analyser, which is essentially a sampling oscilloscope for high frequencies, with 50 Ω probes. Synchronisation must be provided by a reference microwave signal, usually the input signal. Two directional couplers are inserted at each side of the DUT, and active or passive loads terminate the chain. Therefore, the instrument is equivalent to a harmonic vector source/load-pulling set-up [19, 22–25]. An example is shown in Figure 2.15.

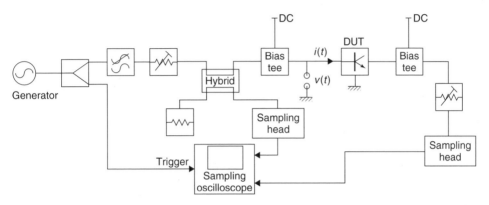

Figure 2.15 A nonlinear vector network analyser based on a fast-sampling oscilloscope

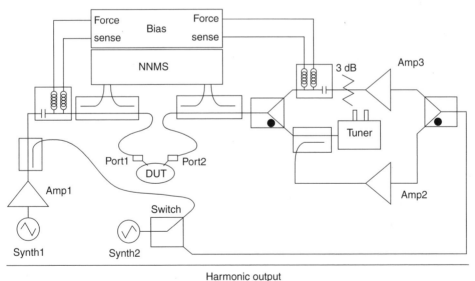

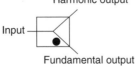

Figure 2.16 Set-up for the measurement of the nonlinear scattering functions

A VNNA can also be used for the modelling of nonlinear devices under large-signal operations. For instance, the device behaviour can be represented within a range of input power and loadings by a black-box equivalent model, whose behaviour is linearised around a range of large-signal working points [26], and then modelled by means of a neural network or any similar approximating and interpolating system. The corresponding experimental set-up is shown in Figure 2.16.

2.4 PULSED MEASUREMENTS

In this paragraph, the currently available techniques for DC and RF pulsed measurements are described. Examples of measured data are also given.

An active device is characterised for linear applications in a small neighbourhood of an operating point, corresponding to a given value of quiescent voltages and currents. The quiescent point determines the state of some 'slow' phenomena within the device, that is, phenomena with long time constant [27–31]; they are mainly the thermal processes, determining the temperature of the device, and the carrier generation and recombination processes, determining the trap occupancy within the semiconductor. The time constants are in the order of seconds to milliseconds for thermal phenomena and down to microseconds for trapping and de-trapping phenomena. A superimposed microwave signal, be it small or large, does not have the time within a microwave period to affect any of these phenomena. The microwave properties of the device are however affected by the 'quiescent' state of the device: by way of example, it is obvious that a high temperature affects the microwave gain of a device. Therefore, temperature and bias voltages and currents must be specified when a microwave measurement is performed.

These considerations are obvious for small-signal characterisation; they are less so for large-signal measurements. Let us illustrate this with a few examples.

Let us consider the output (drain–source) $I–V$ characteristics of a power FET measured in DC conditions. A high-current characteristic curve will show a negative slope with respect to drain–source voltage (Figure 2.17); this implies a negative output (drain–source) conductance. In fact, the decrease in current is due to an increase in the temperature of the device, which reduces the carrier mobility within the device, which in turn decreases the current in the channel.

If the output conductance around a high-current bias point is measured with a small microwave signal, as for instance with a linear VNA, it is always found to be positive, except for very special cases. This is because the microwave small signal does not have enough time within its period to warm up or to cool the device; it is therefore a constant-temperature measurement. The same can be said for trapping or de-trapping phenomena that do not simply happen during a microwave cycle. The small-signal measurement is therefore an isothermal and 'isotrap' measurement.

Let us consider now the same transistor pinched off by a negative DC gate–source voltage (for instance −3 V), while the DC drain–source voltage is well beyond the knee voltage (for instance 6 V). The drain current is zero, and the device is cold. When a large

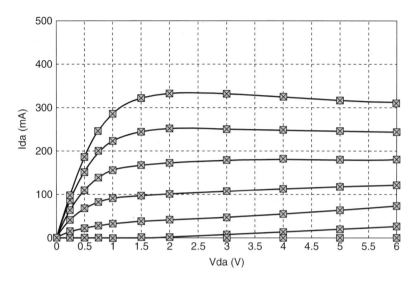

Figure 2.17 Output I/V curves of a power transistor

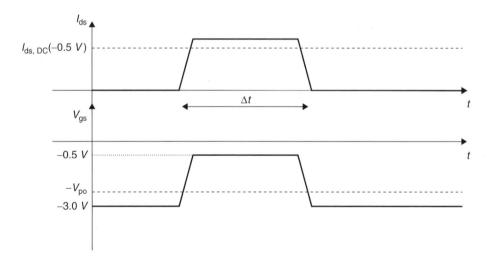

Figure 2.18 Gate-voltage pulse and pulsed drain current

gate–source voltage step is applied (for instance a 2.5 V pulse), the current flows in the channel, to be pinched off again when the input pulse goes off (Figure 2.18).

If the pulse is short enough, say below 1µs, the transistor does not warm up, and no trapping or de-trapping takes place within it. The current is therefore higher than in the case when the same total gate–source voltage is applied statically, that is, a −0.5 V DC gate–source voltage in the case of our example.

Let us now consider a transistor biased at two different quiescent points, dissipating the same power, because $P_{\text{diss}} = V_{\text{ds,bias\_1}} \cdot I_{\text{ds,bias\_1}} = V_{\text{ds,bias\_2}} \cdot I_{\text{ds,bias\_2}}$ is the same; in our case, it is $V_{gs,bias\_1} = -0.75$ V, $V_{ds,bias\_1} = 3$ V and $V_{gs,bias\_2} = 0$ V, $V_{ds,bias\_2} = 1$ V. The temperature is the same at both the bias points, but trap occupancy is probably different because of the different depths of the depleted regions within the semiconductor. We can now apply two simultaneous pulses to the gate and drain electrodes, and measure the current during the short pulses; the temperature and trap states will not change during the pulses. By varying the amplitude of the pulses, all the output I/V characteristics are measured from each bias point. Comparison of the two sets of curves (Figure 2.19) shows that the trap state also plays a role in determining the characteristics and therefore the performances of the device.

A consequence of what has been seen above is that the output current must be measured with short pulses, starting from the actual static condition that will be present during large-signal operations. This means that both DC bias voltages and device temperature must be the same as in large-signal operations. The device temperature does not depend on bias voltages and currents only, that is, from the DC power dissipated in the device, since heat removal and external environment temperature can actually differ from case to case. The instantaneous drain current can therefore be written as follows:

$$I_{ds}(t) = I_{ds}(V_{gs,DC}, V_{ds,DC}, T, V_{gs}(t), V_{ds}(t)) \tag{2.2}$$

The instantaneous current is a function of the static DC voltages and average temperature, and of the instantaneous 'fast' voltages [31, 32].

Several possibilities are available for pulsed I/V curve measurements. A possible scheme is shown in Figure 2.20.

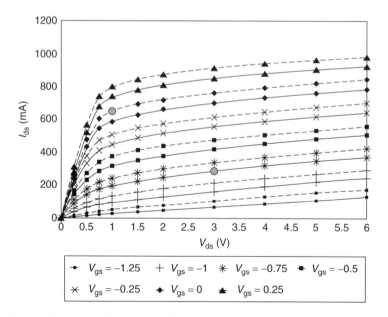

Figure 2.19 Pulsed output I/V curves of a power transistor from two different bias points, marked with circles ($V_{gs} = -0.75$ V, $V_{ds} = 3$ V and $V_{gs} = 0$ V, $V_{ds} = 1$ V)

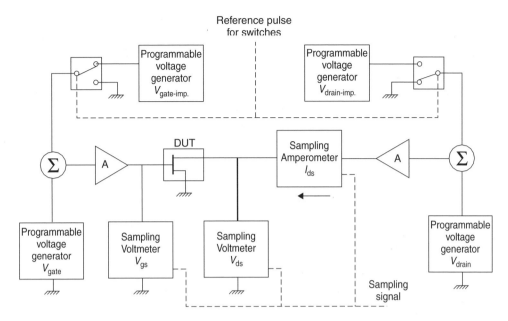

Figure 2.20   A possible scheme for measuring the pulsed output I/V characteristics of an FET

The set-up is based on two dual bias supplies and on electronic switches. The gate–source voltage is switched between a static value and a pulsed value by means of a standard CMOS switch. Since the gate pulse must not provide any current, a standard CMOS amplifier is fast enough to supply a fast pulse to the gate of the device. The voltage is sampled by means of a sampling oscilloscope. The drain–source voltage is switched by means of fast transistors (for example, two complementary HEXFETs with high current capability) between two bias supplies, a Schottky diode and a load resistance. The voltage is sampled at both ends of a current-viewing resistor that also prevents oscillations in the device; drain–source voltage and drain current are sampled by means of other channels of the digital-sampling oscilloscope. A typical sequence of pulsed voltages and currents is shown in Figure 2.21.

Another scheme, including temperature control, is shown in Figure 2.22 [31].

It must be remarked that the DC voltages are not always *a priori* known in large-signal operations, because of the rectification phenomena that are actually often present. Operating temperature is not always *a priori* known either, as remarked above. A complete characterisation of an active device for nonlinear applications must therefore include measurements in several different static conditions for a complete characterisation even if the application (e.g. Class-B power amplifier) is known.

Pulsed I/V curves are isothermal and 'isotrap', when correctly performed. Their partial derivatives with respect to gate and drain voltages in the bias point must coincide with the transconductance and output conductance of the small-signal equivalent circuit when measured at the same bias point by means of a dynamic small-signal measurement,

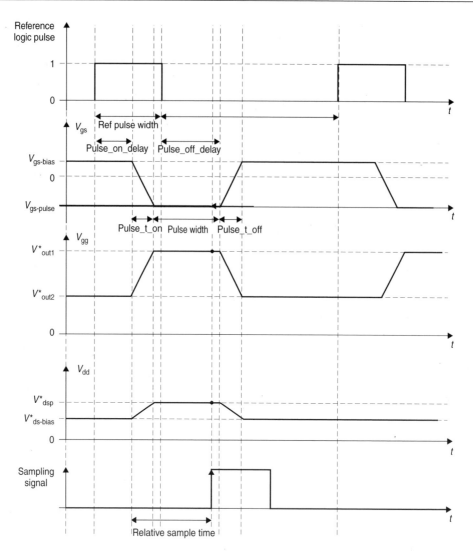

Figure 2.21 Typical voltages and currents for the measurement of pulsed I/V characteristics

for example linear S-parameters:

$$\frac{\partial I_{ds}}{\partial V_{gs}}\bigg|_{V_{ds}=\text{const}} = g_m \qquad \frac{\partial I_{ds}}{\partial V_{ds}}\bigg|_{V_{gs}=\text{const}} = g_{ds} \qquad (2.3)$$

Similarly, the integral of transconductance and output conductance around a closed contour in the $V_{gs} - V_{ds}$ plane should sum up to zero:

$$\oint (g_m \cdot dV_{gs}(t) + g_{ds} \cdot dV_{ds}(t)) = \oint \frac{dI_{ds}}{dV(t)} \cdot \frac{dV(t)}{dt} \cdot dt = \oint_T \frac{dI_{ds}}{dt} \cdot dt = 0 \qquad (2.4)$$

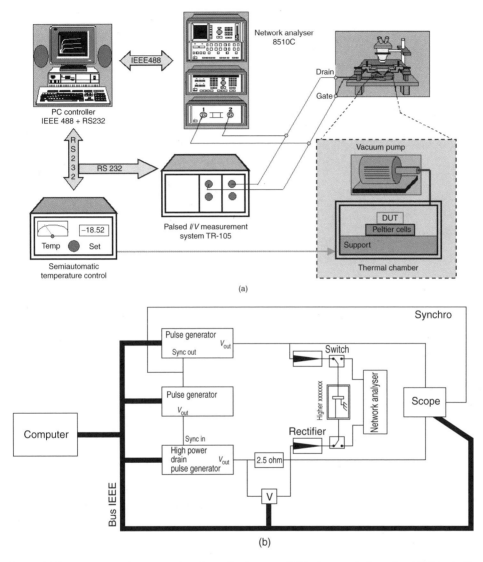

Figure 2.22 A scheme for measuring the pulsed output I/V characteristics of an FET including temperature control: general arrangement (a) and pulse generator (b)

or, for a periodic circuit after a cycle, the current must have the same value as at the beginning of the cycle (Figure 2.23).

However, the transconductance and the output conductance, as evaluated from S-parameters, are not the derivatives of the isothermal and 'isotrap' I/V characteristic, because they are measured in different static conditions along the closed contour. In fact, every small-signal S-parameter measurement is made at a specific bias point, while the closed contour is followed by the dynamic operating point in isothermal and 'isotrap' conditions corresponding to the quiescent point of the large-signal operations.

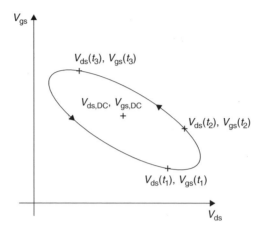

Figure 2.23 Isothermal cycle in the $V_{gs} - V_{ds}$ plane

This has deep implications in the extraction of a large-signal model, as will be described in Chapter 3.

Similar considerations can be made for capacitances in an active device as partial derivatives with respect to gate–source and drain–source voltages of the charges stored in the device itself. Capacitances are usually evaluated from the small-signal S-parameters measured at many bias points, and therefore suffer from the same limitations as described for conductances. It is, however, not as straightforward and easy to measure charges in pulsed conditions as it is for currents. A solution consists of performing pulsed S-parameters measurement, that is, small-signal dynamic measurements taken in a very short time during a pulsed step of gate–source and drain–source voltage. The instrumentation is very complex, requiring the superposition of a sinusoidal test signal on a bias voltage step during less than a microsecond and the measurement of S-parameters in the same short time [33].

The conductances and capacitances extracted from pulsed S-parameters measurements prove to be actual partial derivatives of single-valued functions, that is, current and charge respectively. A very large amount of data must be measured by means of a costly equipment requiring a corresponding effort for data processing, making the total cost of this approach very high.

2.5 BIBLIOGRAPHY

[1] J.M. Cusack, S.M. Perlow, B.S. Perlman, 'Automatic load contour mapping for microwave power transistors', *IEEE Trans. Microwave Theory Tech.*, **MTT-22**, 1146–1152, 1974.
[2] F. Sechi, R. Paglione, B. Perlman, J. Brown, 'A computer-controlled microwave tuner for automated load pull', *RCA Rev.*, **44**, 566–572, 1983.
[3] J. Fitzpatrick, 'Error models for systems measurement', *Microwave J.*, **21**, 63–66, 1978.
[4] R.S. Tucker, P.B. Bradley, 'Computer-aided error correction of large-signal load-pull measurements', *IEEE Trans. Microwave Theory Tech.*, **MTT-32**, 296–300, 1984.

[5] I. Hecht, 'Improved error-correction technique for large-signal load-pull measurement', *IEEE Trans. Microwave Theory Tech.*, **MTT-35**, 1060–1062, 1987.

[6] A. Ferrero, U. Pisani, 'An improved calibration technique for on-wafer large-signal transistor characterisation', *IEEE Trans. Instrum. Meas.*, **42**(2), 360–364, 1993.

[7] C. Tsironis, 'Two-tone intermodulation measurements using a computer-controlled microwave tuner', *Microwave J.*, **32**, 156–161, 1989.

[8] C. Tsironis, 'A novel design method of wideband power amplifiers', *Microwave J.*, **35**, 303–304, 1992.

[9] G. Madonna, A. Ferrero, M. Pirola, U. Pisani, 'Testing microwave devices under different source impedance values – a novel technique for on-line measurement of source and device reflection coefficients', *IEEE Trans. Instrum. Meas.*, **49**(2), 285–289, 2000.

[10] R. Stancliff, D. Poulin, 'Harmonic load-pull', *IEEE MTT-S Int. Microwave Symp. Dig.*, 1979, pp. 185–187.

[11] Y. Takayama, 'A new load-pull characterisation method for microwave power transistors', *IEEE MTT-S Int. Microwave Symp. Dig.*, June 1976, pp. 218–220.

[12] G.P. Bava, U. Pisani, V. Pozzolo, 'Active load technique for load-pull characterisation at microwave frequencies', *Electron. Lett.*, **18**(4), 178–180, 1982.

[13] R.B. Stancliff, D.D Poulin, 'Harmonic load pull', *Proc. IEEE Int. Microwave Symp.*, 1979, pp. 185–187.

[14] F.M. Ghannouchi, R. Larose, R.G. Bosisio, 'A new multiharmonic loading method for large-signal microwave and millimetre-wave transistor characterisation', *IEEE Trans. Microwave Theory Tech.*, **MTT-39**(6), 986–992, 1991.

[15] G. Berghoff, E. Bergeault, B. Huyart, L. Jallet, 'Automated characterisation of HF power transistors by source-pull and multiharmonic load-pull measurements based on six-port techniques', *IEEE Trans. Microwave Theory Tech.*, **MTT-46**(12), 2068–2073, 1998.

[16] D.-L Lê, F.M. Ghannouchi, 'Multitone characterisation and design of FET resistive mixers based on combined active source-pull/load-pull techniques', *IEEE Trans. Microwave Theory Tech.*, **MTT-46**(9), 1201–1208, 1998.

[17] P. Heymann, R. Doerner, M. Rudolph, 'Multiharmonic generators for relative phase calibration of nonlinear network analysers', *IEEE Trans. Instrum. Meas.*, **50**(1), 129–134, 2001.

[18] A. Ferrero, U. Pisani, 'Large-signal 2nd harmonic on-wafer MESFET characterisation', *36th ARFTG Conf. Dig.*, Monterey (CA), Nov. 1990, pp. 101–106.

[19] D. Barataud, F. Blache, A. Mallet, P. Bouisse, J.-M. Nebus, J.P. Illotte, J. Obregon, J. Verspecht, Ph. Auxemery, 'Measurement and control of current/voltage waveforms of microwave transistors using a harmonic load-pull system for the optimum design of high efficiency power amplifiers', *IEEE Trans. Instrum. Meas.*, **48**(4), 835–842, 1999.

[20] B. Hughes, A. Ferrero, A. Cognata, 'Accurate on-wafer power and harmonic measurements of mm-wave amplifiers and devices', *IEEE MTT-S Int. Symp. Dig.*, Albuquerque (NM), June 1992, pp. 1019–1022.

[21] D.D. Poulin, J.R. Mahon, J.P. Lantieri, 'A high power on-wafer active load-pull system', *IEEE MTT-S Int. Symp. Dig.*, Albuquerque (NM), 1992, pp. 1431–1433.

[22] M. Sipilä, K. Lehtinen, V. Porra, 'High-frequency periodic time-domain waveform measurement system', *IEEE Trans. Microwave Theory Tech.*, **MTT-36**(10), 1397–1405, 1988.

[23] G. Kompa, F. Van Raay, 'Error-corrected large-signal waveform measurement system combining network analyser and sampling oscilloscope capabilities', *IEEE Trans. Microwave Theory Tech.*, **MTT-38**(4), 358–365, 1990.

[24] M. Demmler, P.J. Tasker, M. Schlechtweg, 'A vector corrected high power on-wafer measurement system with a frequency range for higher harmonics up to 40 GHz', *Proc. EuMC*, 1994.

[25] J. Verspecht, P. Sebie, A. Barel, L. Martens, 'Accurate on-wafer measurement of phase and amplitude of the spectral components of incident and scattered voltage waves at the signal

ports of a onlinear microwave device', *IEEE MTT-S Int. Symp. Dig.*, Orlando (FL), 1995, pp. 1029–1032

[26] J. Verspecht, P. Van Esch, 'Accurately characterizing of hard nonlinear behavior of microwave components by the nonlinear network measurement system: introducing the nonlinear scattering functions', *Proc. INNMC'98*, Duisburg (Germany), Oct. 1998, pp. 17–26.

[27] M. Paggi, P.H. Williams, J.M. Borrego, 'Nonlinear GaAs MESFET modelling using pulsed gate measurements', *IEEE Trans. Microwave Theory Tech.*, **MTT-36**(12), 1593–1597, 1988.

[28] C. Camacho-Peñalosa, C. Aitchison, 'Modelling frequency dependence of output impedance of a microwave MESFET at low frequencies', *Electron. Lett.*, **21**, 528–529, 1985.

[29] P. Ladbrooke, S. Blight, 'Low-field low-frequency dispersion of rate-dependent anomalies', *IEEE Trans. Electron Devices*, **35**, 257–267, 1988.

[30] J.M. Golio, M.G. Miller, G.N. Maracas, D.A. Johnson, 'Frequency-dependent electrical characteristics of GaAs MESFETs', *IEEE Trans. Electron Devices*, **ED-37**, 1217–1227, 1990.

[31] F. Filicori, G. Vannini, A. Mediavilla, A. Tazon, 'Modelling of the deviations between static and dynamic drain characteristics in GaAs FET's', *Proc. EuMC Conf.*, Madrid (Spain), 1993, pp. 454–457.

[32] J.F. Vidalou, F. Grossier, M. Camiade, J. Obregon, 'On-wafer large-signal pulsed measurements', *IEEE MTT-S Int. Symp. Dig.*, 1989, pp. 831–834.

[33] J.-P. Teyssier, Ph. Bouisse, Z. Ouarch, D. Barataud, Th. Peyretaillade, R. Quéré, 40-GHz/150-ns versatile pulsed measurement system for microwave transistor isothermal characterisation', *IEEE Trans. Microwave Theory Tech.*', **MTT-46**(12), 2043–2052, 1998.

3

Nonlinear Models

3.1 INTRODUCTION

In this introduction, some general concepts are introduced, together with the main types of models available both in the literature and in commercial simulators.

We have seen in Chapter 1 that a nonlinear active device is represented by a nonlinear model for nonlinear circuit analysis. The model must reproduce the electrical behaviour of the device in large-signal operating conditions for the accurate prediction of circuit performances. In general, it is pointless to reduce the simulation error below the reproducibility of the technology, both for the active and for the passive elements. In fact, the currently available nonlinear analysis algorithms are usually accurate enough to make the numerical or truncation error small enough for practical purposes. The current limit to the accuracy of the simulation lies in the limited capability to accurately model the electrical behaviour of the elements of the circuit. For passive elements, this may depend on the simplified electromagnetic representation of the single elements or on the limited capability to take into account the electromagnetic interactions among different elements within the circuit. For active devices, the picture is more complex. In part, an oversimplified structure of the model with respect to the actual device may account for a loss of accuracy. However, another serious shortcoming usually comes from the lack of suitable measurements.

In general, models for active devices belong to two categories: physical and empirical models. Physical models describe the device in terms of its physical structure and predict its performances by means of electromagnetic and charge transport equations. In principle, the behaviour of the device can be predicted *a priori*, without the need for actually fabricating the device itself. In practice, some parameters must be adjusted *a posteriori*, because not everything is known of the actual phenomena taking place inside the device, and because of the tolerances of the fabrication process. Moreover, the physical equations are usually simplified in order to keep the numerical burden to a manageable level; as a consequence, some empirical parameters must do for the missing terms in the equations. Anyhow, physical models tend to be computationally heavy, and their accuracy is usually below acceptable levels for circuit design. Their use lies essentially in the

Nonlinear Microwave Circuit Design   F. Giannini and G. Leuzzi
© 2004 John Wiley & Sons, Ltd   ISBN: 0-470-84701-8

possibility to optimise a device before fabrication, at least preliminarily, and in a better comprehension of the device behaviour and possible causes of misfunctioning. They are also useful for yield optimisation when their computational cost is low enough.

Empirical models are extracted from data measured on the fabricated device. They may include some *a priori* knowledge of the physical structure of the device, or they may be a powerful and flexible interpolating scheme: in the latter case they are usually referred to as black-box models, while in the former case they are referred to as equivalent-circuit models. Empirical models vary greatly in the required amount of measured data, extraction procedure, operating regime and, of course, accuracy. While they may differ in computational burden, they are orders of magnitude faster than physical models, and usually fast enough for interactive nonlinear analysis. When properly defined and extracted, they are also accurate enough for practical circuit design. What can actually be burdensome in an empirical model is the amount of measurements required for its extraction, and usually also the extraction procedure itself. Quite often, the extraction procedure requires a good deal of skill; it is sometimes wise not to try to extract a general model valid for all operating regimes, but it is better to limit oneself to the extraction of a model valid for specific operating regimes or circuit applications.

A separate category is constituted by simple or simplified models. Their structure is simple enough to avoid the need for the cumbersome analysis algorithms described in Chapter 1 and to allow for intuitive reasoning. On the other hand, their accuracy is very limited, but nonetheless sufficient to understand the main design topics. Typical examples are piecewise-linear or simple polynomial models. Obviously, the simplification is done in such a way that the main effects due to nonlinearities are preserved, while minor effects are neglected. A great deal of *a priori* knowledge of the behaviour of the device is required, and usually the model is tailored to a specific application (e.g. power amplifiers).

The three main types of nonlinear models will be described in the following paragraphs, with the main emphasis being put on empirical models, by far the most commonly used in nonlinear CAD.

3.2 PHYSICAL MODELS

In this paragraph, the main types of physical models are described. The models are classified on the basis of their basic equations, and their performances are described.

3.2.1 Introduction

Physical models cover a wide range of active devices and differ very widely for the formulation of the equations they are based upon. The motion equation for the charged particles (electrons and/or holes) can be the quantistic Schrödinger's equation, or the semiclassical Boltzmann's equation, depending on the scale of the microscopic phenomena. In electronic devices, the electric field always appears among the causes of the motion of the particles inside the semiconductor; therefore, Poisson's equation must also be added since it links electric field and charged particles inside the semiconductor. No realistic device has been so far demonstrated where Poisson's equation must be replaced by Maxwell's equations, although full electromagnetic/transport models have been developed [1].

The equations are written in three dimension for a full description of the effects of actual devices (3D models). Normally, however, only a section of the device is taken into account, assuming that the device is uniform in the lateral direction, saving a lot of computational effort (2D models). In some cases, only the main motion direction of the charged particles is considered (e.g. from source to drain, or from emitter to collector), resulting in a one-dimensional model; in this case, the results are even less accurate, but the computational effort is drastically reduced (1D models). Often, the second direction is somehow taken into account by means of additional equations, yielding a quasi-two-dimensional model (quasi-2D models).

The mathematical description strongly depends on the formulation of the model. Sometimes, the device is described in terms of Schrödinger's and Poisson's equations only. More often, quantistic effects take place only in a limited part of the device, for example, at a heterojunction or in a resonant superlattice structure; the rest of the device is well described by semiclassical equations. In other cases, only a semiclassical motion equation and Poisson's equation are required for sufficient accuracy. The semiclassical Boltzmann's equation is often expanded in moments, by suitable integration in the momentum space, and then only the first ones are retained. The first moment is the particle conservation equation or continuity equation; the second is the momentum conservation equation or current density equation; the third is the energy conservation equation, and so on.

Once the type of formulation is defined, the main forces driving the motion of the charged particles must be written into the equations in such a way that only the meaningful physical effects are retained and the solution remains reasonably simple and accurate. It is often not easy to describe the forces with sufficient detail and accuracy, given their mathematically complicated formulation and the uncertainty on parameters related to fabrication process. A compromise must be sought, sometimes by means of semi-empirical parameters derived from measurements or practical evidence.

All equations are differential with respect to space and time. In the special case of steady-state models, the dependence on time is removed, and in very simple models, the dependence on space can be made to be non-differential; in any case, the equations require boundary conditions. For Poisson's equation, in most cases the boundary conditions are given by the applied external voltage. For the transport equation, they must be a physically meaningful condition: for example, neutrality or equilibrium conditions very far from the junctions or channel. The solution of the differential equations then usually requires a numerical solving scheme, and the model is said to be numerical. Solving schemes can be stochastic or deterministic, that is, incorporating or not incorporating the statistical properties of the microscopic behaviour of particles in a semiconductor. When the mathematical formulation is simple enough to allow for an analytical, explicit solution in terms of external voltages and currents, the model is said to be analytical. Only very simplified models belong to this category, and their utility lies essentially in their simplicity and clearness of description; their accuracy is often below acceptable limits for practical circuit simulation.

Physical models are normally used to predict the behaviour of the intrinsic part of an active device; for a realistic evaluation of its performances, it is often necessary to add parasitic elements as contact resistances, pad capacitances or line inductances; these can hardly be theoretically predicted, and are usually evaluated by means of empirical or semi-empirical expressions. The evaluation of the effects due to the layout of the device is also difficult and has been recently addressed by coupling an electromagnetic analysis of the connecting parts of the device to the physical study of the intrinsic part. This approach is probably going to gain importance as the operating frequency increases in the millimetre-wave region and beyond.

3.2.2 Basic Equations

We write here the basic equations in one dimension for simplicity; Schrödinger time-dependent equation reads [2]

$$i\frac{\partial \varphi}{\partial t} = -\hbar^2 \frac{\partial^2 \varphi}{\partial x^2} + q(V - V_0) \tag{3.1}$$

where $\varphi(x, t)$ is the probability function or wave function, $h = 2\pi \cdot \hbar$ is Planck's constant, i is the imaginary unit and V is the electrical potential. The derivative of the wave function with respect to time corresponds quantistically to the energy of the particle. The derivative of the wave function with respect to the space variable times Planck's constant corresponds to the classical momentum. The equation states that the energy of the particle equals the sum of the square of the momentum (the kinetic energy) and of the potential energy. If the potential is time invariant, the time-invariant Schrödinger equation is obtained as

$$w = -\hbar^2 \frac{\partial^2 \varphi}{\partial x^2} + q(V - V_0) \tag{3.2}$$

where w is the time-invariant energy of the particle, and its solution is a time-invariant quantistic state, or level, for the electron. A time-dependent solution of Schrödinger's equation (3.2) is obtained as a linear superposition of time-invariant solutions or states.

Boltzmann's semiclassical equation reads [3]

$$\frac{dF}{dt} = \frac{\partial k}{\partial t}\frac{\partial F}{\partial k} + \frac{\partial x}{\partial t}\frac{\partial F}{\partial x} + \frac{\partial F}{\partial t} \tag{3.3}$$

where $F(x, k, t)$ is the time-dependent distribution function of the particle in the real space (x) and momentum space (k). The derivative of the momentum k with respect to time is the externally applied force, while the derivative of the space variable x with respect to time is the velocity of the particle; the derivative of the distribution function with respect to the momentum k is the effective mass of the particle. The equation states that the particle changes its momentum or position if an externally applied force is present, because of its inertial motion, or if the system is time-variant $\left(\dfrac{\partial F}{\partial t}\right)$. By multiplication by the powers of the momentum k^0, k^1, k^2, and so on and integration with respect to k, that is, by saturation of the momentum space, the moments of Boltzmann's equation are obtained. By integration over the momentum space, the distribution function reduces to the particle density in the real space only. In the case of electrons, the zeroth-order moment is

$$\frac{dn}{dt} = -v \cdot \frac{\partial n}{\partial x} + \left(\frac{\partial n}{\partial t}\right)_{\text{coll.}} \tag{3.4}$$

where n is the electron density and v is the electron velocity; this is the particle conservation equation or continuity equation. The last term is the recombination term due to collisions (with holes). The velocity is obtained from the first-order moment that reads

$$\frac{d(n \cdot v)}{dt} = \frac{n \cdot qE}{m_{\text{eff}}} - \frac{\partial (n \cdot v^2)}{\partial x} - \frac{2}{3} \cdot \frac{\partial \left(\dfrac{n \cdot w}{m_{\text{eff}}} - \dfrac{n \cdot v^2}{2}\right)}{\partial x} - \left(\frac{\partial (n \cdot v)}{\partial t}\right)_{\text{coll.}} \tag{3.5}$$

where qE is the external (Coulomb) force applied to the electron, m_{eff} is the effective mass of the electron, w is the electron energy and the last term is the contribution of collisions; this is the momentum conservation equation or the current density equation. This equation states that the momentum of a particle changes if there is an external force, if there are diffusion or inertial phenomena or because of collisions. The energy is obtained by the second-order moment that reads

$$\frac{d(n \cdot w)}{dt} = n \cdot qE \cdot v - \frac{\partial (n \cdot (w + k_{\text{B}}T) \cdot v + n \cdot Q)}{\partial x} - \left(\frac{\partial (n \cdot w)}{\partial t}\right)_{\text{coll.}} \tag{3.6}$$

where k_{B} is Boltzmann's constant, Q is the heat flux and the last term is the contribution of collisions. The equation states that the energy of a particle changes if there is an external power, if there are diffusion or inertial phenomena or because of collisions. The heat flux is obtained by the third-order moment; however, it is usually neglected or approximated. In this way, the expansion is truncated; the higher-order moments could nevertheless be obtained in a similar way.

Poisson's equation in one dimension is

$$\frac{\partial E}{\partial x} = -\frac{\rho}{\varepsilon} \tag{3.7}$$

where E is the electric field, ρ is the charge density and ε is the dielectric constant.

It is worth remarking that Boltzmann's equation takes into account the statistical characteristics of particle motion in the semiconductor through the collision terms; on the other hand, Schrödinger's equation does not account for statistical information. Thus, the quantum equivalent of the Boltzmann's equation is not directly the Schrödinger equation but the Liouville equation for the density matrix. In many cases, however, statistical information can be easily coupled to Schrödinger equation, and this will be sufficient for a correct description of the system [2].

3.2.3 Numerical Models

When the equations together with their boundary conditions cannot be solved explicitly, and no analytical solution is available, a numerical solving scheme must be implemented. The solving schemes belong to two main categories: deterministic and stochastic. In the first approach, the charged particles in the semiconductor are approximated by a single-particle gas, where the motion of each particle is driven by deterministic forces that are the same for each particle in the gas; all particles in the gas are identical. In some cases, the charged particles are approximated by two gases, one for high-energy and one for low-energy particles. The equations are solved once by exact calculations (in numerical sense), and the macroscopic quantities of interest are computed for the whole gas. In the second approach, the forces and events that determine the motion of a particle retain their statistical nature so that the trajectory of each particle is different from that of any other in the gas. In this second approach, at the basis of the so-called Monte Carlo models, macroscopic quantities like current or gas temperature are computed as an average of the behaviour of the population of the gas; a large number of particles must be simulated in order to get statistically meaningful results.

We first describe deterministic models. The models are classified by the type of equations (quantistic, semiclassical, mixed; Poisson's or Maxwell's equation); if semi-classical, by the number and type of included moments of Boltzmann's equation (hydro-dynamic, energy-balance, drift-diffusion, etc.); by the number of dimensions (3D, 2D, quasi-2D, 1D); by the dependence on time (steady-state, time-dependent); and by the regime of validity (small-signal, large-signal, noise). The detailed description of the physical effects included in the different formulations, with the related accuracy, is beyond the scope of this book; therefore, not all possible combinations are described in the following but only some examples that are particularly meaningful for a circuit designer.

Let us now describe the semiclassical models. The first three moments of Boltzmann's equation are relative to the conservation of the number of particles, the conservation of their momentum and the conservation of their energy; each equation includes derivatives with respect to time and space. If the time or space scale of the microscopic events is much smaller than that of the characteristic times or distances of the device, the importance of the corresponding terms is very small, and they can be neglected. In this way, simpler expressions are derived. Moreover, phenomena related to the energy of the particles are sometimes neglected, allowing the suppression of the energy conservation equation; phenomena like the velocity overshoot cannot be included in this case, but an additional simplification is obtained.

The equations are discretised in the space domain and in the time domain, if time dependence is included, so that the system of partial differential equations is replaced by a finite-difference equation system [4]. The system is usually solved in the time domain, with similar schemes as those described for nonlinear circuits (Section 1.2). The choice of the space step in one or more dimensions (the mesh) and of the time step is criti-cal, and many schemes have been devised for automatic adjustment ensuring numerical convergence. Recently, a Fourier series approach has also been proposed, making use of harmonic balance or waveform balance techniques [5], that has demonstrated computation speed advantages especially for intermodulation prediction, while being obviously limited to steady-state analysis.

Models including the first three moments are usually referred to as hydrodynamic models. Several dynamic, 2D formulations have been proposed in the literature, most of them very powerful [6–11]; hot electrons effects, velocity overshoot, penetration into the substrate, channel dynamics and other effects are predicted quite accurately. Unfortunately, these models tend to have critical numerical convergence properties, and not always are a solution of the equations obtained. Typical applications are the analysis of MESFETs, HEMTs and HBTs on GaAs-based material. A typical result for electron temperature distribution in the section of a high electron mobility transistor (HEMT) is shown in Figure 3.1.

If the energy conservation equation is neglected, many of the stiffest numerical convergence problems are removed; however, phenomena related to the heating of carriers in the semiconductor cannot be accounted for [12–15]. For example, velocity overshoot phenomena in a short channel of an FET or in a very short base of a heterojunction bipolar transistor (HBT) are not accounted for, somehow restricting the accuracy especially for very high-frequency devices. However, the relative simplicity of the equations and the reliability of use make this model very popular, especially for commercial software. Several commercial companies actually offer flexible and powerful simulation tools.

As a comparison between the hydrodynamic and the drift-diffusion approaches, in Figure 3.2, the drain current of an HEMT is plotted as a function of gate–source voltage for a constant drain–source voltage.

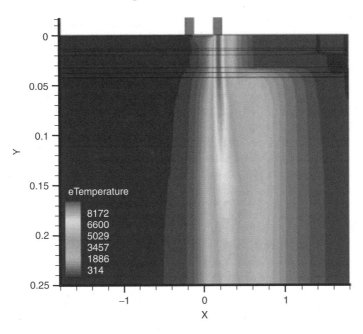

Figure 3.1 Electron temperature distribution in the section of an HEMT, computed with a 2D drift-diffusion model

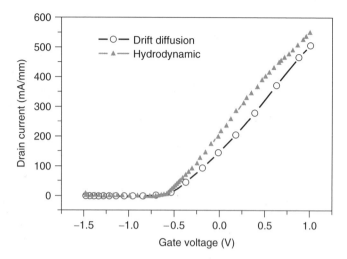

Figure 3.2   Drain current computed with a 2D drift-diffusion model and a hydrodynamic model

Several quasi-2D hydrodynamic models have been developed for both MESFETs and HEMTs. The control of channel depletion (for MESFETs) or electron density at the heterojunction (for HEMTs) is computed in the vertical dimension by the solution of Poisson's equation (for MESFETs) by an approximation of a self-consistent solution of Schrödinger's and Poisson's equations (for HEMTs); the charge transport in the source-to-drain direction is analysed by means of a semiclassical approach (eqs. (3.4)–(3.7)) with various degrees of approximations. The channel control and the charge transport are then coupled in a self-consistent solution. The quasi-2D approach combines a reduced computational burden with a reasonably complete set of equations and an often-acceptable accuracy. Several formulations have been proposed for small-signal [16, 17], large-signal [18–23] and noise applications [16].

Stochastic models are often referred to as Monte Carlo models. In this approach, the physics of the semiconducting material are easily taken into account in a very detailed way: for instance, an accurate description of the band structure, or detailed models of the scattering phenomena, can be included in the equations. The model typically includes Boltzmann's equation coupled to Poisson's equation [24–27]. An example of detailed physical data obtained by a Monte Carlo approach is shown in Figure 3.3: the evolution of charges within the section of an HEMT at the startup of breakdown.

Physical models are traditionally used for device optimisation; however, many applications within circuit solvers have been proposed. The first applications made use of algorithms based on Kirchhoff's equations [11, 22, 23, 28, 29]. A goal of this approach consists of an evaluation of the effects of fabrication tolerances on circuit performances towards a statistical analysis and design for yield maximisation [29]. However, analytical models (see below) are better used for this application, given the large computational effort required by this approach. Recently, the physical models have been included in electro-magnetic field solvers, in order to better evaluate the effect of the environment on the performances of the device, especially at very high frequencies [30–33]. As an example,

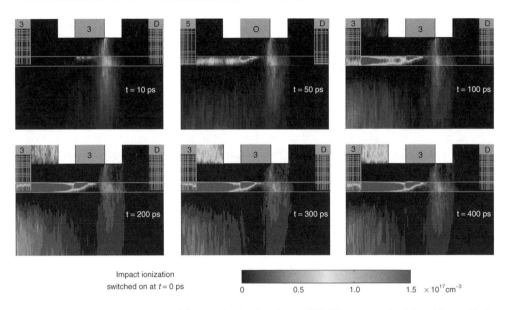

Impact ionization
switched on at $t = 0$ ps

0   0.5   1.0   1.5 $\times 10^{17}$cm^{-3}

Figure 3.3 Evolution in time of impact ionisation in an HEMT computed with a Monte Carlo method

a dynamic, large-signal quasi-2D model of an HEMT coupled to an FDTD numerical electromagnetic field solver for the prediction of the effects of layout at millimetre-wave frequencies is shown in Figure 3.4 [33]: the active region is divided into several individual quasi-2D devices, then coupled to the electromagnetic field by means of equivalent nonlinear controlled current sources.

A different approach for the exploitation of physical models has also been proposed. The model is used for the generation of data, from which an equivalent circuit is extracted; the latter is actually used for the circuit analysis. Many examples have been demonstrated: a 2D energy-balance model [34], a standard 2D drift-diffusion model [29] and a 2D drift-diffusion model Fourier-transformed for efficient small-signal frequency-domain application [35], a quasi-2D static model [19, 20], have been used as measurement

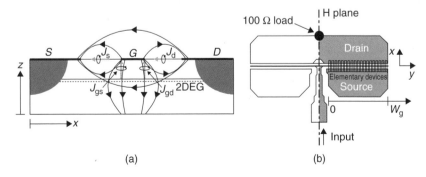

(a)   (b)

Figure 3.4 Cross section and top view of an HEMT showing the intrinsic and extrinsic regions

data simulators for the successive extraction of an equivalent-circuit model of GaAs MES-FETs. This approach has a great potential for providing a better insight into the physical meaningfulness of the elements of the equivalent circuit.

3.2.4 Analytical Models

The transport equations can be solved analytically if some simplifications are made. A well-known approach includes the first two moments of Boltzmann's equation (continuity and current equations, eqs. (3.4) and (3.5)), and Poisson's equation (eq. (3.7)), in one dimension and in steady state; a simplified expression for energy is adopted, and its derivatives with respect to space are neglected. For electrons we have (see Appendix A.9)

$$\frac{dn}{dt} = 0 = \frac{\partial j_n}{\partial x} - \frac{(n - n_0)}{\tau_n} \tag{3.8}$$

$$j_n = q D_n \cdot \frac{\partial n}{\partial x} + q \mu_n \cdot E \tag{3.9}$$

$$\frac{\partial E}{\partial x} = -\frac{q \cdot (n - n_0)}{\varepsilon} \tag{3.10}$$

where j_n is the electron current density, τ_n is the generation/recombination time constant for electrons, D_n is the diffusion constant, μ_n is the electron mobility, n_0 is the equilibrium density of electrons, E is the electric field and ε is the dielectric constant.

A very simple analytical solution of these equations for the bipolar junction transistor (BJT) is the Ebers–Moll model [36]. The physical data describing the material are the mobility, the diffusion constants, the generation/recombination time constants and the dielectric constant; the data describing the structure of the device are the junction area, the emitter, base and collector width and the doping densities. The boundary conditions are the voltages applied at the external terminals and the junction law at the emitter/base and base/collector junctions. An empirical parameter is the ideality factor of the two junctions. After analytical integration, the global parameters of the device are found: the reverse saturation currents and the transport factors α_e and α_c. The model then reads

$$I_c = \alpha_e \cdot I_{0e} \cdot (e^{\frac{V_{be}}{n_e \cdot V_T}} - 1) + I_{0c} \cdot (e^{\frac{V_{bc}}{n_c \cdot V_T}} - 1)$$

$$I_e = \alpha_c \cdot I_{0c} \cdot (e^{\frac{V_{bc}}{n_c \cdot V_T}} - 1) + I_{0e} \cdot (e^{\frac{V_{be}}{n_e \cdot V_T}} - 1) \tag{3.11}$$

The Ebers–Moll model is a particularly simple one that does not take into account many important effects in a microwave device, as for instance the transients or the inertial (ballistic) phenomena. It has been mentioned here essentially for illustration purposes.

We remark that the Ebers–Moll model is seen here as a physical model; however, if its parameters are extracted from measurements, it can be seen as an empirical model. In particular, since a simple equivalent circuit can represent eq. (3.10), it can be seen as an equivalent-circuit empirical model (Figure 3.5).

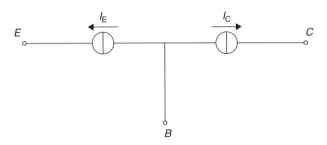

Figure 3.5 The equivalent circuit of the Ebers–Moll model

A more complex example is the analytical model for the metal semiconductor field-effect transistor (MESFET) with gradual-channel approximation [29, 37–45]. The basic equations are again the first two moments of Boltzmann's equation, that is, the continuity and current equations, and Poisson's equation (eqs. (3.8)–(3.10)), in steady-state conditions or including transient behaviour. The electron motion is assumed to be unidirectional from source to drain; the forcing term in the current equation is only the electric field–induced drift. The transport equations are therefore solved in one dimension; moreover, the physical quantities in the channel are assumed to vary slowly along the propagation direction of the charged particles. The channel can be assumed to be either totally depleted, that is, with zero electron density, or having constant electron density equal to the doping concentration; sometimes, a transition region between depleted and conductive regions is introduced. Poisson's equation is usually solved in two dimensions. In some models, however, the solutions in the vertical and horizontal directions are decoupled: the depleted region is evaluated by integrating from the gate down to the conducting region of the channel in the vertical direction; the conducting channel is the residual channel depth. The material data are the dielectric constant, the doping concentration and the electron velocity as a function of the electric field in the semiconductor. For low fields, the mobility is a constant, while for high fields the particle velocity becomes saturated and the incremental mobility is zero; in the intermediate region, the curve has a negative slope corresponding to negative differential resistance. A velocity-field plot for electrons in bulk GaAs at room temperature is shown in Figure 3.6 for several doping concentrations, as resulting from analytical approximation.

The geometrical data are the length, depth and width of the channel, and the gate–source and gate–drain spacings. The channel is divided in more than one region with simplified equivalent geometry, as shown in Figure 3.7; the transport equations are solved in each region and then piecewise-connected. The boundary conditions for Poisson's equation are the voltage values at the source, gate and drain electrodes and the neutrality condition at the source and drain ends of the channel for the transport equations. The solution is usually only semi-analytical, since a numerical solution of the resulting current and charge (or capacitance) equations is required for most models. However, no discretisation of either space or time is required, and the numerical effort is extremely limited. Sometimes a simpler, explicit model can be extracted, with acceptable loss of accuracy [41].

The gate current is computed as the displacement current through the depleted capacitive region below the gate. It is expressed as the derivative with respect to time of

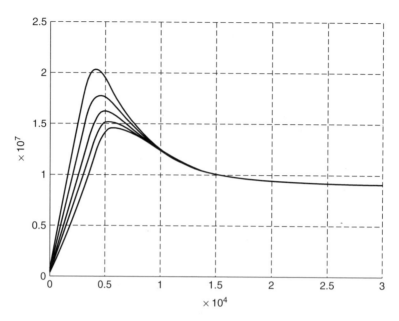

Figure 3.6 Velocity-field plot in GaAs for several doping concentrations

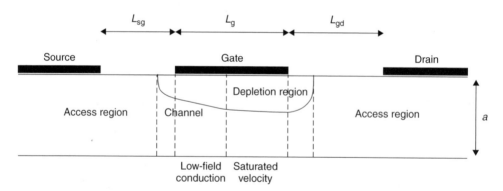

Figure 3.7 The MESFET channel in the gradual-channel approximation

the charge stored in the depletion region [29, 42, 45]; this is obtained by integration of the field or by means of analytical formulae derived by the simple approximated geometry as shown in Figure 3.7.

Similar to the MESFET, the HEMT also has been given analytical solutions to the Poisson-transport equations [46–49]. In this case, the vertical control of the channel is modelled through the dependence on gate-channel voltage of the equivalent sheet-carrier density of the two-dimensional electron gas at the heterointerface (Figure 3.8). Several simple formulae have been given from simple linear approximations [46, 47], usually valid only in the low-field conduction region, to more sophisticated ones, valid also in the saturated velocity region [48, 49]. Derivation of the current–voltage and charge–voltage characteristics follow the same guidelines of the MESFET; conduction in

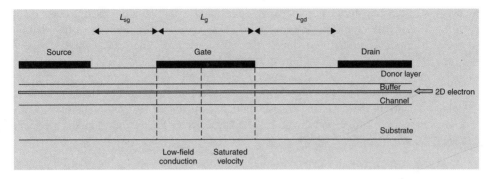

Figure 3.8 The simplified structure of an HEMT for analytical model calculation

the parasitic MESFET is also usually included. Recently, analytical models of MESFET and HEMT devices based on GaN have been developed, aimed at device optimisation for low intermodulation distortion applications [50, 51].

The analytical models are very simple and fast, but their accuracy is limited; as already mentioned, their most useful feature is the possibility to predict and possibly optimise the yield of a circuit, once the fabrication tolerances of the physical parameters and their statistical properties are known. In fact, an analytical model can also be treated as an equivalent-circuit model, where the parameters are not directly the values of the elements, but rather the physical parameters in the analytical formulation, which must therefore be intended as equivalent parameters [29, 52]. Fit-to-measured data proves to be almost as good as in the case of the equivalent circuit, but it is usually more stable, and the extracted values show better correlation to the technological process. Nonetheless, the accuracy is sometimes not sufficient for an accurate circuit design.

3.3 EQUIVALENT-CIRCUIT MODELS

In this paragraph, the basic characteristics of the main types of equivalent-circuit models are described, with special emphasis on the requirements for physical consistency in large-signal regime. An example of extraction procedure is also given.

3.3.1 Introduction

Empirical models are the solution of choice for practical and accurate CAD of non-linear circuits. The effort required by long measurements and sometimes troublesome extraction procedures pays off in terms of flexibility and accuracy for computer-aided design applications.

Equivalent-circuit models are the most successful empirical models for CAD applications so far. The active device is modelled within the simulation programme by means of an electrical circuit, whose elements usually are lumped, with frequency-independent but voltage-dependent values. The distributed-element equivalent circuits have not been successful for nonlinear design so far for two main reasons: first, distributed nonlinear elements are not easily dealt with by nonlinear CAD algorithms; second, active devices are usually fabricated in such a way that distributed effects within the intrinsic device are negligible in the operating frequency band. Of course, distributed access elements are not a problem, at least for frequency- or mixed-domain analysis methods; nonetheless, they are usually replaced by lumped elements without a significant loss of accuracy, given the small dimensions of the devices. Frequency-dependent nonlinear elements pose problems to nonlinear analysis algorithms, unless a purely frequency-domain approach is adopted; if this is the case, then these models turn out to be quite general and practical. However, no commercial CAD programmes are so far available in the frequency domain.

All equivalent-circuit models are made of two subnetworks: an extrinsic, linear one corresponding to parasitic access elements like contact resistances, pad capacitances or line or wire inductances and an intrinsic, nonlinear one corresponding to the inner device. The topology of both parts should be *a priori* known before the extraction, although it can be adjusted during the extraction if the measurements suggest it.

Black-box models in principle require no *a priori* knowledge of the physical structure of the device; however, many of them require either the preliminary extraction of the extrinsic elements or some hypothesis on the behaviour of the inner device. We will call these models quasi-black-box models, reserving the traditional 'black-box' term only to purely behavioural models.

Before discussing nonlinear models, a brief recapitulation of linear models is presented in the following for a better understanding. Then, the procedures and constraints for the extension from a linear to a nonlinear model are described. Finally, some nonlinear models are introduced, together with an example of extraction procedure.

3.3.2 Linear Models

For linear analysis, a frequency-dependent model is acceptable, since each frequency component of a signal is independent of the others in the frequency domain (see Section 1.1); therefore, any general frequency-dependent complex four-parameter equivalent representation of a two-port device can be used. In practice, at microwave and millimetre-wave frequencies the scattering parameters are commonly used. As an example, a Y-matrix equivalent network with four complex frequency-dependent admittance parameters is shown in Figure 3.9.

If the device is reciprocal, the network has only three complex parameters, as shown in Figure 3.10.

If the reactive behaviour of the device is capacitive, an equivalent-circuit representation of this network is shown in Figure 3.11, where the admittance parameters are written as

$$
\begin{aligned}
Y_{11} &= g_i(\omega) + j\omega C_i(\omega) \\
Y_{12} &= g_{mr}(\omega) + j\omega C_{mr}(\omega) \\
Y_{21} &= g_{mf}(\omega) + j\omega C_{mf}(\omega) \\
Y_{22} &= g_o(\omega) + j\omega C_o(\omega)
\end{aligned}
\tag{3.12}
$$

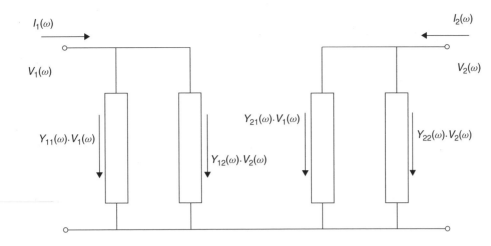

Figure 3.9   An admittance-parameters equivalent network

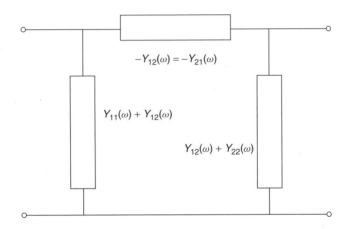

Figure 3.10   A reciprocal equivalent-pi network

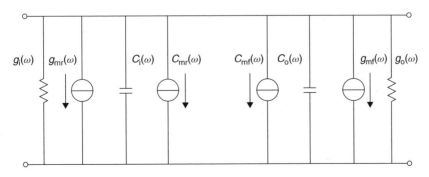

Figure 3.11   An equivalent-circuit representation of the admittance-parameter network for a resistive–capacitive device

The real parts of the admittance parameters are conductances and transconductances; the imaginary parts are capacitances and transcapacitances.

This circuit is easily rewritten in terms of conduction and displacement currents, instead of resistances and capacitances (Figure 3.12) where conduction currents are written as

$$I_{c1} = I_{c1}(V_1, V_2) = g_i \cdot V_1 + g_{mr} \cdot V_2$$
$$I_{c2} = I_{c1}(V_1, V_2) = g_{mf} \cdot V_1 + g_o \cdot V_2 \tag{3.13}$$

because of the linearity of the device. Similarly, the displacement currents are written as

$$I_{q1} = \frac{dQ_1(V_1, V_2)}{dt} = \frac{d}{dt} \cdot (C_i \cdot V_1 + C_{mr} \cdot V_2) = C_i \cdot \frac{dV_1}{dt} + C_{mr} \cdot \frac{dV_2}{dt}$$
$$I_{q2} = \frac{dQ_2(V_1, V_2)}{dt} = \frac{d}{dt} \cdot (C_{mf} \cdot V_1 + C_o \cdot V_2) = C_{mf} \cdot \frac{dV_1}{dt} + C_o \cdot \frac{dV_2}{dt} \tag{3.14}$$

where the input and output charges have been introduced. The correspondence between the integral description in terms of currents and charges and the differential small-signal circuit in terms of conductances and capacitances is defined by

$$
\begin{aligned}
g_i &= \frac{\partial I_{c1}}{\partial V_1} & g_{mr} &= \frac{\partial I_{c1}}{\partial V_2} & g_{mf} &= \frac{\partial I_{c2}}{\partial V_1} & g_o &= \frac{\partial I_{c2}}{\partial V_2} \\
C_i &= \frac{\partial Q_1}{\partial V_1} & C_{mr} &= \frac{\partial Q_1}{\partial V_2} & C_{mf} &= \frac{\partial Q_2}{\partial V_1} & C_o &= \frac{\partial Q_2}{\partial V_2}
\end{aligned} \tag{3.15}
$$

If the device is reciprocal, we have

$$g_{mr} = g_{mf} \quad C_{mr} = C_{mf} \tag{3.16}$$

or

$$\frac{\partial I_{c1}}{\partial V_2} = \frac{\partial I_{c2}}{\partial V_1} \quad \frac{\partial Q_1}{\partial V_2} = \frac{\partial Q_2}{\partial V_1} \tag{3.17}$$

and the equivalent circuit becomes (Figure 3.13).

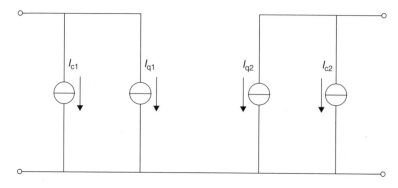

Figure 3.12   A current-charge equivalent network

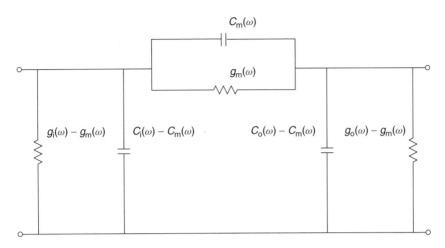

Figure 3.13   A reciprocal resistive–capacitive equivalent circuit

In fact, conduction currents are non-reciprocal in all active devices of interest.

Sometimes it is possible or desirable that the conduction currents also be expressed in terms of the charges inside the device. If this is the case, the conduction currents are expressed as

$$
\begin{aligned}
I_{c1} &= I_{c1}(V_1, V_2) = k_i \cdot Q_1(V_1, V_2) + k_{mr} \cdot Q_2(V_1, V_2) \\
I_{c2} &= I_{c1}(V_1, V_2) = k_{mf} \cdot Q_1(V_1, V_2) + k_o \cdot Q_2(V_1, V_2)
\end{aligned}
\tag{3.18}
$$

while the displacement currents have the same expression as above. This is called a charge-control model. For example, the conduction current in a forward-biased pn junction can be written as the diffusion charge divided by the recombination time [36]:

$$
i = \frac{Q_n}{\tau_n} + \frac{Q_p}{\tau_p}
\tag{3.19}
$$

Since the diffusion charge is expressed as

$$
Q_n = A \cdot q n_0 \cdot L_n \cdot \left(e^{\frac{V}{\eta V_T}} - 1 \right) \quad Q_p = A \cdot q p_0 \cdot L_p \cdot \left(e^{\frac{V}{\eta V_T}} - 1 \right)
\tag{3.20}
$$

this is equivalent to the usual expression for the conduction current:

$$
i = \frac{A \cdot q n_0 \cdot L_n \cdot \left(e^{\frac{V}{\eta V_T}} - 1 \right)}{\tau_n} + \frac{A \cdot q p_0 \cdot L_p \cdot \left(e^{\frac{V}{\eta V_T}} - 1 \right)}{\tau_p} = i_0 \cdot \left(e^{\frac{V}{\eta V_T}} - 1 \right)
\tag{3.21}
$$

The number of parameters of the model is obviously the same.

If the device is such that the eight equivalent-circuit parameters are frequency-independent, then the device is quasi-static, because the currents and charges at a given time instant depend on voltages present at the ports of the device at the same time instant. In other words, the device responds instantaneously to the input signal. Naturally, the equivalent circuit is not necessarily an admittance network, but it can be any equivalent representation; for instance, a T network corresponding to an impedance-matrix representation or a hybrid network. For example, the Giacoletto model for the bipolar transistor is a quasi-static admittance-matrix network if we neglect the base resistance $r_{bb'}$ and the base-collector resistance $r_{b'c}$, as shown in Figure 3.14.

Since the reactances in the network are reciprocal, the circuit can be redrawn as in Figure 3.15.

If the device is non-quasi-static, either frequency-dependent elements or time-delayed responses must be present in a circuit. For example, if the base resistance $r_{bb'}$ is not neglected in the Giacoletto equivalent circuit, the current in the collector–emitter current source is controlled by the voltage $V_{b'e}$ delayed with respect to the external base-emitter voltage V_{be} because of the low-pass behaviour of the input mesh; the charges in the two junctions are also proportional to the delayed voltage $V_{b'e}$. The circuit becomes non-quasi-static (Figure 3.16).

Another example is a simple quasi-static equivalent circuit of a field-effect transistor shown in Figure 3.17.

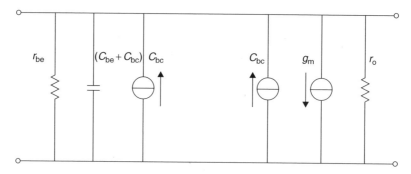

Figure 3.14   A quasi-static, simplified Giacoletto equivalent network for a bipolar transistor with non-reciprocal capacitances

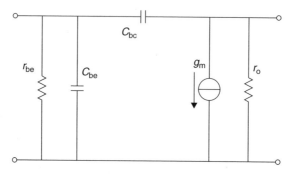

Figure 3.15   A quasi-static, simplified Giacoletto equivalent network for a bipolar transistor

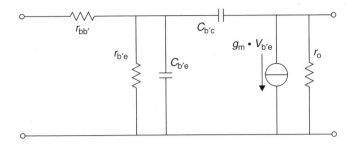

Figure 3.16   The non-quasi-static Giacoletto equivalent network for a bipolar transistor

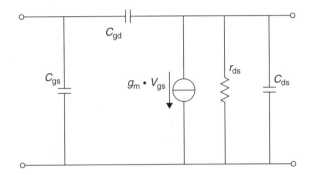

Figure 3.17   A quasi-static equivalent network for a field-effect transistor

The circuit becomes non-quasi-static if one or more modifications are made: for example, a delay RC circuit is introduced in the input mesh, and the drain–source current source is controlled by the voltage across the gate–source capacitance or by a voltage delayed by a constant value τ. All of these modifications are introduced in Figure 3.18.

In the frequency domain, the drain–source current source can be written as

$$I(\omega) = g_{\mathrm{m}} \cdot e^{j\omega\tau} \cdot V_{\mathrm{i}}(\omega) \tag{3.22}$$

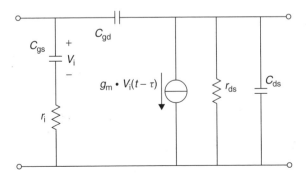

Figure 3.18   A non-quasi-static equivalent network for a field-effect transistor

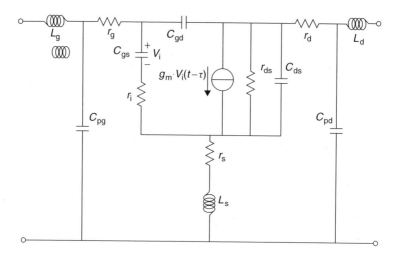

Figure 3.19   A complete equivalent circuit for a field-effect transistor

More equivalent-circuit elements are usually added to the equivalent circuit in order to accurately represent the actual behaviour of the device with only frequency-independent elements. For example, parasitic effects such as contact resistances, pad capacitances or line inductances are modelled by separate elements; a complete equivalent circuit may look like that shown in Figure 3.19.

3.3.3 From Linear to Nonlinear

In this paragraph, the extension of linear equivalent circuits to nonlinear regime is descri-bed. Some concepts from Section 2.4 are repeated for the sake of clarity.

Whatever type of linear model is concerned, there is an apparently simple path from a linear to a nonlinear representation. It has been assumed above for a linear model that the conduction currents and the charges are linear with respect to voltages; for simplicity, let us take a linear one-port device with conduction and displacement current components as a working example (Figure 3.20).

$$I = I_c + I_q \tag{3.23a}$$

$$I_c(V) = g \cdot V \tag{3.23b}$$

$$Q(V) = C \cdot V \quad I_q = \frac{dQ(V)}{dt} = C \cdot \frac{dV}{dt} \tag{3.23c}$$

In actual devices, the linearity relations hold only for small perturbations around a quiescent point or bias point (Figure 3.21) [53]:

$$V = V_0 + v(t) \tag{3.24}$$

$$I_c(V) = I_c(V_0) + \left.\frac{dI_c}{dV}\right|_{V=V_0} (V - V_0) + \cdots = I_{c0} + g \cdot v(t) + \cdots \tag{3.25a}$$

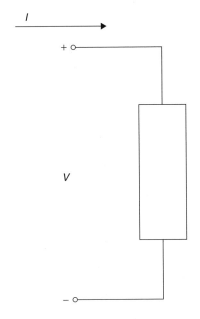

Figure 3.20 A one-port element

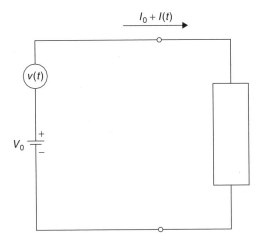

Figure 3.21 A one-port nonlinear element with a combined DC and RF excitation

$$Q(V) = Q(V_0) + \left.\frac{\mathrm{d}Q}{\mathrm{d}V}\right|_{V=V_0} (V - V_0) + \cdots = Q_0 + C \cdot v(t) + \cdots \quad (3.25\text{b})$$

$$I_\mathrm{q}(V) = \frac{\mathrm{d}Q(V)}{\mathrm{d}t} = C \cdot \frac{\mathrm{d}v(t)}{\mathrm{d}t} + \cdots \quad (3.25\text{c})$$

where the Taylor series is truncated after the first-order term; this corresponds to a simple equivalent network for the small perturbation (Figure 3.22).

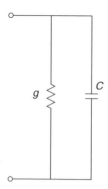

Figure 3.22   A one-port resistive–capacitive element

The conductance and the capacitance can be evaluated at a given bias point from an incremental (small-signal) measurement, for example an admittance measurement:

$$\overline{V}(\omega) = \Im(v(t)) \quad \overline{I}(\omega) = \Im(i(t)) \quad Y(\omega) = \frac{\overline{I}(\omega)}{\overline{V}(\omega)} = g + j\omega C \qquad (3.26)$$

where the symbol $\Im$ denotes a Fourier transform. If the device is quasi-static, the conductance and the capacitance are constant with frequency. For actual devices, extraction procedures for more realistic but similar equivalent circuits are available, and will be described in Section 3.3.4 below.

If we now want to extract a nonlinear model of our one-port device from the linear small-signal model just defined, we can try to repeat the measurements at many bias points, that is, in this example for many values of the bias voltage V_0. The linearised conductance and capacitance are evaluated at each bias point, and their dependence on the applied bias voltage is found.

$$g(V_0) \quad C(V_0) \qquad (3.27)$$

In principle, we have found the nonlinear dependence of the values of the elements on the applied voltage. We can now compute the dependence on applied voltage of the integral parameters, that is, current and charge, by simple integration:

$$I(V_0) = \int_0^{V_0} g(\beta) \cdot \mathrm{d}\beta \quad Q(V_0) = \int_0^{V_0} C(\beta) \cdot \mathrm{d}\beta \qquad (3.28)$$

For the conduction current, there is also an alternative possibility: a direct DC measurement; this is obviously not possible for the charge.

In a nonlinear circuit, however, the large-signal voltage applied to the nonlinear element is not a DC quantity as above, but a fast signal sweeping a large voltage range in a very short time, that is, in the period of the microwave signal: a value ranging from a nanosecond to tens of picoseconds. In a semiconductor device, there are phenomena that require some time to build up, and that affect the conduction phenomena or charge behaviour in the device itself [54–62]. One of these phenomena is the temperature rise or fall because of the power dissipated in the device by the bias voltage and current and by the electrical signal; another one is the trapping or detrapping of the carriers in the semiconducting material (see Section 2.4). Both the phenomena have characteristic time constants typically above a hundred nanoseconds. In static conditions, these effects have plenty of time to build up, changing the characteristics of conduction phenomena or charge state, while in fast (microwave) large-signal operations the short time during which the voltage is applied does not allow them to take place, and they respond only to the average, quiescent point voltages. This is a well-known phenomenon; a typical example is the hysteresis in the measurement of swept I/V characteristics, when the measurement is fast enough not to allow heating or trapping in the device during the way on but not fast enough to prevent heating or trapping during the way back (Figure 3.23).

This behaviour of the current curves at low frequencies is usually called the low-frequency dispersion of the characteristics.

Let us resume the situation of the three possible measurements for our one-port device and for the conduction current only. In the first one, the conduction current is measured in static conditions, that is, by applying a DC voltage and measuring the DC

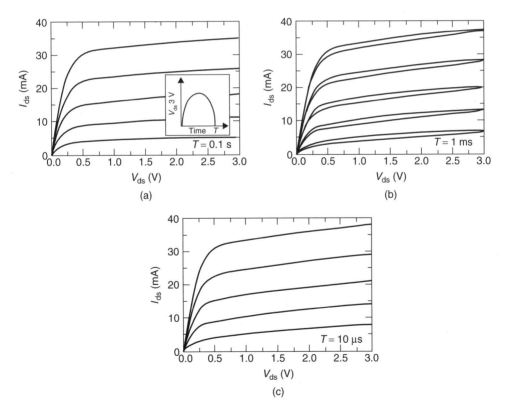

Figure 3.23 (a) Slow swept measurement of the I/V output characteristics of an MESFET showing the DC behaviour of the device; (b) medium-speed swept measurement showing hysteresis; (c) fast swept measurement showing no hysteresis

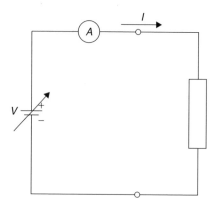

Figure 3.24   A DC current–voltage measurement scheme

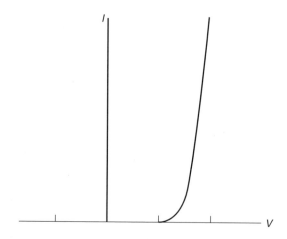

Figure 3.25   A DC I/V curve for a diode

current at many points along the curve (Figure 3.24). We get a curve where every point has a different temperature and traps state (Figure 3.25).

In the second one, a fast swept measurement is performed, that is, the voltage is swept and the current is measured all along the I/V curve with a period shorter than the time constants of slow phenomena; typically, below 100 ns, which is an affordable speed (Figure 3.26).

In this figure, the current is computed as

$$I(t) = \frac{V_{\text{sense},1}(t) - V_{\text{sense},2}(t)}{R_{\text{sense}}} \tag{3.29}$$

where the sense voltages are measured, for instance, by a sampling oscilloscope, and

$$V_{\text{sweep}}(t) = \overline{V_{\text{sweep}}} \cdot \sin(\omega t) \tag{3.30}$$

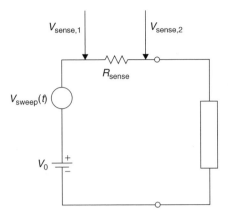

Figure 3.26   A fast current–voltage measurement scheme

with a sufficiently high ω (e.g. above 10 MHz), and an amplitude of V_{sweep} large enough for covering the whole curve.

In fact, many measurements of this type can be performed by starting the voltage sweep each time from a different bias point. A different curve is obtained for each bias point V_0, since the temperature and traps state of the device depends on the quiescent point only (Figure 3.27).

In fact, the same curves can be measured by pulsing the voltage from the bias point for a short time to all points of the corresponding curve, one after the other, and measuring the current during the pulses. A typical set-up and a sequence of pulsed voltages and currents in the case of an FET are shown in Section 2.4, Figures 3.20 and 3.21. This arrangement is in fact the most common; this is why fast, isothermal and 'isotrap' current measurements are also called pulsed measurements.

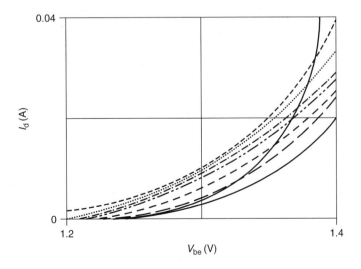

Figure 3.27   A DC I/V curve with several fast I/V curves from different bias points for a diode

In the third one, the device is biased at many points along the curve; at each point, the differential (small-signal) conductance is measured by a microwave admittance measurement. In practice, small-signal S-parameter measurements are performed, and the reflection coefficients are converted to admittances. The small-signal conductance corresponds to the tangent of the fast measurement performed from that bias point (Figure 3.28).

This is easily seen by considering the measurement set-up for large-signal fast measurements, and comparing it to a standard vector network analyser. Once the small-signal conductance is evaluated for all bias points, the current curve is computed by integration with respect to voltage.

Let us now discuss the three measurements. The DC curve must be used for the simulation of slow or constant phenomena, as the DC bias or rectified voltages and currents, the low-frequency second-order intermodulation signal in multi-tone systems or the down-converted phase-noise in oscillators, when their frequency is very low (below approximately 100 KHz). The pulsed curves must be used for the simulation of microwave large signals; in this case, the curve must be measured from the same quiescent point as that of the large-signal operation. This is however, in general, not predictable, since it includes rectified terms (see Chapter 1); the model must then include the curves measured from all quiescent points and must be able to adjust to the actual quiescent point obtained in the analysis.

The curve obtained by the integration of the small-signal measurements, finally, is not physically correct and should never be used. Unfortunately, this is a popular method to extract nonlinear models, because it is also the most practical and traditional from the point of view of both instrumentation and extraction procedure. A complete measurement set-up for DC, small-signal and pulsed S-parameter measurement has been demonstrated [63], allowing consistent modelling; however, it is not usually available to modellers and designers.

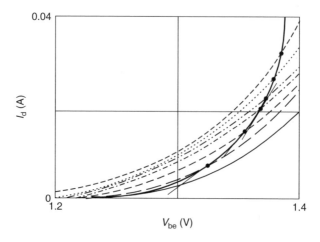

Figure 3.28 A DC I/V curve with the differential (incremental) conductances at several bias points

If the measurements in Figure 3.27 are available, then the I/V characteristics have the following general form, equivalent to eq. (2.2) in Chapter 2:

$$i(t) = i(v_{DC}, T, v(t)) \tag{3.31}$$

where the dependence on the instantaneous voltage, on the DC voltage and on temperature is a consequence of the described phenomena. The DC curve is obviously a particular case of the above formulation, obtained when the instantaneous voltage coincides with the DC voltage:

$$i_{DC}(t) = i(v_{DC}, T, v_{DC}) \tag{3.32}$$

A procedure for the correction of small-signal data in order to account for temperature and trap effect has been proposed, under certain hypotheses, allowing the use of small-signal parameters for consistent large-signal modelling [64–67]. Let us assume that we are working in isothermal conditions and that the current for a given instantaneous voltage $v(t)$ when the DC voltage v_{DC} is equal to the current obtained from the DC curve for a slightly different voltage (Figure 3.29):

$$i(t) = i_{DC}(u(t)) = i_{DC}(v(t) - \Delta v) \tag{3.33}$$

The correction can be written as a fraction of the difference between the static and actual voltage value:

$$\Delta v = \alpha \cdot (v(t) - v_{DC}) \tag{3.34}$$

If the same instantaneous voltage is reached from a different bias point, a similar expression holds (Figure 3.30):

$$i'(t) = i_{DC}(u'(t)) = i_{DC}(v(t) - \Delta v') \tag{3.35}$$

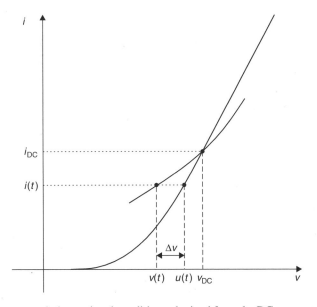

Figure 3.29 The current in large-signal conditions obtained from the DC curve at an offset voltage

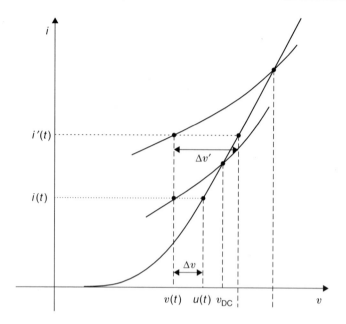

Figure 3.30   The current in large-signal conditions from two different bias points

The correction can be expressed in a similar way as

$$\Delta v' = \alpha' \cdot (v(t) - v'_{DC}) \tag{3.36}$$

If the constant $\alpha = \alpha'$ is the same in both cases, that is, if the correction is the same fraction of the distance of the actual instantaneous voltage from the bias point, then it is easily evaluated from a single fast (pulsed) measurement with $v(t) \neq v_{DC}$ and from the DC curve; however, it is also evaluated from a single small-signal measurement at $v_{DC} \cong v(t)$ and from the DC curve. It is easy to see that, for a $v(t)$ very close to v_{DC}, that is, for a small incremental instantaneous signal (Figure 3.30)

$$g_{ss} = \frac{di}{dv}\bigg|_{ss} \cong \frac{i(t) - i_{DC}}{v(t) - v_{DC}} = \frac{1}{(1-\alpha)} \cdot \frac{i(t) - i_{DC}}{(u(t) - v_{DC})} \cong \frac{1}{1-\alpha} \cdot \frac{di}{dv}\bigg|_{DC} = \frac{g_{DC}}{1-\alpha} \tag{3.37}$$

where g_{ss} is the small-signal incremental conductance, tangent to the pulsed curve (see above), and g_{DC} is the tangent to the DC curve. The parameter α is immediately evaluated. It is reasonable however that the parameter α is a function of the instantaneous voltage v; therefore, the evaluation must be repeated for all voltages along the I/V curve.

The above assumption can be given a physical interpretation in the case of an FET: the charge due to free carriers in the channel responds to the instantaneous voltage, while the trapped charge responds to the static DC voltage because of the long trapping and detrapping times. Therefore, if the instantaneous voltage is different from the DC one, the effective DC potential corresponding to the same amount of free charge in the channel is in between the two values. It is also a reasonable assumption that this effect be linear.

Physically different effects can be treated in the same way if they have a linear effect on the current. For instance, mobility is inversely proportional to temperature; therefore, a change in temperature (e.g. increase in ambient temperature) for a fixed DC voltage can be reduced to a change in voltage proportional to the difference in temperature (Figure 3.31):

$$i_{DC}(v_{DC}, T') = i_{DC}(u_{DC}, T) = i_{DC}(v_{DC} + \Delta v, T) \tag{3.38}$$

$$\Delta v = \beta \cdot (T' - T) \tag{3.39}$$

Let us allow the device to increase its temperature as a consequence of increased dissipated power within the device itself, for example, as a consequence of increased applied DC voltage (Figure 3.32). We can write

$$i_{DC}(v'_{DC}, T') = i_{DC}(u_{DC}, T) = i_{DC}(v'_{DC} + \Delta v, T) \tag{3.40}$$

$$\Delta v = \beta \cdot (T' - T) = \beta \cdot R_{th} \cdot (i_{DC}(v'_{DC}, T') \cdot v'_{DC} - i_{DC}(v_{DC}, T) \cdot v_{DC}) \tag{3.41}$$

Again, a single isothermal measurement and a DC I/V curve are enough to identify the parameter β.

An alternative procedure can be described as follows [53, 68–75]. Let DC measurements of the I/V curve and also the differential conductance measurements, be performed the measured data are

$$i_{DC}(v) \text{ and } g_{ss}(v) \tag{3.42}$$

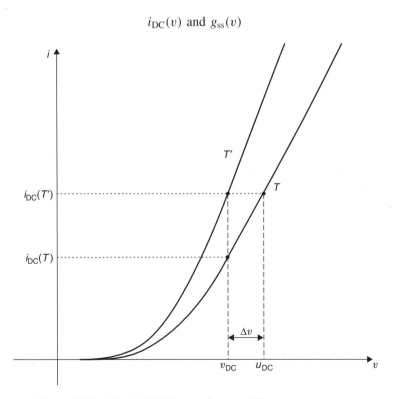

Figure 3.31   The DC I/V curve for two different temperatures

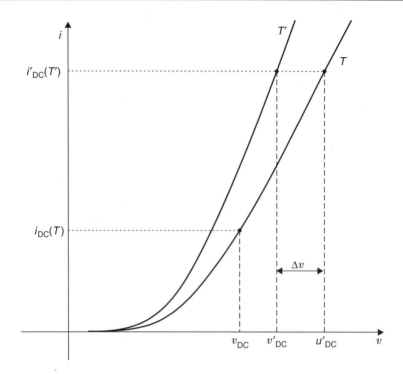

Figure 3.32 The DC I/V curve for two different temperatures and applied DC voltages

that do not fulfil the condition

$$i_{DC}(v_0) + \int_{v=v_0}^{v=v_1} g_{ss}(y) \cdot dy = i_{DC}(v_1) \tag{3.43}$$

for any pair of voltages (v_0, v_1), because of the problems mentioned. The DC curve is now fitted by any fitting function f that depends on a number of parameters p_i: a set of values for the parameters $p_{DC,i}$ is determined as a consequence of the fitting procedure:

$$i_{DC}(v) = f(p_{DC,i}, v) = f_{DC}(v) \tag{3.44}$$

Then, the difference between the measured small-signal conductance and the derivative of the DC curve is computed for all voltages and fitted to the derivative of another fitting function:

$$g_{DC}(v) = \frac{\partial i_{DC}(v)}{\partial v} = \frac{\partial f_{DC}(v)}{\partial v} \qquad g_{diff}(v) = g_{ss}(v) - g_{DC}(v) \tag{3.45}$$

$$g_{diff}(v) = \frac{\partial f_{diff}(v)}{\partial v} \tag{3.46}$$

Then, when a large-signal voltage $v(t) = v_{DC} + v_{RF}(t)$ is applied, the current is computed as

$$i(t) = f_{DC}(v(t)) + f_{diff}(v_{RF}(t)) \tag{3.47}$$

Alternatively, the small-signal conductance is fitted to the derivative of a fitting function:

$$g_{ss}(v) = \frac{\partial f_{RF}(v)}{\partial v} \qquad (3.48)$$

Then, when a large-signal voltage $v(t) = v_{DC} + v_{RF}(t)$ is applied, the current is computed as (Figure 3.33):

$$i(t) = f_{RF}(v(t)) + f_{DC}(v_{DC}) - f_{RF}(v_{DC}) \qquad (3.49)$$

The expression in eq. (3.47) can be implemented as an equivalent circuit (Figure 3.34), where the capacitor C_{RF} is large enough to allow all signals above the frequency where dispersion is present.

So far for the conduction current; when the displacement current contribution is considered, things are more complex. First of all, the DC measurements are not possible. Second, the fast measurements require that complex data (amplitude and phase) be measured in a short time. In fact, this requires pulsed S-parameter measurements: during a short bias pulse, a microwave small-signal excitation is fed to the device and the scattering

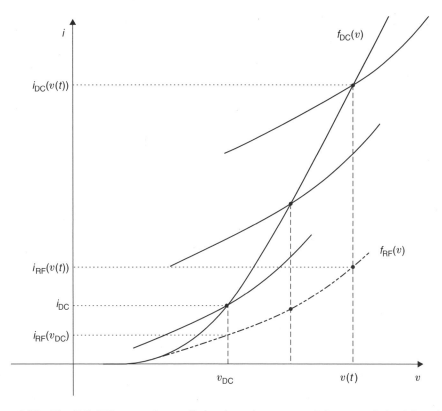

Figure 3.33　The DC I/V curve, the small-signal conductances and the curve derived from integration of the latter

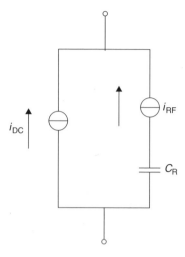

Figure 3.34   The equivalent circuit of a two-terminal device with low-frequency dispersion

parameters are evaluated. However, this technique involves expensive instrumentation that is not always available (see Section 2.4). Moreover, the extraction procedure becomes very burdensome. Therefore, this technique is so far limited to special cases, as for instance very high power transistors used for pulsed operations. On the other hand, the reactive part of an active device is less sensitive to low-frequency dispersion effects, and the consequences on the performances of a device are less pronounced. The loss of accuracy caused by the imperfect model is therefore limited.

We now rewrite the above considerations for a two-port device [53]. The currents in a linear device are

$$
\begin{aligned}
I_1 &= I_{c1} + I_{q1} \\
I_2 &= I_{c2} + I_{q2}
\end{aligned}
\tag{3.50}
$$

$$
\begin{aligned}
I_{c1}(V_1, V_2) &= g_{11} \cdot V_1 + g_{12} \cdot V_2 \\
I_{c2}(V_1, V_2) &= g_{21} \cdot V_1 + g_{22} \cdot V_2
\end{aligned}
\tag{3.51}
$$

$$
\begin{aligned}
Q_1(V_1, V_2) &= C_{11} \cdot V_1 + C_{12} \cdot V_2 \\
Q_2(V_1, V_2) &= C_{21} \cdot V_1 + C_{22} \cdot V_2
\end{aligned}
$$

$$
\begin{aligned}
I_{q1}(V_1, V_2) &= \frac{dQ_1(V_1, V_2)}{dt} \\
&= C_{11} \cdot \frac{dV_1}{dt} + C_{12} \cdot \frac{dV_2}{dt} \\
I_{q2}(V_1, V_2) &= \frac{dQ_2(V_1, V_2)}{dt} \\
&= C_{21} \cdot \frac{dV_1}{dt} + C_{22} \cdot \frac{dV_2}{dt}
\end{aligned}
\tag{3.52}
$$

The small-signal incremental expressions in the neighbourhood of a bias point are

$$
\begin{aligned}
V_1 &= V_{10} + v_1(t) \\
V_2 &= V_{20} + v_2(t)
\end{aligned}
\tag{3.53}
$$

$$I_{c1}(V_1, V_2) = I_{c1}(V_{10}, V_{20}) + \left.\frac{dI_{c1}}{dV_1}\right|_{\substack{V_1 = V_{10} \\ V_2 = V_{20}}} (V_1 - V_{10}) + \left.\frac{dI_{c1}}{dV_2}\right|_{\substack{V_1 = V_{10} \\ V_2 = V_{20}}} (V_2 - V_{20})$$

$$+ \cdots = I_{c10} + g_{11} \cdot v_1(t) + g_{12} \cdot v_2(t) + \cdots$$

$$I_{c2}(V_1, V_2) = I_{c2}(V_{10}, V_{20}) + \left.\frac{dI_{c2}}{dV_1}\right|_{\substack{V_1 = V_{10} \\ V_2 = V_{20}}} (V_1 - V_{10}) + \left.\frac{dI_{c2}}{dV_2}\right|_{\substack{V_1 = V_{10} \\ V_2 = V_{20}}} (V_2 - V_{20})$$

$$+ \cdots = I_{c20} + g_{21} \cdot v_1(t) + g_{22} \cdot v_2(t) + \cdots \tag{3.54}$$

$$Q_1(V_1, V_2) = Q_1(V_{10}, V_{20}) + \left.\frac{dQ_1}{dV_1}\right|_{\substack{V_1 = V_{10} \\ V_2 = V_{20}}} (V_1 - V_{10}) + \left.\frac{dQ_1}{dV_2}\right|_{\substack{V_1 = V_{10} \\ V_2 = V_{20}}} (V_2 - V_{20})$$

$$+ \cdots = Q_{10} + C_{11} \cdot v_1(t) + C_{12} \cdot v_2(t) + \cdots$$

$$Q_2(V_1, V_2) = Q_2(V_{10}, V_{20}) + \left.\frac{dQ_2}{dV_1}\right|_{\substack{V_1 = V_{10} \\ V_2 = V_{20}}} (V_1 - V_{10}) + \left.\frac{dQ_2}{dV_2}\right|_{\substack{V_1 = V_{10} \\ V_2 = V_{20}}} (V_2 - V_{20})$$

$$+ \cdots = Q_{20} + C_{21} \cdot v_1(t) + C_{22} \cdot v_2(t) + \cdots \tag{3.55}$$

$$I_{q1}(V_1, V_2) = \frac{dQ_1(V_1, V_2)}{dt} = C_{11} \cdot \frac{dv_1(t)}{dt} + C_{12} \cdot \frac{dv_2(t)}{dt} + \cdots$$

$$I_{q2}(V_1, V_2) = \frac{dQ_2(V_1, V_2)}{dt} = C_{21} \cdot \frac{dv_1(t)}{dt} + C_{22} \cdot \frac{dv_2(t)}{dt} + \cdots \tag{3.56}$$

where the bi-dimensional Taylor series is truncated after the first-order terms. A simple equivalent network can be introduced in this case too (Figure 3.35), where the $g_{11}, C_{11}, g_{22}, C_{22}$ terms are conductances and capacitances and the $g_{12}, C_{12}, g_{21}, C_{21}$ terms are transconductances and transcapacitances. They can be evaluated at any given bias point from an incremental (small-signal) measurement, as for the one-port device.

If the reactive part of the device is reciprocal, the circuit is simplified ($C_{12} = C_{21}$) (Figure 3.36).

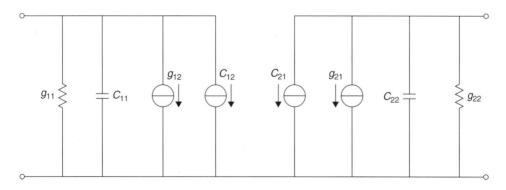

Figure 3.35   An equivalent circuit of the model as in eqs. (3.53)–(3.56)

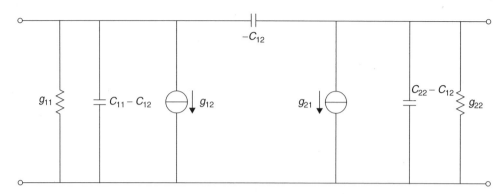

Figure 3.36 An equivalent circuit of the model as in eqs. (3.37)–(3.40) with reciprocal capacitances

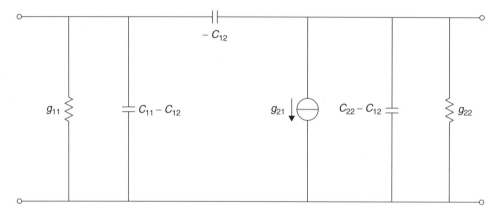

Figure 3.37 An equivalent circuit with reciprocal capacitances and unilateral transconductances

Often, the conduction currents are unilateral in actual devices ($g_{12} = 0$); the equivalent circuit then becomes (Figure 3.37).

For an FET, the quasi-static equivalent circuit is shown in Figure 3.38, where

$$g_{21} = g_m \quad g_{22} = g_{ds} \tag{3.57a}$$

$$C_{gs} = C_{11} + C_{12} \quad C_{gd} = -C_{12} \quad C_{ds} = C_{22} + C_{12} \tag{3.57b}$$

A non-quasi-static symmetric equivalent circuit of an FET is shown in Figure 3.39, where the capacitive region of the device is included within the dashed line.

The reciprocity of the reactive part of the device can be derived from an independent principle [71]. We first define the reactive energy stored in the capacitive region of the device:

$$E(V_1, V_2) = Q_1(V_1, V_2) \cdot V_1 + Q_2(V_1, V_2) \cdot V_2 \tag{3.58}$$

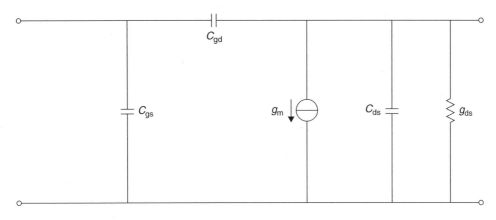

Figure 3.38   A quasi-static equivalent circuit of an FET

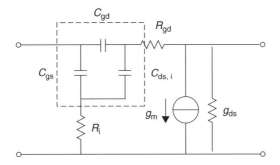

Figure 3.39   A non-quasi-static symmetric equivalent circuit of the intrinsic FET

The reactive energy is also conserved after one cycle, because there is no dissipation within a purely reactive region. Therefore, the energy is a single-valued function of voltages. Its first-order derivatives are

$$Q_1(V_1, V_2) = \frac{\partial E(V_1, V_2)}{\partial V_1} \qquad Q_2(V_1, V_2) = \frac{\partial E(V_1, V_2)}{\partial V_2} \tag{3.59}$$

The second-order partial derivatives of the single-valued energy function E must fulfil the following relations:

$$C_{11}(V_1, V_2) = \frac{\partial Q_1(V_1, V_2)}{\partial V_1} = \frac{\partial^2 E(V_1, V_2)}{\partial V_1^2}$$

$$C_{12}(V_1, V_2) = \frac{\partial Q_1(V_1, V_2)}{\partial V_2} = \frac{\partial^2 E(V_1, V_2)}{\partial V_1 \partial V_2} = \frac{\partial Q_2(V_1, V_2)}{\partial V_1} = C_{21}(V_1, V_2) \tag{3.60}$$

$$C_{22}(V_1, V_2) = \frac{\partial Q_2(V_1, V_2)}{\partial V_2} = \frac{\partial^2 E(V_1, V_2)}{\partial V_2^2}$$

and the capacitive part of the device then is reciprocal.

A nonlinear model of the two-port device can be extracted from the linear small-signal model by using the small-signal measurements at many bias points, that is, for many values of the bias voltages $V_{gs,0}$ and $V_{ds,0}$. The linearised conductances and capacitances are evaluated at each bias point, and their dependence on the applied bias voltages is found:

$$\begin{array}{ll} g_m(V_{gs,0}, V_{ds,0}) & C_{gs}(V_{gs,0}, V_{ds,0}) \\ g_{ds}(V_{gs,0}, V_{ds,0}) & C_{gd}(V_{gs,0}, V_{ds,0}) \\ & C_{ds}(V_{gs,0}, V_{ds,0}) \end{array} \tag{3.61}$$

We can now compute the dependence on applied voltages of the integral parameters, that is, currents and charges, by integration; however, this time the integration must be performed along a line in the V_1/V_2 plane:

$$I_{ds}(V_{gs}, V_{ds}) = \int_0^{V_{gs}, V_{ds}} (g_m(\beta_1, \beta_2) \cdot d\beta_1 + g_{ds}(\beta_1, \beta_2) \cdot d\beta_2) \tag{3.62a}$$

$$Q_g(V_{gs}, V_{ds}) = \int_0^{V_{gs}, V_{ds}} ([C_{gs}(\beta_1, \beta_2) + C_{gd}(\beta_1, \beta_2)] \cdot d\beta_1 - C_{gd}(\beta_1, \beta_2) \cdot d\beta_2)$$

$$\tag{3.62b}$$

$$Q_d(V_{gs}, V_{ds}) = \int_0^{V_{gs}, V_{ds}} (-C_{gd}(\beta_1, \beta_2) \cdot d\beta_1 + [C_{ds}(\beta_1, \beta_2) + C_{gd}(\beta_1\beta_2)] \cdot d\beta_2)$$

The results of these integrations are independent of the integration path only if the conductances on the one hand and the capacitances on the other hand are the partial derivatives of single-valued functions:

$$g_m = \frac{\partial I_{ds}(V_{gs}, V_{ds})}{\partial V_{gs}} \qquad g_{ds} = \frac{\partial I_{ds}(V_{gs}, V_{ds})}{\partial V_{ds}} \tag{3.63a}$$

$$C_{gs} = \frac{\partial Q_g(V_{gs}, V_{ds})}{dV_{gs}} + \frac{\partial Q_g(V_{gs}, V_{ds})}{\partial V_{ds}} \qquad C_{ds} = \frac{\partial Q_d(V_{gs}, V_{ds})}{dV_{gs}} + \frac{\partial Q_d(V_{gs}, V_{ds})}{\partial V_{ds}}$$

$$C_{gd} = -\frac{dQ_g(V_{gs}, V_{ds})}{dV_{ds}} = -\frac{dQ_d(V_{gs}, V_{ds})}{dV_{gs}} \tag{3.63b}$$

In other terms, a line integral along a closed curve on the V_1/V_2 plane must sum up to zero; from an electrical point of view, this corresponds to the charges and currents coming back to the initial value after one period in periodic large-signal regime (Section 2.4) [53] (Figure 3.40):

$$\oint (g_m \cdot dV_{gs}(t) + g_{ds} \cdot dV_{ds}(t)) = \oint \frac{dI_{ds}}{dV(t)} \cdot \frac{dV(t)}{dt} \cdot dt$$

$$= \oint_T \frac{dI_{ds}}{dt} \cdot dt = 0 \tag{3.64a}$$

$$\oint ([C_{gs} + C_{gd}] \cdot dV_{gs}(t) - C_{gd} \cdot dV_{ds}(t)) = \oint \frac{dQ_g}{dV(t)} \cdot \frac{dV(t)}{dt} \cdot dt$$

$$= \oint_T \frac{dQ_g}{dt} \cdot dt = 0$$

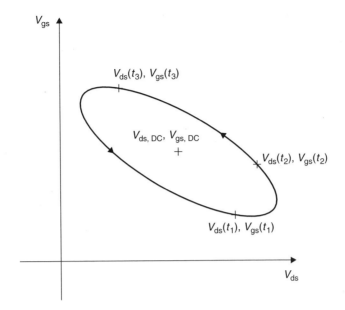

Figure 3.40   Microwave load curve in the $V_{gs} - V_{ds}$ plane

$$\oint \left(-C_{gd} \cdot dV_{gs}(t) + [C_{gs} + C_{gd}] \cdot dV_{ds}(t) \right) = \oint \frac{dQ_d}{dV(t)} \cdot \frac{dV(t)}{dt} \cdot dt$$

$$= \oint_T \frac{dQ_d}{dt} \cdot dt = 0 \qquad (3.64b)$$

However, if the device exhibits low-frequency dispersion, the measured small-signal conductances and capacitances do not fulfil the above conditions: they are valid for isothermal and 'isotrap' contours in the V_1/V_2 plane, while the small-signal measurements are affected by changes in temperature and trap state in different positions in the plane due to different bias voltages. Once more, only DC and fast or pulsed data should be used, unless an identification procedure similar to that described above is carried out.

From what has been said above, it is apparent that a good nonlinear model must account for frequency dispersion of the output I/V characteristics. DC data for the I/V characteristics is easily available; high-frequency data is available from pulsed measurement or corrected extraction from S-parameters, as described above. A suitable arrangement for the equivalent circuit is shown in Figure 3.41 [72, 75].

If a smoother transition from DC to RF frequency is desired, low-pass and high-pass filters with transition frequency in the MHz range must be replaced in the DC and RF branches respectively.

For the sake of illustration of the charge expressions above, a particular, simple expression for the charges is described, that is, the case of a two-port device, for example an FET, with C_{gs} function of V_{gs} only, C_{gd} function of $(V_{gs} - V_{ds})$ only and constant C_{ds}. This is the case of an FET where the gate–source and gate–drain capacitances are

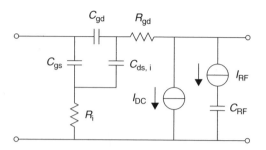

Figure 3.41 An equivalent circuit of an FET including dispersion effects

modelled as the depletion capacitances of the gate–source and gate–drain diodes, and the drain–source capacitance as a constant parasitic, that is, the substrate capacitance. This circuit automatically fulfils the nonlinear constraints just described. Let the diode capacitances have an expression of the type

$$C_{gs}(V_{gs}) = \frac{C_{g0}}{\sqrt{1 - \dfrac{V_{gs}}{V_{bi}}}} \qquad C_{gd}(V_{gd}) = \frac{C_{d0}}{\sqrt{1 - \dfrac{V_{gd}}{V_{bi}}}} \tag{3.65}$$

where

$$V_{gd} = V_{gs} - V_{ds} \tag{3.66}$$

From eq. (3.63b), we have

$$\frac{\partial Q_g}{\partial V_{gs}} = C_{gs} + C_{gd} \qquad \frac{\partial Q_g}{\partial V_{ds}} = \frac{\partial Q_d}{\partial V_{gs}} = C_{gd} \qquad \frac{\partial Q_d}{\partial V_{ds}} = C_{ds} + C_{gd} \tag{3.67}$$

By integration, the charge expressions are found to be

$$Q_g(V_{gs}, V_{ds}) = \int \left(C_{gs}(V_{gs}, V_{ds}) + C_{gd}(V_{gs}, V_{ds}) \right) \cdot dV_{gs}$$

$$= -2V_{bi} \cdot C_{g0} \cdot \sqrt{1 - \frac{V_{gs}}{V_{bi}}} - 2V_{bi} \cdot C_{d0} \cdot \sqrt{1 - \frac{V_{gd}}{V_{bi}}} \tag{3.68a}$$

$$Q_d(V_{gs}, V_{ds}) = \int \left(C_{gd}(V_{gs}, V_{ds}) + C_{ds}(V_{gs}, V_{ds}) \right) \cdot dV_{ds}$$

$$= 2V_{bi} \cdot C_{d0} \cdot \sqrt{1 - \frac{V_{gd}}{V_{bi}}} + C_{ds} \cdot V_{ds} \tag{3.68b}$$

In the following paragraph, a typical multi-bias extraction of a linear equivalent circuit that does not take into account the above considerations explicitly is described; therefore, it is rigorously valid only if the data are pulsed *S*-parameters measurements.

3.3.4 Extraction of an Equivalent Circuit from Multi-bias Small-signal Measurements

The behaviour of a real device is distributed by nature; therefore, a good equivalent circuit can at best be an approximation. In general, the higher the number of elements in the equivalent circuit, the better is the approximation; however, the number of elements should be kept as low as possible, both for practical model extraction and for physical meaningfulness of the circuit elements. On the one hand, the evaluation of the element values should be as easy and straightforward as possible, and this is seriously hampered by an excessive number of elements in the circuit. On the other hand, the behaviour of the elements must satisfy the physical constraints (see previous paragraph) that are best fulfilled by elements with a clear correspondence to actual physical effects inside the device. Moreover, when physically meaningful, the equivalent circuit gives interesting information on the structure of the device, both as a feedback to technology and for a qualitative evaluation of the device performances by the designer.

As an example, the correspondence between a simple equivalent circuit of an MESFET and its physical structure is shown in Figure 3.42.

Several similar topologies are available for most active devices at microwave and millimetre-wave frequencies, like MESFETs, HEMTs, MOSFETs, BJTs and HBTs. In

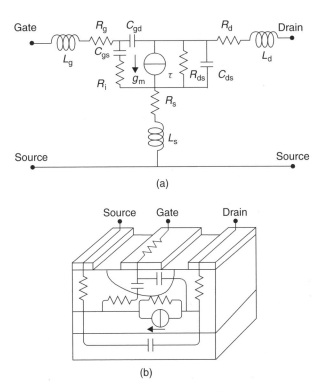

Figure 3.42 The physical structure of an MESFET with the equivalent-circuit elements

general, they fit the wide-band small-signal (linear) parameters of the device for a given bias point; when this changes, the values of the intrinsic elements change too, while parasitics are unchanged. If this is true and the fit to wide-band small-signal parameters is still good, the equivalent circuit is a valid candidate for nonlinear applications. Then, the dependence of the values of the intrinsic elements on the applied voltages is modelled by some fitting functions, with the limitations described in the previous paragraph; if this is true, the model is a good nonlinear model.

Let us illustrate this procedure with an example. The circuit in Figure 3.43 is an equivalent circuit suitable for fitting MESFETs and HEMTs in a wide-frequency band. As an example, the measured S-parameters from 0.1 GHz to 40.1 GHz are shown in Figure 3.44, together with the S-parameters computed from the equivalent circuit, in a range of bias points ($V_{ds} = 2.5$ V, $V_{gs} = -1.8 \div 0.5$ V).

In Figure 3.45, the values of the intrinsic elements are plotted as a function of the gate–source voltage V_{gs} and of the drain–source voltage V_{ds}, together with a fitting function, in this case a neural network model [76].

The topology of the equivalent circuit together with the fitting functions identifies a large-signal equivalent-circuit model.

Let us now describe how the values of the elements of the equivalent circuit are extracted from small-signal data for a practical device. Two main approaches are available: a wide-band fit of the equivalent circuit to the small-signal measured data by means of numerical optimisation routines and the selective identification of groups of parameters by analytical means at special bias points.

A wide-band fit of the equivalent circuit to small-signal data is performed by means of any commercial CAD package. The optimisation variables are the values of

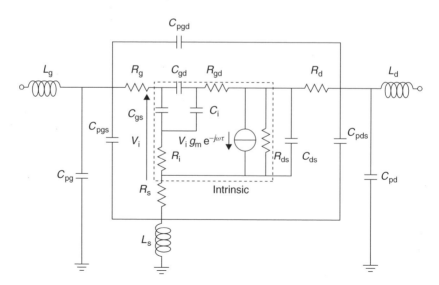

Figure 3.43   The equivalent circuit of an FET

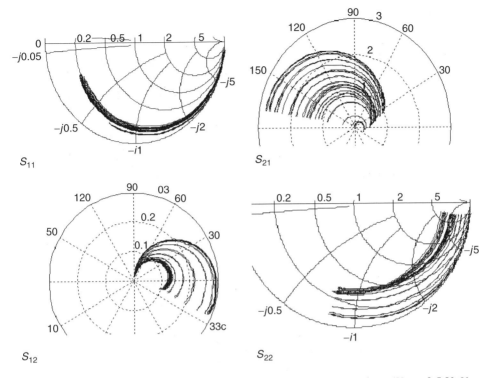

Figure 3.44 Measured and modelled S-parameters at many bias points ($V_{ds} = 2.5$ V, $V_{gs} = -1.8 \div 0.5$ V)

the elements of the equivalent circuit at a given bias point; the optimisation routine varies them until the S-parameters of the equivalent circuit are as close as possible to the measured ones, in the whole frequency band of interest. Alternatively, Y-parameters or any other linear equivalent parameters can be fitted. The optimisation can be performed for each bias point separately or for all the data from all bias points of interest at once [29]. In the former case, the risk is that the optimised values of the parasitic elements vary from bias point to bias point, contrary to the assumption: this is an indication of bad topology or bad optimisation. In the latter case, the parasitics are forced to have the same values at all bias points; however, the numerical burden greatly increases. In both the cases, the optimisation algorithm risks to get trapped in local minima, never reaching the absolute minimum. The goal function is usually not very sensitive to some elements, whose values are therefore rather uncertain. This can be a problem for some applications: for example, the gate resistance in an FET is difficult to extract from normal, operating-point S-parameters, but its value is meaningful for the evaluation of the noise performances of the device. If this is the case, it is wise to adopt the global fitting procedure, also including special bias point as in the 'two-tier' procedure (see below). On the other hand, this approach has a remarkable advantage: it is very easy to change or adjust the topology of the equivalent circuit and have a fast feedback on its fitting accuracy. In addition, it is not restricted or dedicated to any topology or device, and there is no need to develop dedicated software.

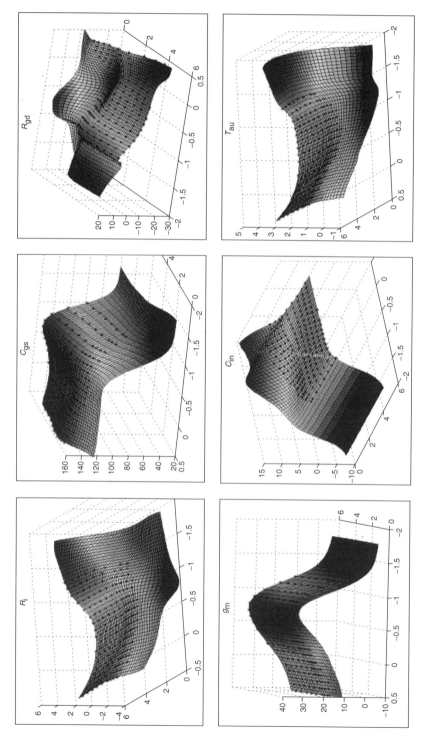

Figure 3.45 Extracted values of the intrinsic elements of the circuit in Figure 3.34 as a function of V_{gs} and V_{ds}

The alternative approach, that is, the selective identification of the elements of the equivalent circuit, is based on the bias-independence of the parasitics. A 'two-tier' extraction procedure is performed: parasitics are first evaluated at special, suitable bias conditions, and their values are not changed afterwards; then, the intrinsic elements are evaluated at each normal, operating bias point within the region of interest. This approach has a clear advantage: the values of the elements are extracted by means of simple, analytical formulae without any optimisation. Usually, a better understanding of the structure of the equivalent circuit is also gained.

A great variety of bias conditions has been proposed for parasitics evaluation [77–87], but almost all require some measurements on a 'cold' device, that is, zero drain or collector voltage. This condition greatly simplifies the behaviour of the inner device, and parasitics are better evaluated. Basically, the measured small-signal parameters of the 'cold' device are equated to the corresponding analytic expressions of the small-signal parameters of the model; the equations are then explicitly solved for the values of the elements. Measurements at a single frequency can be used for the evaluation of the parameters, but averaging over frequency in a suitable band allows for reduction of random measurement errors. In general, diversity in frequency is a useful tool for improving the meaningfulness of the extraction.

A two-port S-parameter measurement at a 'cold' bias condition provides three complex equations, yielding six real values; if the parasitic elements are more than six, more than one 'cold' bias condition must be used. Moreover, not all equations yield reliable results, and it is usually better to have redundant data.

Let us illustrate the procedure by an example concerning an FET [83, 87]; the procedure described hereafter is by no means the only possible one and not necessarily the best: it is only one of the many proposed so far. In fact, it often turns out that a specific device may require modifications of the procedure, because of minor but important differences in the structure of the device; however, the approach is usually similar.

Three 'cold' bias conditions are used in this example: depleted channel, that is, pinched-off FET ($V_{ds} = 0$, $V_{gs} < V_{po}$), open channel ($V_{ds} = 0$, $V_{po} < V_{gs} < V_{bi}$) and gate-channel junction in weak forward conduction ($V_{ds} = 0$, $V_{bi} < V_{gs}$). The intrinsic device behaves very differently in the three conditions in such a way that different parasitic elements are relevant in each bias condition. Somehow, this is similar to what is required from different calibration standards; in fact, in this case also, access elements of a device under investigation (in this case the intrinsic device) must be identified and removed. The equivalent circuits of the device in the three 'cold' conditions are shown in Figure 3.46.

In the case of pinched-off device, the measured S-parameters are converted to admittance Y-parameters, which are easier to express analytically for the equivalent circuit. If we limit ourselves to a suitably low-frequency range, the inductances can be neglected, and the simplified equations read as follows:

$$\text{Im}(Y_{11})_{po} = \omega(C_{pg} + C_{pgs} + C_{gs} + C_{pgd} + C_{gd}) \tag{3.69a}$$

$$\text{Im}(Y_{12})_{po} = \omega(C_{pgd} + C_{gd}) \tag{3.69b}$$

$$\text{Im}(Y_{22})_{po} = \omega(C_{in} + C_{ds} + C_{pds} + C_{pd} + C_{pgd} + C_{gd}) \tag{3.69c}$$

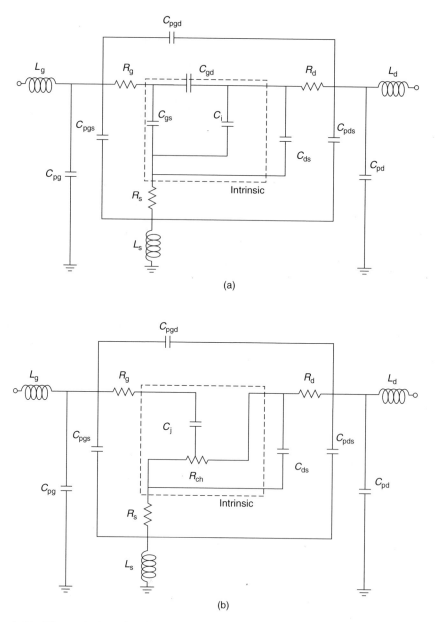

Figure 3.46 The equivalent circuits of a 'cold' FET: pinch-off (a), open channel (b) and weak forward conduction of the gate junction (c)

We consider only the imaginary part of the parameters for the extraction, since the real parts are not always found to give reliable results in this bias condition. When plotted against frequency, the measured parameters exhibit a linear behaviour at low frequencies, confirming the negligible effect of inductances in that frequency range; in Figure 3.47, the imaginary part of the Y_{11} parameter versus frequency is shown for $V_{gs} = -2$ V.

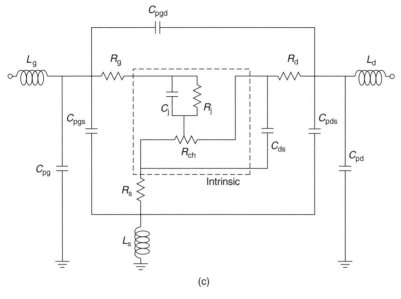

(c)

Figure 3.46   (*continued*)

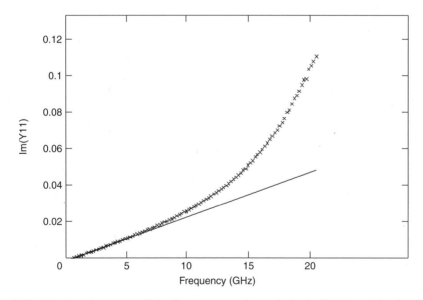

Figure 3.47   The imaginary part of the Y_{11} parameter for a pinched-off FET and the fitted capacitive susceptance

The slope of the measured curve in the low-frequency range yields the values of the capacitances. If the measurement is repeated for decreasing gate–source voltages, the channel is more and more depleted; the values of the intrinsic capacitances tend to zero, and a plot of the extracted capacitance value versus gate–source voltage V_{gs} can be extrapolated to $V_{gs} = -\infty$ to eliminate their contribution (Figure 3.48).

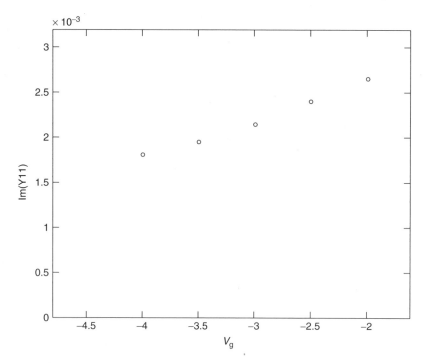

Figure 3.48 Extrapolation of the extracted capacitance to $V_{gs} = -\infty$

In this way, the parasitic capacitances are found. In fact, the model shown in Figure 3.43 has five parasitic capacitances, while we have used only three measurements. The additional informations required for complete identification are usually obtained from measurement of transistors of the same type but with different gate periphery. This approach allows not only a more reliable extraction, but also the availability of a scalable model. However, in the case that the gate–source and drain–source parasitic capacitances C_{pgs} and C_{pds} are negligible, the measurements are sufficient.

When pinched off, the intrinsic device behaves as a high impedance. Now, the different condition of weak forward-biased gate-channel junction is considered: the intrinsic device behaves as a low impedance. It is neither necessary nor advisable to bias the junction in full conduction in order to avoid junction degradation: a small gate current I_g is sufficient to open a conductive path from the gate to the channel. The expressions of the impedance Z-parameters of the equivalent circuit read as follows:

$$\text{Re}(Z_{12})_{wfb} = R_s + \frac{R_{ch}}{2} - \frac{R_{ch}^2}{12nV_T} \cdot I_g \tag{3.70a}$$

$$\text{Im}(Z_{12})_{wfb} = \omega\left(L_s + K_1 + \frac{K_2}{I_g}\right) \tag{3.70b}$$

$$\text{Re}(Z_{22})_{wfb} = R_s + R_d + R_{ch} \tag{3.70c}$$

$$\text{Im}(Z_{22})_{wfb} = \omega(L_s + L_d + K_3) \tag{3.70d}$$

where I_g is the gate current, and K_1, K_2 and K_3 are functions of resistances and capacitances [87, 88]; their expressions are given in Appendix A.2. As in the previous case, not all equations have been considered: the Z_{11} parameter does not usually yield reliable results. Plotting the measured data versus frequency, a suitable frequency range is identified for the extraction of the parameter; for example, the real part of the Z_{12} parameter is better evaluated at relatively high frequency (Figure 3.49).

The extracted value depends on the value of the gate current I_g: the corresponding term can be eliminated by repeating the measurement for several current values and extrapolating to $I_g \to 0$ (Figure 3.50).

From this bias condition, the source and drain inductances are found, and two equations in the three unknown resistances R_s, R_d and $R_{channel}$ are found.

The third 'cold' bias condition is somehow in between the other two: the channel is open, and controlled by the gate–source voltage. The equations relative to the impedance Z-parameters are

$$\mathrm{Re}(Z_{11})_{oc} = \frac{C_g^2\left(R_g + R_s + \dfrac{\gamma R_{ch}}{3}\right) + C_p^2(R_s + R_d + \gamma R_{ch}) + 2C_g C_p\left(R_s + \dfrac{R_{ch}}{2}\right)}{(C_g + C_{pg} + C_p)^2} \tag{3.71a}$$

$$\mathrm{Im}(Z_{11})_{oc} = -\frac{1}{\omega(C_g + C_p + C_{pg})} + \omega(L_g + K_4 L_s + K_5) \tag{3.71b}$$

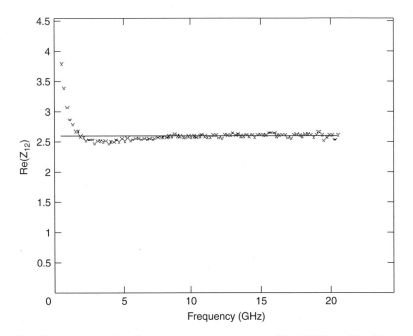

Figure 3.49 The real part of the Z_{12} parameter for a pinched-off cold FET and the fitted resistance

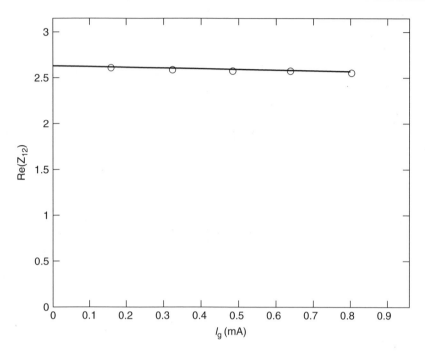

Figure 3.50 Extrapolation of the extracted resistance to $I_g = 0$

$$\mathrm{Re}(Z_{22})_{oc} = R_s + R_d + \gamma R_{ch} \tag{3.71c}$$

$$\gamma = \left(1 - \sqrt{\frac{V_b - V_{gs}}{V_b - V_p}}\right)^{-1} \tag{3.71d}$$

where all capacitances except the gate-channel capacitance C_g are known, and K_4 and K_5 are functions of the resistances and capacitance [87, 88]; their expressions are given in Appendix A.2. Once more, a plot of the measured data versus frequency indicates the best frequency range for the extraction. For example, the real part of the Z_{11} parameter is shown in Figure 3.51, suggesting a moderate- to high-frequency range for the evaluation.

The extracted value depends on the value of the gate–source voltage via a linear dependence on the γ parameter (Figure 3.52): the extrapolated value to $\gamma \rightarrow 0$ and the slope versus γ yield two equations.

Similarly, three other equations are given by the other parameters. In all, five equations in eight unknowns R_g, R_s, R_d, $R_{channel}$, L_g, L_s, L_d, and C_g are obtained. Together with the weak forward-biased junction condition, we have now eight equations in as many unknowns: the system is easily solved, and the parasitics are extracted. Their values will not be changed by successive steps of the extraction.

Now the intrinsic, bias-dependent elements must be evaluated at normal operating points. At each bias point, the S-parameters are measured and then de-embedded from the already evaluated parasitics. This is accomplished by successive transformations

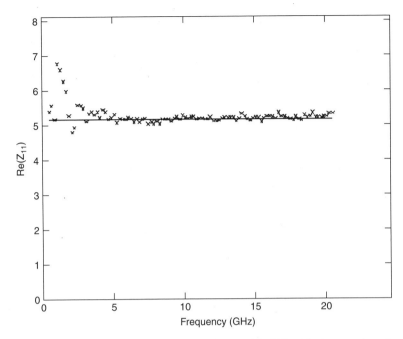

Figure 3.51   The real part of the Z_{11} parameter for a cold FET with the gate junction in weak forward conduction and the fitted resistance

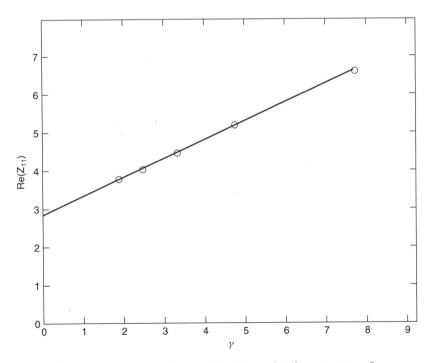

Figure 3.52   Extrapolation of the extracted resistance to $\gamma = 0$

to Z- and Y-matrices, from which series and shunt parasitic elements respectively are subtracted [83] (Figure 3.53); the de-embedded data correspond to the equivalent circuit shown in Figure 3.54.

The analytical expressions of the admittance Y-parameters of this intrinsic equivalent network read

$$Y_{11} = \omega^2(R_{gd}C_{gd}^2 + R_iC_{gs}^2) + i\omega(C_{gd} + C_{gs})$$
$$Y_{12} = \omega^2[R_iC_iC_{gs} - R_{gd}C_{gd}(C_{gd} + C_i)] - i\omega C_{gd}$$
$$Y_{21} = g_m - i\omega[C_{gd} + g_m(\tau + R_iC_{gs})]$$
$$Y_{22} = g_{ds} + i(C_i + C_{gd} - g_mR_iC_i) \tag{3.72}$$

$$[S]_{measured} \rightarrow [Z]_{measured} \rightarrow \begin{bmatrix} Z_{11}-j\omega L_g & Z_{12} \\ Z_{21} & Z_{22}-j\omega L_d \end{bmatrix} \rightarrow [Y] \rightarrow$$

$$\begin{bmatrix} Y_{11}-j\omega(C_{pg}+C_{pgd}) & Z_{12}-j\omega C_{pgd} \\ Z_{21}-j\omega C_{pgd} & Z_{22}+j\omega(C_{pd}-C_{pgd}) \end{bmatrix} \rightarrow [Z] \rightarrow$$

$$\begin{bmatrix} Z_{11}-L_s & Z_{12}-L_s \\ Z_{21}-L_s & Z_{22}-L_s \end{bmatrix} \rightarrow [Y] \rightarrow$$

$$\begin{bmatrix} Y_{11}-j\omega C_{pgs} & Y_{12} \\ Y_{21} & Y_{22}-j\omega C_{pds} \end{bmatrix} \rightarrow [Z] \rightarrow$$

$$\begin{bmatrix} Z_{11}-R_g-R_s & Z_{12}-R_s \\ Z_{21}-R_s & Z_{22}-R_d-R_s \end{bmatrix} \rightarrow [Z]_{intrinsic} \rightarrow [Y]_{intrinsic}$$

Figure 3.53 Procedure for the de-embedding of the parasitic elements from measured S-parameters

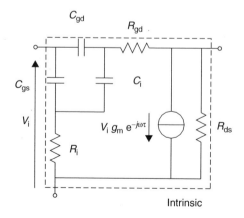

Figure 3.54 Equivalent circuit of the intrinsic FET

and must be equated to the de-embedded measured data. By analytically solving for the intrinsic elements,

$$C_{gs} \cong \frac{\text{Im}(Y_{11} + Y_{12})}{\omega}$$

$$C_{gd} \cong -\frac{\text{Im}(Y_{12})}{\omega}$$

$$C_{in} \cong \frac{\text{Im}(Y_{12} + Y_{22})}{\omega}$$

$$R_{gd} \cong \frac{1}{\omega^2} \cdot \frac{C_{in} \cdot \text{Re}(Y_{11}) - C_{gs} \cdot \text{Re}(Y_{12})}{C_{in} \cdot C_{gd}^2 - C_{gs} \cdot C_{gd} \cdot (C_{gd} + C_{in})}$$

$$g_m = \sqrt{(\text{Re}(Y_{21} - Y_{12}))^2 + (\text{Im}(Y_{21} - Y_{12}))^2}$$

$$\tau = \frac{a \tanh\left(-\dfrac{\text{Re}(Y_{21} - Y_{12})}{\text{Im}(Y_{21} - Y_{12})}\right)}{\omega}$$

$$R_i \cong \frac{\text{Re}(Y_{11}) - R_{gd}\omega^2 C_{gd}^2}{\omega^2 C_{gs}^2}$$

$$R_{ds} \cong \frac{1}{\text{Re}(Y_{22})} \tag{3.73}$$

The values of the intrinsic elements are found by averaging the above expressions over frequency in a low-to-medium frequency band. After evaluation, the measured and de-embedded intrinsic Y-parameters can be plotted versus frequency and compared to the extracted equivalent-circuit data for verification of the extraction. As an example, the real and imaginary parts of the Y_{21} parameter are shown in Figure 3.55 for the operating point $V_{gs} = -0.6$ V, $V_{ds} = 5$ V.

Alternatively, the extracted values of the elements are plotted versus frequency; as an example, the transconductance g_m and the transit time τ are plotted versus frequency in Figure 3.56.

From the plotted data, it is apparent that the elements of the intrinsic equivalent circuit are constant with frequency, fulfilling the basic hypothesis and confirming the validity of the topology and extraction procedure of the equivalent circuit for this device.

The extraction of the intrinsic elements is now repeated for all operating points of interest, and a table with the extracted values is available for nonlinear model construction, as explained in the previous paragraph.

3.3.5 Nonlinear Models

A nonlinear model consists of the topology of the equivalent circuit together with specific fitting functions for the extracted values of the equivalent-circuit elements. A large variety

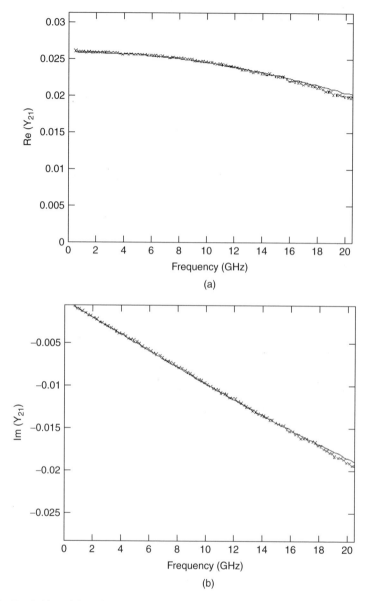

Figure 3.55 Real (a) and imaginary parts (b) of the measured and extracted Y_{21} of the intrinsic FET for $V_{gs} = -0.6$ V, $V_{ds} = 5$ V

of models are available in the literature for practically all microwave devices, and many of them are implemented in commercial CAD programmes. Some of them are described in this paragraph for ease of use by the design engineer.

A simple function for fitting the I/V output characteristics is defined in [89]:

$$I_d(V_{gs}, V_{ds}) = (A_0 + A_1 \cdot V_1 + A_2 \cdot V_1^2 + A_3 \cdot V_1^3) \cdot \tanh(\alpha \cdot V_{ds}) \tag{3.74}$$

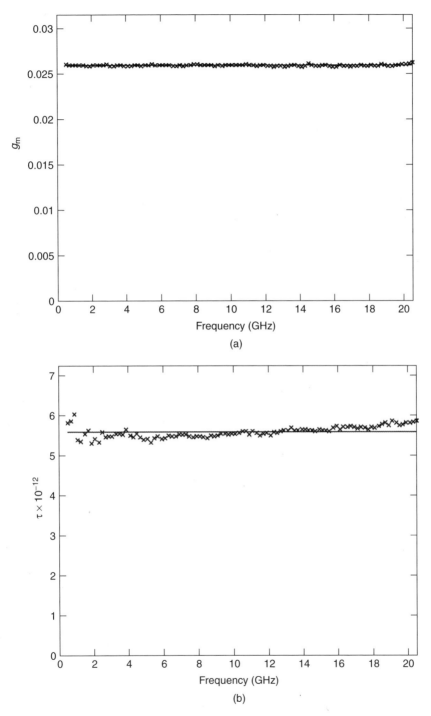

Figure 3.56 Measured and extracted transconductance g_m (a) and delay time τ (b) of the intrinsic FET for $V_{gs} = -0.6$ V, $V_{ds} = 5$ V

where

$$V_1 = V_{gs}(t - \tau) \cdot (1 + \beta \cdot (V_{d0} - V_d)) \qquad (3.75)$$

The drain current is modelled as a third-order polynomial in the delayed gate–source voltage, with an additional term that accounts for the shift in pinch-off voltage for high drain voltages. The current is set to zero when eq. (3.74) becomes negative. While accurate enough for quasi-linear operations in a region not too far from a Class-A bias point, this model does not fit sufficiently well for the pinch-off behaviour and the knee of the I/V characteristics of the device.

A more flexible model has been developed on the basis of the one proposed in [90].

$$I_{ds} = I_{dss} \cdot \left(1 - \frac{V_{gi}}{V_{po} + \gamma V_{ds}}\right)^{E + K_e \cdot V_{gi}} \cdot \left(1 + S_s \cdot \frac{V_{ds}}{I_{dss}}\right) \cdot tgh\left[S_1 \cdot \frac{V_{ds}}{I_{dss}} \cdot \left(\frac{1}{1 - K_g V_{gi}}\right)\right]$$

$$(3.76a)$$

where

$$V_{gi} = V_{gs}(t - \tau) \qquad (3.76b)$$

The nine parameters $(I_{dss}, V_{po}, \gamma, E, K_e, S_s, S_1, K_g, \tau)$ allow a good fitting in all parts of the characteristic. The function is again set to zero when eq. (3.76a) becomes negative.

A fitting function developed for HEMT, with very good properties, has been proposed in [91]:

$$I_{ds} = I_{pk} \cdot (1 + \tanh \psi) \cdot (1 + \lambda \cdot V_{ds}) \tanh(\alpha \cdot V_{ds}) \qquad (3.77)$$

where I_{pk} is the current where the transconductance has the maximum (peak) value and ψ is a polynomial function of V_{gs} centred around the value corresponding to maximum transconductance:

$$\psi = P_1(V_{gs} - V_{pk}) + P_2(V_{gs} - V_{pk})^2 + \cdots \qquad (3.78)$$

and

$$V_{pk} = V_{pk0} + \gamma \cdot V_{ds} \qquad (3.79)$$

accounts for a shift with drain voltage of the value corresponding to peak transconductance. This model is very flexible, and reproduces with good accuracy also the derivatives of the drain current with respect to the gate voltage.

Special attention must be paid to the model when intermodulation distortion or mixing performances must be analysed. It is clear from Section 1.3.1 that the intermodulation depends very much on the derivatives of the drain current with respect to the controlling voltages. Therefore, models that have a better capability to model the derivatives of the drain current, especially with respect to the gate voltage, have been developed [92–101]. A typical behaviour of the drain current and the first two derivatives with respect to the gate voltage in the saturation region ($V_{ds} = 2$ V) for an HEMT device is shown in Figure 3.57.

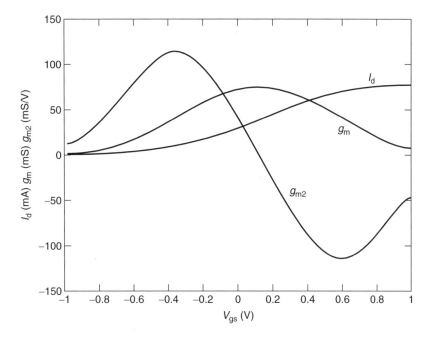

Figure 3.57   Typical drain current with its first two derivatives versus gate voltage for an HEMT device

A function for fitting the drain current and its derivatives has been proposed in [94] as

$$I_{ds} = (A_0 \cdot x + A_1 \cdot \sin x + A_2 \cdot \sin 2x + A_3 \cdot \sin 3x) \cdot \tanh(\alpha \cdot V_{ds}) \tag{3.80}$$

where

$$x = \pi \cdot \left(\frac{V_{gs} - V_{po}}{V_{bi} - V_{po}} \right) \tag{3.81}$$

and A_0, A_1.A_2, A_3, α, V_{po}, V_{bi} are the model parameters. Similar functions with accurate modelling of the derivatives with respect to the gate voltage [92, 93, 97, 99, 100] or both gate and drain voltages [95, 98, 101] have been proposed. Some models are valid in the saturation region of the FET [92–94, 97, 99–101], while others are especially intended for the application to the analysis of low-intermodulation, resistive FET mixers (see Chapter 7) [95]. A symmetric model valid in both regions and also in the inverse saturated region has also been proposed [98]. The drain current is expressed as a function of gate–source and gate–drain voltages for symmetry:

$$I_{ds} = g[f_1(U_{gd}^+, U_{gs}^+) \cdot f_2(V_{gs} - V_{gd}) - f_1(U_{gs}^-, U_{gd}^-) \cdot f_2(V_{gd} - V_{gs})] \tag{3.82}$$

where

$$f_1(x, y) = (1 + a \cdot x) \cdot \left[1 - \tanh(e^{-b \cdot (y+c)}) \right] \quad f_2(z) = 1 - \tanh(e^{-d \cdot z}) \tag{3.83}$$

and

$$
\begin{bmatrix} U_{gd}^+ \\ U_{gs}^+ \end{bmatrix} = \begin{bmatrix} \cos\varphi & \sin\varphi \\ -\sin\varphi & \cos\varphi \end{bmatrix} \cdot \begin{bmatrix} V_{gs} \\ V_{gs} \end{bmatrix}
\qquad
\begin{bmatrix} U_{gd}^- \\ U_{gs}^- \end{bmatrix} = \begin{bmatrix} \cos\varphi & -\sin\varphi \\ \sin\varphi & \cos\varphi \end{bmatrix} \cdot \begin{bmatrix} V_{gd} \\ V_{gs} \end{bmatrix}
$$
(3.84)

and a, b, c, d, φ are the model parameters.

However, when higher-order derivatives of the drain current must be modelled, it is not trivial to get accurate data. Differentiation of DC or RF I/V data can, in principle, be attempted, but the accuracy is poor. A simple but effective approach consists of accurately fitting the first-order derivatives of the drain current, that is, the small-signal transconductance and output conductance g_m and g_{ds} as functions of the gate–source and drain–source voltages [75, 91, 98]. If the small-signal equivalent circuit has been accurately extracted, the model can actually predict intermodulation distortion with good accuracy. A more sophisticated approach requires the measurement of the harmonics of a low-frequency (in the MHz range) single-tone excitation with standard terminations [92–95] or with different output loads for increased accuracy [99–101].

A different approach is adopted in [96, 102–104]. The dependence of the drain current on the two controlling voltages (e.g. gate–source and gate–drain voltages for an FET) is locally expressed in the neighbourhood of the bias point as a bi-dimensional Taylor expansion up to the third order:

$$
\begin{aligned}
I_{ds} = {} & g_m V_{gs} + g_{ds} V_{ds} + \\
& + g_{m2} V_{gs}^2 + g_{md} V_{gs} V_{ds} + g_{d2} V_{ds}^2 + \\
& + g_{m3} V_{gs}^3 + g_{m2d} V_{gs}^2 V_{ds} + g_{md2} V_{gs} V_{ds}^2 + g_{d3} V_{ds}^3 + \\
& + \cdots
\end{aligned}
$$
(3.85)

Clearly, the coefficients of the expansion are functions of the bias point. The values of the coefficients are found by intermodulation measurements with low-frequency (in the 100-MHz range) two-tone test voltages: one sinusoidal voltage at the first test frequency is applied to the gate (or base) of the transistor, and a second sinusoidal signal at a second, very close test frequency is applied to the drain (or collector), as in Figure 3.58.

The amplitudes of the test signals are small enough for a Volterra series representation of the intermodulation phenomena to be valid. By selectively shorting some harmonic and intermodulation frequencies at the ports of the device and by expressing the spectral powers for the extraction as functions of the Taylor series expansion, it is possible to use the power of the intermodulation signals coming out of the device at intermodulation frequencies for the selective and successive identification of all the coefficients of the Taylor series expansions. This is obtained by inversion of the Volterra series expressions for the intermodulation terms that include the coefficients; the technique is similar to the nonlinear current method (see Section 1.3.1). Since the measurements are absolute values (powers), some care is required for the identification of the sign of the coefficients. The instrumentation required for the measurements is very limited: the characterisation frequency (in the 100-MHz range) is high enough to avoid low-frequency dispersion, but

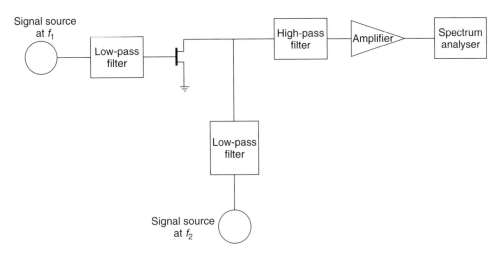

Figure 3.58 The measurement set-up for evaluation of the coefficients of the Taylor series expansion of the drain current

low enough to allow for a very cheap and simple measurement set-up and to make the contribution of capacitive nonlinearities negligible.

Once the local data are extracted, their bias dependence can be modelled by means of neural networks in order to be able to predict intermodulation distortion for any bias point within the measured region [105]. A typical application has been the identification of the so-called 'sweet spots' of a transistor, that is, the bias points where the intermodulation contributions from the various nonlinearities inside an active device cancel, so that a very low distortion is obtained from the device for a significant input power range. Another application has been the analysis of resistive mixers that are typically used for low-intermodulation applications: in this case, the extracted data have been fitted by an empirical function [104]. The procedure has also been extended to capacitances [103].

3.3.6 Packages

Packaged transistors are often used for high-power applications. Usually, suppliers suggest the optimum bias and loading and also the layout of a circuit for the most common applications. However, the availability of a nonlinear model is often useful, when special non-standard applications are envisaged, in terms of frequency band or input signal or bias point. In this case, a model of the package must be evaluated prior to the extraction of the transistor model.

The package is a distributed network; however, it can often be approximated with a lumped or semi-lumped equivalent circuit. The topology of the circuit depends on the specific package, but a typical structure is shown in Figures 3.59 and 3.60 in which pre-matching structures are also present.

In-package pre-matching is often used when large transistors, or even several large transistors in parallel, are packaged for high-power applications. Such an arrangement

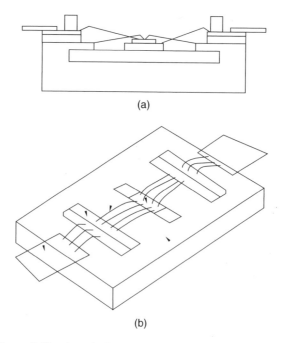

(a)

(b)

Figure 3.59   A typical package for a high-power transistor

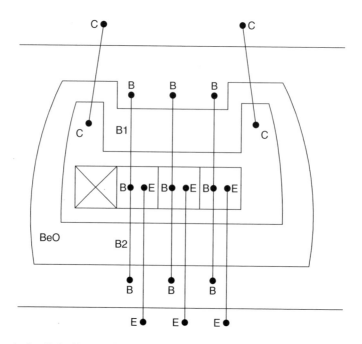

Figure 3.60   A detailed picture of the bonding arrangement for a bipolar transistor within the package

obviously makes the modelling procedure practically impossible, unless the package is accurately modelled.

Sometimes, the model of the package is available from the supplier; however, this is not the typical case. Contrariwise, it is very advisable that at least some knowledge of the structure and dimensions of the package be gained from the supplier for a sufficiently reliable modelling. If this is the case, the definition and extraction of a segmented model can be attempted, as shown in the Figure 3.61.

The single sections of the package can be modelled by means of electrical lumped or distributed elements, as shown in Figure 3.62. The extraction of the values of the

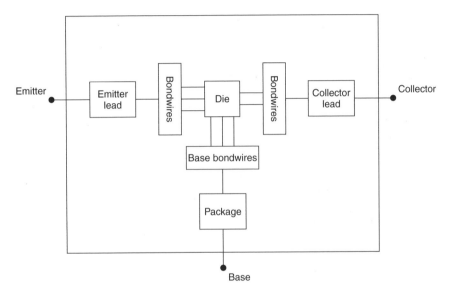

Figure 3.61   A possible structure of the equivalent network of the package

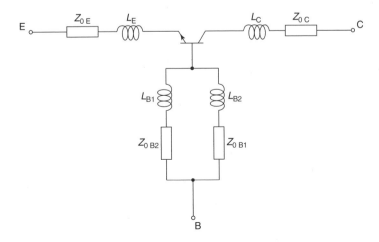

Figure 3.62   The equivalent circuit of the segmented package structure as in Figure 3.61

elements can be performed by means of full-wave electromagnetic simulators or approximated with suitable equivalent expressions. It is normally useful to measure an empty package and fit the extracted model of the package to the measurements for final trimming of the element values.

3.4 BLACK-BOX MODELS

In this paragraph, the main black-box approaches are described: an example of technology-independent model for time-domain or harmonic balance analysis, a model for nonlinear frequency-domain analysis (spectral balance) and behavioural models, mostly based on a describing-function approach.

3.4.1 Table-based Models

Several approaches have been proposed for the construction of a nonlinear model based on measured data only without any assumption on the structure of the device [106–113]. They allow the automatic construction of the model, avoiding troublesome extraction procedures and inaccuracies due to topology approximation of equivalent circuits (see Section 3.3). Many of them have been developed for use in conjunction with harmonic balance analysis algorithms; in the following, the mathematical formulation is justified as in [108, 112], showing that this approach is a generalisation of eq. (3.47).

The mathematical formulation can be demonstrated by means of a Volterra series representation of a nonlinear device (Section 1.3.1), and by remarking that the 'memory' of past inputs on the nonlinear behaviour of a semiconductor device is usually limited to a very short time compared to a microwave period. Assuming a voltage-controlled current nonlinearity, the standard Volterra series expansion can be rewritten as

$$i(t) = F_{DC}(v(t)) + \sum_n \int \ldots \int g_n(v(t), \tau_1, \ldots \tau_n) \cdot e(\tau_1, t) \ldots e(\tau_n, t) d\tau_n \ldots d\tau_1$$

$$(3.86)$$

where $F_{DC}(v)$ is the DC I/V characteristic of the device,

$$e(v(t), \tau) = v(\tau) - v(t)$$

$$(3.87)$$

is the difference between the input at the current time instant and its past values, and

$$g_n(v(t), \tau_1, \ldots \tau_n) = \sum_{n=k}^{\infty} v^{(n-k)}(t) \cdot \binom{n}{k} \int \ldots \int h_n(t - \tau_1, \ldots t - \tau_n) d\tau_n \ldots d\tau_1$$

$$(3.88)$$

are dynamic nonlinear impulse responses that account for the response of the nonlinear current to the difference between the current value of the input voltage and its past values. In practice, the first term in the right-hand side of eq. (3.86) gives the nonlinear response of the device to the input voltage, as if it had kept constant at the current value for an infinite time in the past. The second term accounts for the memory effects within the device due to the time evolution of the input voltage in the past; these effects are nonlinear, and require a Volterra-like formulation including higher-order nonlinear terms. Unlike the classical Volterra series formulation, the first term in eq. (3.86) accounts for the largest part of the nonlinear response due to 'instantaneous' nonlinearities, while the additional terms account only for the deviations due to memory. From another point of view, eq. (3.86) is equivalent to a Volterra series expansion in the neighbourhood of a bias point, which in this case is dynamic (the voltage $v(t)$). If the memory of the device is short enough, that is, if the dynamic impulse responses decay to zero in a time sufficiently shorter that the microwave period, the first-order response dominates the term, and it is the only one to be retained:

$$i(t) = F_{\mathrm{DC}}(v(t)) + \int_{t-\tau_m}^{t} g_1(v(t), t - \tau) \cdot (v(\tau) - v(t)) \cdot \mathrm{d}\tau \qquad (3.89)$$

The convolution integral in eq. (3.89) accounts for the effect of time-varying input, linearised in the neighbourhood of the instantaneous voltage $v(t)$. Fourier transformation of eq. (3.89) yields [108]

$$\Im[g_1(v, t)] = \hat{Y}(v, \omega) = Y(v, \omega) - \frac{\partial F_{\mathrm{DC}}}{\partial v} \qquad (3.90)$$

where $Y(v, \omega)$ are the standard admittance parameters measured at a bias point v and $\dfrac{\partial F_{\mathrm{DC}}}{\partial v}$ is the conductance computed from the DC characteristics of the device. The formulation is a generalisation of the eq. (3.47), where only the drain conduction current had been considered. The convolution integral in eq. (3.88) is easily included in a harmonic balance algorithm [112]. The result is a table-based model, where integration of measured data along the load line is performed during large-signal analysis.

3.4.2 Quasi-static Model Identified from Time-domain Data

Under the assumption that the device is quasi-static, a direct extraction procedure can be performed on the basis of time-domain large-signal data, as obtained from a time-domain large-signal measurement set-up (see Section 2.3). Let us consider a two-port nonlinear element; a quasi-static representation is as shown in Figure 3.63.

The voltages and currents at the ports of the device are

$$i_1(t) = i_{gs}(v_{gs}(t), v_{ds}(t)) + C_{gs}(v_{gs}(t), v_{ds}(t)) \cdot \frac{\mathrm{d}v_{gs}(t)}{\mathrm{d}t}$$

$$+ C_{gd}(v_{gs}(t), v_{ds}(t)) \cdot \frac{\mathrm{d}v_{ds}(t)}{\mathrm{d}t} \qquad (3.90a)$$

$$i_2(t) = i_{ds}(v_{gs}(t), v_{ds}(t)) + C_{dg}(v_{gs}(t), v_{ds}(t)) \cdot \frac{\mathrm{d}v_{gs}(t)}{\mathrm{d}t}$$

$$+ C_{ds}(v_{gs}(t), v_{ds}(t)) \cdot \frac{\mathrm{d}v_{ds}(t)}{\mathrm{d}t} \qquad (3.90b)$$

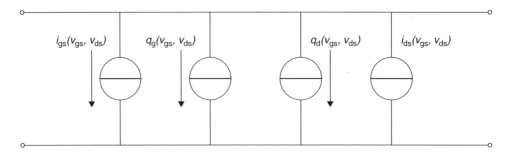

Figure 3.63   The quasi-static model of a two-port active device

where the capacitances and transcapacitances have been defined above (Section 3.3). When a large-signal excitation is applied, the port currents, voltages and voltage derivatives are measured in the time domain. The values of the elements of the model are directly evaluated for a pair of port voltages v_{gs}, v_{ds} when three independent measurements are available for each port for the same voltage pair but different time derivatives [114–117]. The model is identified for all voltage pairs scanned by the waveform measurement, that is, for all the voltage pairs assumed by the voltage waveforms during their time-domain evolution. The time-domain identification can be associated with the de-embedding of parasitic elements in order to make the quasi-static assumption more realistic.

3.4.3 Frequency-domain Models

Nonlinear black-box models can be constructed in the frequency domain; they find their natural application within nonlinear analysis algorithms of the spectral balance type (Chapter 1) [118, 119]. The basic principle requires the definition of the current in a nonlinear element as an extension of the quasi-static formula:

$$i(t) = i_g(v(t)) + \frac{dq^{(1)}(v(t))}{dt} + \frac{d^2q^{(2)}(v(t))}{dt^2} + \frac{d^3q^{(3)}(v(t))}{dt^3} + \cdots \qquad (3.91)$$

The equivalent circuit is as shown in Figure 3.64.

Let us assume that the static current and the generalised charges do not depend explicitly on time and on voltage derivatives, but only on voltage. If a small signal is

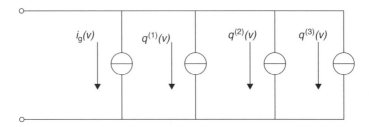

Figure 3.64   A frequency-domain black-box equivalent circuit

superimposed to a static large voltage

$$v(t) = V_0 + \tilde{v}(t) \quad i(t) = I_0 + \tilde{i}(t) \tag{3.92}$$

then the current in the element can be written as

$$I_0 + \tilde{i}(t) = i_g(V_0) + \left.\frac{di_g(v)}{dv}\right|_{v=V_0} \cdot \tilde{v}(t) + \left.\frac{dq^{(1)}(v)}{dv}\right|_{v=V_0} \cdot \frac{d\tilde{v}(t)}{dt} + \left.\frac{dq^{(2)}(v)}{dv}\right|_{v=V_0}$$

$$\cdot \frac{d^2\tilde{v}(t)}{dt^2} + \left.\frac{dq^{(3)}(v)}{dv}\right|_{v=V_0} \cdot \frac{d^3\tilde{v}(t)}{dt^3} + \cdots \tag{3.93}$$

In the frequency domain

$$\tilde{I}(\omega) = g(V_0) \cdot \tilde{V}(\omega) + C^{(1)}(V_0) \cdot j\omega \cdot \tilde{V}(\omega) + C^{(2)}(V_0) \cdot (j\omega)^2 \cdot \tilde{V}(\omega)$$

$$+ C^{(3)}(V_0) \cdot (j\omega)^3 \cdot \tilde{V}(\omega) + \cdots \tag{3.94}$$

The ratio between small-signal current and voltage, that is, the small-signal admittance is

$$Y(V_0, \omega) = g(V_0) + j\omega \cdot C^{(1)}(V_0) - \omega^2 \cdot C^{(2)}(V_0) - j\omega^3 \cdot C^{(3)}(V_0) + \cdots \tag{3.95}$$

therefore, the frequency dependence of the small-signal admittance at a given bias point is represented by a polynomial in the frequency itself, allowing a much better fit to experimental data than the usual quasi-static formulation. Since the model is implemented in a frequency-domain algorithm, the dependence on higher-order time derivatives does not introduce any additional difficulty. Moreover, any frequency dependence of the parameters of the model is possible, including very low-frequency or very high-frequency dispersion. The large-signal model is constructed by integration of the small-signal conductance and generalised charges, as seen above:

$$i_g(v) = i_g(v_0) + \int_{v_0}^{v} g(\beta) \cdot d\beta \qquad q^{(1)}(v) = q^{(1)}(v_0) + \int_{v_0}^{v} C^{(1)}(\beta) \cdot d\beta$$

$$\tag{3.96}$$

$$q^{(2)}(v) = q^{(2)}(v_0) + \int_{v_0}^{v} C^{(2)}(\beta) \cdot d\beta \qquad q^{(3)}(v) = q^{(3)}(v_0) + \int_{v_0}^{v} C^{(3)}(\beta) \cdot d\beta$$

In order to be used within a frequency-domain spectral balance algorithm, the dependence of the nonlinear current and charge functions on voltage must be modelled as a polynomial function of the controlling voltage itself (see Chapter 1). In practice, polynomials of the Chebyshef type provide a good fit up to a high order [118].

Even if, in principle, the model can fit any frequency dependence of the measured data, in practice it is convenient to de-embed passive, parasitic elements in order to reduce the number of higher-order generalised charges. Quite naturally, the model is easily generalised to an n-port network.

3.4.4 Behavioural Models

Behavioural models are based on the assumption that no information on the internal structure of the device is known. The behaviour of the device or subsystem is measured, and then it is approximated by means of suitable fitting functions, usually of the class of artificial neural networks (ANN). This assumption seriously limits the predictive properties of the model, which are valid only in a neighbourhood of the measuring conditions. However, in some cases this neighbourhood includes very interesting situations in which big savings in terms of simulation times are obtained.

Let us show this with an example. For a narrowband amplifier or subsystem, the describing-function approach has been described in Chapter 1. The same approach is experimentally very easy to implement, by simply performing source/load-pull measurements with vector capabilities, for a sufficient number of input amplitudes from small signal to saturation (see Chapter 2). In particular, if the device or subsystem is matched, the input and output loads can be simply set to 50 Ω; otherwise, they must be set to the optimum input and output loads for the device or subsystem. The amplitude and phase of the output signal are measured for each value of the amplitude of the input signal, and the results are usually plotted as AM/AM, AM/PM plots [120]. This approach is easily generalised: if some harmonics are non-negligible, as for instance in the case of a frequency multiplier, more than one describing function is defined, one per harmonic, and a multiharmonic measuring set-up must be available. If the model must reproduce the dependence of the output of a device or subsystem not only on a signal at the input port at fundamental frequency but also on signals at other ports (output port, bias ports, local oscillator ports) and other harmonics, the describing functions become functions of the amplitudes and phases of all the input signals except the phase of an arbitrarily selected signal that acts as a reference for the system. It can also happen that the loads are not fixed and can vary in a neighbourhood of the optimum ones, as they usually do in practice because, for example, of fabrication tolerances; in this case, the values of the loads are treated as additional input variables, and actual source/load-pull measurements must be performed in order to get the data [76, 121, 122]. Once the data are obtained, a set of multi-variable fitting functions are trained to the data and implemented in a CAD tool for circuit or system simulation.

So far, the model only reproduces the measurements and acts as a compact description of the experimental data. For example, in the case of complex system simulation, it may be convenient to simulate a single subcircuit by means of a standard nonlinear analysis algorithm, and then train a model to the results of the analysis. The behavioural model will be much faster and reasonably accurate and allows the simultaneous analysis of several nonlinear subcircuits within a large system with a reasonable computational effort. However, in some cases the extrapolation properties of this type of model are also very useful. Let us assume that a system is narrowband and that the input signal is composed of tens or hundreds of closely spaced spectral lines. This is a common case for modern multi-carrier communications systems and corresponds to a carrier-modulated narrowband around its carrier frequency. The total signal can be considered as a centre-frequency carrier signal modulated by a slowly varying envelope signal (see Chapter 1). If the modulating envelope is slow enough with respect to the carrier, it can be assumed that the describing function still reproduces with sufficient accuracy the behaviour of the carrier as it slowly changes with time [121]. With an extension of this approach, the frequency-conversion properties of a nonlinear circuit are also modelled, and the behaviour of a narrowband-modulated signal is predicted.

A different extension of the describing-function approach consists of the linearisation of the function around the large-signal dynamic state of a device or circuit [122]. The large-signal incident and reflected waves are first measured in the large-signal state of interest and then perturbed by a small incremental wave (see Chapter 2). The perturbed wave is written as the unperturbed wave plus a perturbation that is linear with respect to the perturbing wave:

$$b_{k,i} = b_{k,i}^{(0)} + \sum_{j=1,2} \sum_{l=1,N} \left(G_{k,i,l,j} \cdot \mathrm{Re}[a_{l,j}] + H_{k,i,l,j} \cdot \mathrm{Im}[a_{l,j}] \right)$$

$$k = 1, \ldots N, i = 1, 2; \; l = 1, \ldots N, j = 1, 2 \tag{3.97}$$

where $b_{k,j}^{(0)}$ is the unperturbed wave at port i and harmonic frequency k, $b_{k,j}$ is the perturbed wave at the same port and harmonic frequency and $a_{l,j}$ is the small perturbing wave at port j and harmonic frequency l. The complex coefficients $G_{k,i,l,j}$ and $H_{k,i,l,j}$ are called the nonlinear scattering functions; they are a generalisation of the scattering parameters to the case of a time-variant bias voltage (see Chapter 8). The relation is linear but not analytic [122], as already seen for the Jacobian of a harmonic balance algorithm, from which the functions can also be easily computed [123] (see Chapter 1). In Figure 3.65, an example is shown of drain current and voltage around a large-signal state

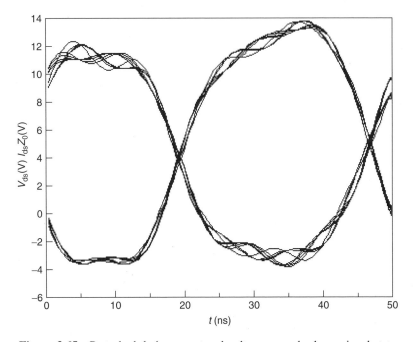

Figure 3.65 Perturbed drain current and voltage around a large-signal state

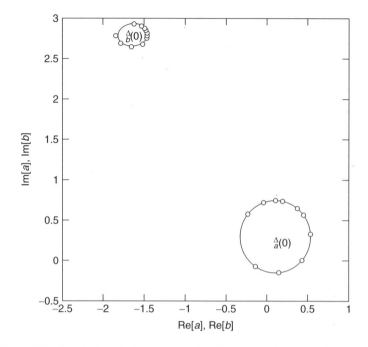

Figure 3.66 Perturbed drain current and voltage around a large-signal state

for a high-efficiency power amplifier perturbed by several small fifth-harmonic waves in the time domain.

In the complex plane of the waves a and b, the waves are represented by vectors; the perturbed waves are therefore represented as constant vectors (the unperturbed wave) to which small perturbing vectors (waves) are added. For better accuracy of the measurement, many perturbing waves a with the same amplitude but different phases are used, describing thus a circle around the unperturbed wave $a^{(0)}$. The perturbed wave vector b correspondingly describes an ellipse around the unperturbed vector $b^{(0)}$, because of the non-analytic nature of eq. (3.97) (Figure 3.66).

The nonlinear scattering parameters find application, for example, when the stability of the large-signal state must be verified or ensured or when the condition of large-signal match is required.

3.5 SIMPLIFIED MODELS

In this paragraph, simplified models are described together with some hints on their main applications.

So far, accuracy has been one of the main desirable features of the described models. In this paragraph, we will describe models that are intentionally not very accurate but that allow for substantial advantages from other points of view. In fact, an accurate model requires an equally accurate nonlinear analysis algorithm, even considering that the limiting factor of the simulation accuracy for the current state of nonlinear CAD is the model itself. However, an accurate analysis algorithm is a numerical algorithm that in itself does not allow a proper insight into the behaviour of the device or circuit. The data are fed into the computer and the results come out. Of course, optimisation is very useful for improving the performances of a circuit; however, numerical problems sometimes do not allow the optimisation algorithm to find the optimum values. Moreover, the definition of a single optimisation goal does not allow for flexibility in the design trade-offs: it is not clear what is gained on one hand if something is lost on the other hand. More importantly, the main mechanisms responsible for good or bad performances of the circuit are not clear, unless a detailed and time-consuming analysis of many simulations is performed by a skilled designer.

A simpler approach consists of the use of a simplified model, including only the main nonlinear characteristics of the active device, and requiring a simplified analysis algorithm. In this way, another advantage of this approach is the much simpler model extraction procedure that can sometimes be performed from data sheets only without actually buying and measuring the device. Obviously, the final design of the circuit will normally be performed by means of a complete model and CAD tool, but a general insight into the performance of a device or circuit will be gained in a short time.

Simple models have been used for a long time for power amplifier design [124–129]. The equivalent circuit can be, for instance, as in Figure 3.67 for the case of an FET where the only nonlinearity is the voltage-controlled drain–source current source. The linear elements are extracted from small-signal parameters at the selected bias point or as an average value over a suitable range of bias voltages. Moreover, the nonlinearity is modelled by a piecewise-linear function, as in Figure 3.68.

In this case, the transconductance is constant with respect to the gate–source voltage V_{gs} within the linear region, and zero outside, unless the operating point reaches the ohmic or breakdown regions. The analysis becomes piecewise-linear as well, and the voltage and current waveforms are computed analytically. For instance, in the case of the

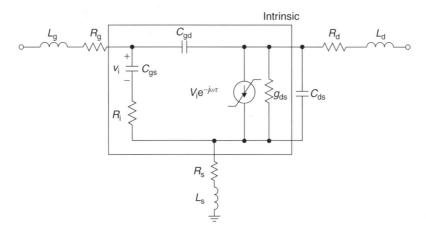

Figure 3.67   Simplified nonlinear equivalent circuit of an FET

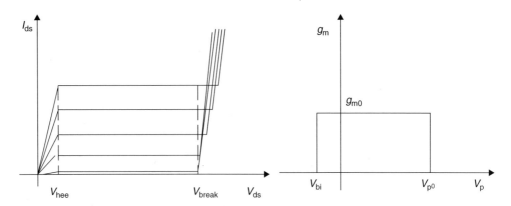

Figure 3.68 Piecewise-linear representation of the drain current and transconductance

current source being considered as a pure transconductance and the input signal being a sinusoid, the drain current is a truncated sinusoid (Figure 3.69).

A simple Fourier analysis yields analytical expressions for the phasors of the harmonics (Figure 3.70).

The output voltage waveform is found by multiplication of the current phasors times the harmonic impedances and time-domain reconstruction. At least for the simplest cases, no iterative analysis is required and explicit expressions are given for voltages and currents.

Piecewise-linear simplified models have been successfully applied to the study and design of nonlinear circuits as power amplifiers, mixers and frequency multipliers; their application will be illustrated in detail in the relevant chapters.

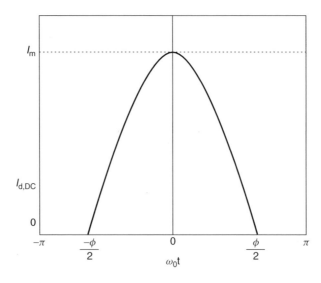

Figure 3.69 Drain current in a simplified piecewise-linear model

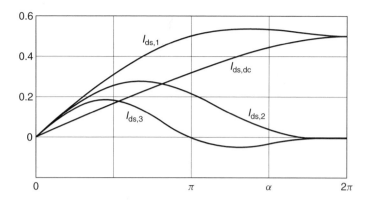

Figure 3.70 Harmonic components of the drain current as a function of the circulation angle as in Figure 3.69

3.6 BIBLIOGRAPHY

[1] M.A. Alsunaidi, S.M. Sohel Imtiaz, S.M. El-Ghazaly, 'Electromagnetic wave effects on microwave transistors using a full-wave time domain model', *IEEE Trans. Microwave Theory Tech.*, **44**(6), 799–807, 1996.

[2] S. Datta, *Electronic Transport in Mesoscopic Systems*, Cambridge University Press, Cambridge (UK), 1995.

[3] K. Bløtekjær, 'Transport equations for electrons in two-valley semiconductors', *IEEE Trans. Electron Devices*, **ED-17**, 38–47, 1970.

[4] S. Selberherr, *Analysis and Simulation of Semiconductor Devices*, Springer-Verlag, Wien, New York (NY), 1984.

[5] B. Troyanovsky, Z.-P. Yu, R.W. Dutton, 'Physics-based simulation of nonlinear distortion in semiconductor devices using the harmonic balance method', *Comput. Methods Appl. Mech. Eng.*, **181**, 467–482, 2000.

[6] T. Shawki, G. Salmer, O. El Sayed, 'MODFET 2-D hydrodynamic energy modelling: optimisation of subquarter-micron-gate structures', *IEEE Trans. Electron Devices*, **37**, 21–29, 1990.

[7] C.M. Snowden, D. Loret, 'Two-dimensional hot-electron models for short-gate length GaAs MESFETs', *IEEE Trans. Electron Devices*, **ED-34**, 212–223, 1987.

[8] D. Loret, 'Two-dimensional numerical model for the high electron mobility transistor', *Solid-State Electron.*, **3**, 1197–1203, 1987.

[9] F.A. Buot, 'Two-dimensional numerical modelling of HEMT using an energy transport model', *International Journal for Computation and Mathematics in Electrical and Electronic Engineering*, **6**, 45–52, 1987.

[10] Y.K. Feng, A. Hintz, 'Simulation of sub-micrometer GaAs MESFETs using a full dynamic transport model', *IEEE Trans. Electron Devices*, **ED-35**, 1419–1431, 1988.

[11] W.R. Curtice, Y.H. Yun, 'A temperature model for the GaAs MESFETs', *IEEE Trans. Electron Devices*, **ED-28**, 954–962, 1981.

[12] R.W. Dutton, B. Troyanovsky, Z. Yu, T. Arnborg, F. Rotella, G. Ma, J. Sato-Iwanaga, 'Device simulation for RF applications', *Proc. IEDM*, 1997, pp. 301–304.

[13] C.M. Snowden, M.J. Hawes, D.V. Morgan, 'Large-signal modelling of GaAs MESFET operations', *IEEE Trans. Electron Devices*, **ED-30**, 1817–1824, 1983.

[14] M. Reiser, 'A two-dimensional numerical FET model for DC, AC and large-signal analysis', *IEEE Trans. Electron Devices*, **ED-26**, 476–490, 1979.

[15] F. Filicori, G. Ghione, C.U. Naldi, 'Physics-based electron device modelling and computer-aided MMIC design', *IEEE Trans. Microwave Theory Tech.*, **MTT-40**(7), 1333–1352, 1992.

[16] B. Carnez, A. Cappy, A. Kaszynski, E. Constant, G. Salmer, 'Modelling of a submicrometer gate field-effect transistor including the effect of nonstationary electron dynamics', *J. Appl. Phys.*, **51**, 784–790, 1980.

[17] P.A. Sandborn, J.R. East, G.I. Haddad, 'Quasi two-dimensional modeling of GaAs MES-FETs', *IEEE Trans. Electron Devices*, **ED-34**(5), 985–981, 1987.

[18] R.R. Pantoja, M.J. Howes, J. Richardson, C.M. Snowden, 'A large signal physical MESFET model for computer aided design and it applications', *IEEE Trans. Microwave Theory Tech.*, **37**(12), 2039–2045, 1989.

[19] C.M. Snowden, R.R. Pantoja, 'Quasi-two-dimensional MESFET simulation for CAD', *IEEE Trans. Electron Devices*, **36**(9), 1564–1574, 1989.

[20] R. Singh, C.M. Snowden, 'A quasi-two-dimensional HEMT model for DC and microwave simulation', *IEEE Trans. Electron Devices*, **45**(6), 1165–1169, 1998.

[21] P.J. Rudge, R.E Miles, M.B. Steer, C.M. Snowden, 'Investigation into intermodulation distortion in HEMTs using a quasi-2-D physical model', *IEEE Trans. Microwave Theory Tech.*, **49**, 2315–2321, 2001.

[22] G. Halkias, H. Gerard, Y. Crosnier, G. Salmer, 'A new approach to RF power operation of MESFETs', *IEEE Trans. Microwave Theory Tech.*, **37**(5), 817–825, 1989.

[23] F. Giannini, G. Leuzzi, M. Kopanski, G. Salmer, 'Large-signal analysis of quasi-2D physical model of MESFETs', *Electron. Lett.*, **29**(21), 1891–1893, 1993.

[24] C. Moglestue, 'A self-consistent Monte Carlo model to analyze semiconductor microcomponents of any geometry', *IEEE Trans. Comput.-Aided Des.*, **CAD-5**(2), 326–345, 1996.

[25] G. Salmer, J. Zimmermann, R. Fauquembergue, 'Modelling of MODFETs', *IEEE Trans. Microwave Theory Tech.*, **MTT-36**, 1124–1140, 1988.

[26] C. Jacoboni, P. Lugli, *The Monte Carlo Method for Semiconductor Device Simulation*, Springer-Verlag, Wien (Austria), 1989.

[27] F. Rossi, A. Di Carlo, P. Lugli, 'Microscopic theory of quantum-transport phenomena in mesoscopic systems: a Monte Carlo approach, *Phys. Rev. Lett.*, **80**, 3348–3352, 1998.

[28] P.J. Rodrigues, M.J. Howes, J.R. Richardson, 'Direct use of a MESFET physical model in non-linear CAD', *Proc. MTT-S Int. Microwave Symp. Dig.*, 1992, pp. 1579–1582.

[29] J.W. Bandler, R.M. Biernacki, Q. Cai, S.H. Chen, S. Ye, Q.-J. Zhang, 'Integrated physics-oriented statistical modelling, simulation and optimisation', *IEEE Trans. Microwave Theory Tech.*, **MTT-40**(7), 1374–1400, 1992.

[30] A. Witzig, Ch. Schluster, P. Regli, W. Fichtner, 'Global modelling of microwave applications by combining the FDTD method and a general semiconductor device and circuit simulator', *IEEE Trans. Microwave Theory Tech.*, **MTT-47**(6), 919–928, 1999.

[31] S. Beaussart, O. Perrin, M.R. Friscourt, C. Dalle, 'Millimetre-wave pulsed oscillator global modelling by means of electromagnetic, thermal, electrical and carrier transport physical coupled models', *IEEE Trans. Microwave Theory Tech.*, **47**, 929–934, 1999.

[32] R.O. Grondin, S. El-Ghazaly, S. Goodnick, 'A review of global modeling of charge transport in semiconductors and full-wave electromagnetics', *IEEE Trans. Microwave Theory Tech.*, **47**(6), 817–829, 1999.

[33] A. Cidronali, G. Leuzzi, G. Manes, F. Giannini, 'Physical/electromagnetic pHEMT modelling', *IEEE Trans. Microwave Theory Tech.*, **MTT-51**(3), 2003, 830–838.

[34] W.R. Curtice, 'Intrinsic GaAs MESFET equivalent circuit models generated from two-dimensional simulation', *IEEE Trans. Comput.-Aided Des.*, **8**, 395–402, 1989.

[35] G. Ghione, U.C. Naldi, F. Filicori, 'Physical modelling of GaAs MESFETs in an integrated CAD environment: from device technology to microwave circuit performance', *IEEE Trans. Microwave Theory Tech.*, **MTT-37**(3), 457–467, 1989.

[36] R.S. Muller, T.I. Kamins, *Device Electronics for Integrated Circuits*, Wiley, New York (NY), 1986.

[37] M.S. Shur, 'Analytical model of GaAs MESFETs', *IEEE Trans. Electron Devices Tech.*, **ED-25**(6), 612–618, 1978.

[38] A. Madjar, F.J. Rosenbaum, 'A large-signal model for the GaAs MESFET', *IEEE Trans. Microwave Theory Tech.*, **MTT-29**, 781–788, 1981.

[39] T.H. Chen, M.S. Shur, 'Analytic models for ion-implanted MESFETs', *IEEE Trans. Electron Devices*, **ED-32**, 18–28, 1985.

[40] T.H. Chen, M.S. Shur, 'A capacitance model for GaAs MESFETs', *IEEE Trans. Electron Devices*, **ED-32**, 883–891, 1985.

[41] A. Madjar, 'A fully analytical AC large-signal model of the GaAs MESFET for nonlinear network analysis and design', *IEEE Trans. Microwave Theory Tech.*, **MTT-36**(1), 61–67, 1988.

[42] M.A. Khatibzadeh, R.J. Trew, 'A large-signal analytic model for the GaAs MESFET', *IEEE Trans. Microwave Theory Tech.*, **MTT-36**, 231–238, 1988.

[43] C.S. Chang, D.Y.S. Day, 'Analytic theory for current-voltage characteristics and field distribution of GaAs MESFET', *IEEE Trans. Electron Devices*, **ED-36**, 269–280, 1989.

[44] S. D'Agostino, G. D'Inzeo, P. marietti, L. Tudini, A. Betti-Berutto, 'Analytic physics-based expressions for the empirical parameters of the Statz-Pucel MESFET model', *IEEE Trans. Microwave Theory Tech.*, **MTT-40**(7), 1576–1581, 1992.

[45] S. D'Agostino, A. betti-Berutto, 'Physics-based expressions for the nonlinear capacitances of the MESFET equivalent circuit', *IEEE Trans. Microwave Theory Tech.*, **MTT-42**(3), 403–406, 1994.

[46] D. Delagebeaudeuf, N.T. Linh, 'Metal-(n) AlGaAs-GaAs two-dimensional electron gas FET', *IEEE Trans. Electron Devices*, **ED-29**, 955–960, 1982.

[47] T.J. Drummond, H. Morkoç, K. Lee, M. Shur, 'Model for modulation doped field-effect transistor', *IEEE Electron Devices Lett.*, **EDL-3**, 338–341, 1982.

[48] A. Shey, W.H. Ku, 'An analytical current-voltage characteristics model for high electron mobility transistor based on nonlinear charge control formulation', *IEEE Trans. Electron Devices*, **36**(10), 2299–2305, 1989.

[49] D.-H. Huang, H.C. Lin, 'Nonlinear charge-control DC and transmission line models for GaAs MODFETs', *Int. MTT-S Symp. Dig.*, 1989, pp. 147–150.

[50] S.S. Islam, A.F.M. Anwar, 'Nonlinear analysis of GaN MESFETs with Volterra series using large-signal models including trapping effects', *IEEE Trans. Microwave Theory Tech.*, **MTT-50**(11), 2474–2479, 2002.

[51] T. Li, R.P. Joshi, R.D. Del Rosario, 'Requirements for low intermodulation distortion in GaN-Al$_x$Ga$_{1-x}$N high electron mobility transistors: a model assessment', *IEEE Trans. Electron Devices*, **ED-49**(9), 1511–1518, 2002.

[52] M.Y. Frankel, D. Pavlidis, 'An analysis of the large-signal characteristics of AlGaAs7GaAs heterojunction bipolar transistor', *IEEE Trans. Microwave Theory Tech.*, **MTT-40**(3), 465–474, 1992.

[53] C. Rauscher, H.A. Willing, 'Simulation of nonlinear microwave FET performance using a quasistatic model', *IEEE Trans. Microwave Theory Tech.*, **MTT-27**(10), 834–840, 1979.

[54] C. Camacho-Peñalosa, C.S. Aitchison, 'Modelling frequency dependence of output impedance of a microwave MESFET at low frequencies', *Electron. Lett.*, **21**(12), 528–529, 1985.

[55] P.H. Ladbrooke, S.R. Blight, 'Low-field, low-frequency dispersion of transconductance in GaAs NESFETs with implications for other rate-dependent anomalies', *IEEE Trans. Microwave Theory Tech.*, **MTT-35**, 257–267, 1988.

[56] M. Paggi, P.H. Williams, J.M. Borrego, 'Nonlinear GaAs MESFET modelling using pulsed gate measurements', *IEEE Trans. Microwave Theory Tech.*, **MTT-36**, 1593–1597, 1988.

[57] N. Scheinber, R. Bayruns, R. Goyal, 'A low-frequency GaAs MESFET circuit model', *IEEE J. Solid-State Circuits*, **23**(2), 605–608, 1988.

[58] J. Reynoso-Hernandez, J. Graffeuil, 'Output conductance frequency dispersion and low-frequency noise in HEMTs and MESFETs', *IEEE Trans. Microwave Theory Tech.*, **MTT-37**(9), 1478–1481, 1989.

[59] J. Golio, M. Miller, G. Maracas, D. Johnson, 'Frequency-dependent electrical characteristics of GaAs MESFETs', *IEEE Trans. Electron Devices*, **ED-37**, 1217–1227, 1990.

[60] F. Filicori, G. Vannini, G. Santarelli, A. Mediavilla, A. Tazón, Y. Newport, 'Empirical modelling of low-frequency dispersive effects due to traps and thermal phenomena in III-V FETs', *IEEE Trans. Microwave Theory Tech.*, **MTT-43**(12), 2972–2981, 1995.

[61] K. Jeon, Y. Kwon, S. Hong, 'A frequency dispersion model of GaAs MESFET for large-signal applications', *IEEE Microwave Guided Wave Lett.*, **7**(3), 78–80, 1997.

[62] T. Roh, Y. Kim, Y. Suh, W. Park, B. Kim, 'A simple and accurate MESFET channel-current model including bias-dependent dispersion and thermal phenomena', *IEEE Trans. Microwave Theory Tech.*, **MTT-45**(8), 1252–1255, 1997.

[63] J.P. Teyssier, J.P. Viaud, J.J. Raoux, R. Quéré, 'Fully nonlinear modelling and characterisation system of microwave transistors with on-wafer pulsed measurements', *IEEE MT-S Int. Symp. Dig.*, Orlando (FL), 1995, pp. 1033–1036.

[64] M. Lee, L. Forbes, 'A self-backgating GaAs MESFET model for low-frequency anomalies', *IEEE Trans. Electron Devices*, **ED-37**, 2148–2157, 1990.

[65] K. Kunihiro, Y. Ohno, 'A large-signal equivalent circuit model for substrate-induced drain-lag phenomena in HJFETs', *IEEE Trans. Electron Devices*, **ED-43**(9), 1336–1342, 1996.

[66] A. Santarelli, F. Filicori, G. Vannini, P. Rinaldi, 'Backgating model including self-heating for low-frequency dispersive effects in III-V FETs', *Electron. Lett.*, **34**(20), 1974–1976, 1998.

[67] A. Santarelli, G. Zucchelli, R. Paganelli, G. Vannini, F. Filicori, 'Equivalent-voltage approach for modelling low-frequency dispersive effects in microwave FETs', *IEEE Microwave Wireless Components Lett.*, **12**(9), 339–341, 2002.

[68] A. Werthof, G. Kompa, 'A unified consistent DC to RF large-signal FET model covering the strong dispersion effects of HEMT devices', *Proc. 22nd EuMC*, Helsinki (Finland), Aug. 1992, pp. 1091–1096.

[69] G. Kompa, 'Modelling of dispersive microwave FET devices using a quasistatic approach', *Int. J. Microwave Millimetre-Wave Comput.-Aided Eng.*, **5**(3), 173–194, 1995.

[70] V. Rizzoli, F. Mastri, A. Neri, A. Lipparini, 'An electrothermal model of the microwave FET suitable for nonlinear simulation', *Int. J. Microwave Millimetre-Wave Comput.-Aided Eng.*, **5**, 104–121, 1995.

[71] V. Rizzoli, A. Costanzo, 'A fully conservative non-linear empirical model of the microwave FET', *Proc. 24th European Microwave Conf.*, Cannes (France), Sept. 1994, pp. 1307–1312.

[72] I. Angelov, L. Bengtsson, M. Garcia, 'Extension of the Chalmers nonlinear HEMT and MESFET model', *IEEE Trans. Microwave Theory Tech.*, **MTT-44**(10), 1664–1674, 1996.

[73] A. Roizes, D. Lazaro, R. Quéré, J.P. Teyssier, 'Low-noise FETs vulnerability prediction under RF pulsed overloads based on nonlinear electrothermal modelling', *IEEE Microwave Guided Wave Lett.*, **9**(7), 280–281, 1999.

[74] T. Fernandez, Y. Newport, J.M. Zamanillo, A. Tazón, A. Mediavilla, 'Extracting a bias-dependent large-signal MESFET model from pulsed I/V measurements', *IEEE Trans. Microwave Theory Tech.*, **MTT-44**(3), 372–378, 1996.

[75] V.I. Cojocaru, T.J. Brazil, 'A scalable general-purpose model for microwave FETs including DC/AC dispersion effects', *IEEE Trans. Microwave Theory Tech.*, **MTT-45**(12), 2248–2255, 1997.

[76] F. Wang, V.K. Devabhaktuni, Q.-J. Zhang, 'A hierarchical neural network approach to the development of a library of neural models for microwave design', *IEEE Trans. Microwave Theory Tech.*, **MTT-46**(12), 2391–2403, 1998.

[77] R.A. Minasian, 'Simplified GaAs MESFET model to 10 GHz', *Electron. Lett.*, **13**(8), 549–541, 1977.

[78] H. Fukui, 'Determination of the basic device parameters of a GaAs MESFET', *Bell Syst. Tech. J.*, **58**(3), 771–797, 1979.

[79] P.L. Hower, N.G. Bechtel, 'Current saturation and small-signal characteristics of GaAs field effect transistors', *IEEE Trans. Electron Devices*, **ED-20**, 213–220, 1973.

[80] F. Diamand, M. Laviron, 'Measurement of the extrinsic series elements of a microwave MESFET under zero current conditions', *Proc. 12th European Microwave Conf.*, Helsinki (Finland), Sept. 1982, pp. 451–453.

[81] K.W. Lee, K. Lee, M. Shur, T.T. Vu, P. Roberts, M.J. Helix, 'Source, drain and gate series resistances and electron saturation velocity in ion-implanted GaAs FETs', *IEEE Trans. Electron Devices*, **ED-32**(5), 987–992, 1985.

[82] W.R. Curtice, R.L. Camisa, 'Self-consistent GaAs FET models for amplifier design and device diagnostics', *IEEE Trans. Microwave Theory Tech.*, **MTT-32**, 1573–1578, 1984.

[83] G. Dambrine, A. Cappy, F. Heliodore, E. Playez, 'A new method for determining the FET small-signal equivalent circuit', *IEEE Trans. Microwave Theory Tech.*, **MTT-36**(7), 1151–1159, 1988.

[84] M. Berroth, R. Bosch, 'Broad-band determination of the FET small-signal equivalent circuit', *IEEE Trans. Microwave Theory Tech.*, **MTT-38**(7), 891–895, 1990.

[85] R. Anholt, S. Swirhun, 'Equivalent circuit parameter extraction fro cold GaAs MESFETs', *IEEE Trans. Microwave Theory Tech.*, **MTT-39**(7), 1243–1247, 1991.

[86] P. White, R. Healy, 'Improved equivalent circuit for determination of MESFET and HEMT parasitic capacitances from cold-FET measurements', *IEEE Trans. Microwave Theory Tech.*, **MTT-41**(7), 453–455, 1993.

[87] G. Leuzzi, A. Serino, F. Giannini, S. Ciorciolini, 'Novel non-linear equivalent-circuit extraction scheme for microwave field-effect transistors', *Proc. 25th European Microwave Conf.*, Bologna (Italy), Sept. 1995, pp. 548–552.

[88] G. Leuzzi, A. Serino, F. Giannini, 'RC-term correction in the evaluation of parasitic inductances for microwave transistor modelling', *Proc. 24th European Microwave Conf.*, Cannes (France), Sept. 1994, pp. 1628–1631.

[89] W.R. Curtice, M. Ettenberg, 'A nonlinear GaAs FET model for use in the design of output circuits for power amplifiers', *IEEE Trans. Microwave Theory Tech.*, **MTT-33**, 1383–1387, 1985.

[90] A. Materka, T. Kacprzak, 'Computer calculation of large-signal GAaAs FET amplifier characteristics', *IEEE Trans. Microwave Theory Tech.*, **MTT-33**, 129–134, 1985.

[91] I. Angelov, H. Zirath, N. Rorsman, 'A new empirical nonlinear model for HEMT and MESFET devices', *IEEE Trans. Microwave Theory Tech.*, **MTT-40**(12), 2258–2266, 1992.

[92] A. Crosmun, S.A. Maas, 'Minimisation of intermodulation distortion in GaAs MESFET small-signal amplifiers', *IEEE Trans. Microwave Theory Tech.*, **MTT-37**(9), 1411–1417, 1989.

[93] S.A. Maas, A.M. Crosmun, 'Modelling the gate I/V characteristics of a GaAs MESFET for Volterra-series analysis', *IEEE Trans. Microwave Theory Tech.*, **MTT-37**(7), 1134–1136, 1989.

[94] S. Maas, D. Neilson, 'Modeling MESFETs for intermodulation analysis of mixers and amplifiers', *IEEE Trans. Microwave Theory Tech.*, **MTT-38**(12), 1964–1971, 1990.

[95] R.S. Virk, S.A. Maas, 'Modelling MESFETs for intermodulation analysis in RF switches', *IEEE Microwave Guided Wave Lett.*, **4**(11), 376–378, 1994.

[96] J.C. Pedro, J. Perez, 'Accurate simulation of GaAs MESFETs intermodulation distortion using a new drain-source current model', *IEEE Trans. Microwave Theory Tech.*, **MTT-42**(1), 25–33, 1994.

[97] S. Peng, P.J. McCleer, G.I. Haddad, 'Simplified nonlinear model for the intermodulation analysis of MESFET mixers', *IEEE MTT-S Int. Symp. Dig.*, 1994, pp. 1575–1578.

[98] K. Yhland, N. Rorsman, M. Garcia, H.F. Merkel, 'A symmetrical HFET/MESFET model suitable for intermodulation analysis of amplifiers and resistive mixers', *IEEE Trans. Microwave Theory Tech.*, **MTT-48**(1), 15–22, 2000.

[99] G. Qu, A.E. Parker, 'New model extraction for predicting distortion in HEMT and MESFET circuits', *IEEE Micrwave Guided Wave Lett.*, **9**, 363–365, 1999.

[100] C.-W. Fan, K.-K.M. Cheng, 'A new method in characterising the nonlinear current model of MESFETs using single-tone excitation', *IEEE MTT-S Int. Symp. Dig.*, 2000, pp. 449–452.

[101] C.-C. Huang, H.-T. Pai, 'A recursive scheme for MESFET nonlinear current coefficient evaluation applied in Volterra-series analysis', *IEEE MTT-S Int. Symp. Dig.*, 2003, pp. 463–466.

[102] R.S. Tucker, 'Third-order intermodulation distortion and gain compression in GaAs FETs', *IEEE Trans. Microwave Theory Tech.*, **MTT-27**(5), 400–408, 1979.

[103] J.A. Garcia, A. Mediavilla, J.C. Pedro, N.B. Carvalho, A. Tazón, J.L. Garcia, 'Characterising the gate-to-source nonlinear capacitor role on GaAs FET IMD performance', *IEEE Trans. Microwave Theory Tech.*, **MTT-46**(12), 2344–2355, 1998.

[104] J.A. Garcia, J.C. Pedro, M.L. de la Fuente, N.B. Carvalho, A. Mediavilla, A. Tazón, 'Resistive FET mixer conversion loss and IMD optimisation by selective drain bias', *IEEE Trans. Microwave Theory Tech.*, **MTT-47**(12), 2382–2392, 1999.

[105] J.A. García, A. Tazón, A. Mediavilla, I. Santamaría, M. Lázaro, C.J. Pantaleón, J.C. Pedro, 'Modeling MESFETs and HEMTs intermodulation distortion behavior using a generalized radial basis function network', *Int. J. RF Microwave Comput.-Aided Eng.*, **9**, 261–276, 1999.

[106] D.E. Root, 'Technology independent large-signal non-quasistatic FET models by direct construction from automatically characterised device data', *Proc. 21st EuMC*, Stuttgart (Germany), 1991, pp. 927–932.

[107] D.E. Root, 'Measurement-based large-signal diode modelling system for circuit and device design', *IEEE Trans. Microwave Theory Tech.*, **MTT-41**, 2211–2217, 1993.

[108] F. Filicori, G. Vannini, V.A. Monaco, 'A nonlinear integral model of electron devices for HB circuit analysis', *IEEE Trans. Microwave Theory Tech.*, **40**(7), 1992.

[109] R.R. Daniels, A.T. Yang, J.P. Harrang, 'A universal large/small-signal 3-terminal FET model using a non-quasistatic charge-based approach', *IEEE Trans. Electron Devices*, **ED-40**, 1723–1729, 1993.

[110] C.-J. Wei, Y.A. Tkachenko, D. Bartle, 'Table-based dynamic FET model assembled from small-signal models', *IEEE Trans. Microwave Theory Tech.*, **MTT-47**(6), 700–705, 1999.

[111] I. Angelov, N. Rorsman, J. Stenarson, M. Garcia, H. Zirath, 'An empirical table-based FET model', *IEEE Trans. Microwave Theory Tech.*, **MTT-47**(12), 2350–2357, 1999.

[112] F. Filicori, A. Santarelli, P. Traverso, G. Vannini, 'Electron device model based on nonlinear discrete convolution for large-signal circuit analysis using commercial CAD packages', *Proc. GAAS '99*, Munchen (Germany), Oct. 1999.

[113] M. Fernández-Barciela, P.J. Tasker, Y. Campos-Roca, M. Demmler, H. Massler, E. Sanchez, M.C. Curras-Francos, M. Schlechtweg, 'A simplified broad-band large-signal non-quasi-static table-based FET model' *IEEE Trans. Microwave Theory Tech.*, **MTT-48**(3), 395–405, 2000.

[114] A. Werthof, F. van Raay, G. Kompa, 'Direct nonlinear FET parameter extraction using large-signal waveform measurements', *IEEE Microwave Guided Wave Lett.*, **3**(5), 130–132, 1993.

[115] C.-J. Wei, Y.E. Lan, J.C.M. Hwang, W.-J. Ho, J.A. Higgins, 'Waveform-based modelling and characterisation of microwave power heterojunction bipolar transistors', *IEEE Trans. Microwave Theory Tech.*, **MTT-43**(12), 2899–2903, 1995.

[116] M.C. Currás-Francos, P.J. Tasker, M. Fernández-Barciela, Y. Campos-Roca, E. Sánchez, 'Direct extraction of nonlinear FET C-V functions from time-domain large-signal measurements', *Electron. Lett.*, **35**(21), 1789–1781, 1999.

[117] D.M. Schreurs, J. Verspecht, S. Vanderberghe, E. Vandamme, 'Straightforward and accurate nonlinear device model parameter-estimation method based on vectorial large-signal measurements', *IEEE Trans. Microwave Theory Tech.*, **MTT-50**(10), 2315–2319, 2002.

[118] T. Närhi, 'Frequency-domain analysis of strongly nonlinear circuits using a consistent large-signal model', *IEEE Trans. Microwave Theory Tech.*, **MTT-44**(2), 182–192, 1996.

[119] L.O. Chua, 'Device modelling via basic nonlinear circuit elements', *IEEE Trans. Circuit Syst.*, **CAS-27**(11), 1014–1044, 1980.

[120] D.D. Weiner, J.F. Spina, *Sinusoidal Analysis and Modelling of Weakly Nonlinear Circuits*, Van Nostrand, New York (NY), 1980.

[121] V. Rizzoli, A. Neri, D. Masotti, A. Lipparini, 'A new family of neural-network based bidirectional and dispersive behavioural models for nonlinear RF/microwave subsystems', *Int. J. RF Microwave Comput.-Aided Eng.*, **12**(5), 51–70, 2002.

[122] J. Verspecht, P. Van Esch, 'Accurately characterizing of hard nonlinear behavior of microwave components by the nonlinear network measurement system: introducing the nonlinear scattering functions', *Proc. INNMC '98*, Duisburg (Germany), Oct. 1998, pp. 17–26.

[123] D. Schreurs, J. Verspecht, G. Acciari, P. Colantonio, F. Giannini, E. Limiti, G. Leuzzi, 'Harmonic-balance simulation of nonlinear scattering functions for computer-aided design of nonlinear microwave circuits', *Int. J. RF Microwave Comput.-Aided Eng.*, **12**(5), 460–468, 2002.

[124] N.O. Sokal, A.D. Sokal, '*Class E* – a new class of high-efficiency tuned single-ended switching power amplifiers', *IEEE J. Solid-State Circuits*, **SC-10**(3), 168–176, 1975.

[125] S.C. Cripps, 'A theory for the prediction of GaAs FET load-pull power contours', *IEEE MTT-S Symp. Dig.*, 221–223, 1983.

[126] L.J. Kushner, 'Output performances of idealised microwave power amplifiers', *Microwave J.*, **32**, 103–110, 1989.

[127] P. Colantonio, F. Giannini, G. Leuzzi, E. Limiti, 'High efficiency low-voltage power amplifier design by second harmonic manipulation', *Int. J. RF and Microwave Comput.-Aided Eng.*, **10**(1), 19–32, 2000.

[128] A.N. Rudiakova, V.G. Krizhanovski, 'Driving waveforms for class-F power amplifiers', *IEEE MTT-S Symp. Dig.*, Boston (MA), June 2000.

[129] D.M. Snider, 'A theoretical analysis and experimental confirmation of the optimally loaded and overdriven RF power amplifiers', *IEEE Trans. Electron Devices*, **ED-14**(6), 851–857, 1967.

4

Power Amplifiers

4.1 INTRODUCTION

In this introduction, the basic concepts and the design quantities of interest are introduced together with their definitions.

Power amplifiers are nonlinear circuits whose main goal is the amplification of a large signal at a given frequency, or rather, in a given frequency band. The signal usually must be amplified to a given power level, and the power gain can also be specified. However, the specification that is at the origin of the nonlinear behaviour of a power amplifier is the request of limited power consumption; contrariwise, an arbitrarily large transistor could be used, working in its linear region for the given signal but consuming a correspondingly large DC power for voltage and current biasing. Therefore, the transistor and the bias source must be large enough to limit the distortion produced by the nonlinearities but not larger than that.

The quantities that characterise a power amplifier are defined in the following. The output power is the power delivered to the load in the specified frequency band:

$$P_{out} = P_{out}(f) \quad f_L \leq f \leq f_U \tag{4.1}$$

The input power is the available power in the same frequency band:

$$P_{in} = P_{in,av}(f) \quad f_L \leq f \leq f_U \tag{4.2}$$

The power gain is the ratio between these two quantities:

$$G(f) = \frac{P_{out}(f)}{P_{in}(f)} \quad f_L \leq f \leq f_U \tag{4.3}$$

The amplifier being nonlinear, the gain depends on the power level of the signal. If the active device is biased in its linear region, for very small signals the amplifier

Nonlinear Microwave Circuit Design   F. Giannini and G. Leuzzi
© 2004 John Wiley & Sons, Ltd   ISBN: 0-470-84701-8

behaves linearly and the power gain reduces to the linear gain:

$$\lim_{P_{\text{in}} \to 0} G = \lim_{P_{\text{in}} \to 0} \left(\frac{P_{\text{out}}}{P_{\text{in}}} \right) = G_{\text{L}} \tag{4.4}$$

For increasing signal amplitude, the output current and voltage tend to be limited by the nonlinearities of the active device and the output power saturates:

$$\lim_{P_{\text{in}} \to \infty} P_{\text{out}} = P_{\text{sat}} \tag{4.5}$$

Correspondingly, the gain tends to zero:

$$\lim_{P_{\text{in}} \to \infty} G = \lim_{P_{\text{in}} \to \infty} \left(\frac{P_{\text{out}}}{P_{\text{in}}} \right) = 0 \tag{4.6}$$

Given the very wide dynamic range of the signal in practical cases, these quantities are usually expressed in a logarithmic scale. The arbitrary power level of 1 mW is commonly used as reference level, and all power levels are expressed in dB with respect to 1 mW or dBm; the conversion formulae between a power level in watt and the same power level in dBm are

$$P_{\text{dBm}} = 10 \cdot \log_{10} \left(\frac{P_{\text{W}}}{10^{-3}} \right) = 10 \cdot \log_{10}(1000 \cdot P_{\text{W}}) \tag{4.7}$$

$$P_{\text{W}} = 10^{-3} \cdot 10^{\frac{P_{\text{dBm}}}{10}} = \frac{10^{\frac{P_{\text{dBm}}}{10}}}{1000} \tag{4.8}$$

The gain is also expressed in logarithmic scale as

$$G_{\text{dB}} = 10 \cdot \log_{10}(G) = 10 \cdot \log_{10} \left(\frac{P_{\text{out,W}}}{P_{\text{in,W}}} \right)$$

$$= 10 \cdot \log_{10}(P_{\text{out,W}}) - 10 \cdot \log_{10}(P_{\text{in,W}}) = P_{\text{out,dBm}} - P_{\text{in,dBm}} \tag{4.9}$$

The power performances of a power amplifier are usually represented graphically on a plot where the x-axis is the input power expressed in dBm and the y-axis is the output power in dBm as well (logarithmic scale).

If the active device is biased in its linear region, for very low power levels the amplifier behaves linearly and the slope of the plot is unitary:

$$P_{\text{out,dBm}} = 10 \cdot \log_{10}(1000 \cdot G_{\text{L}} \cdot P_{\text{in}}) = 10 \cdot \log_{10}(G_{\text{L}}) + 10 \cdot \log_{10}(1000 \cdot P_{\text{in}})$$

$$= P_{\text{in,dBm}} + G_{\text{L,dB}} \tag{4.10}$$

The linear gain of the amplifier is easily found from the plot as the difference between the output power in dBm and the input power in dBm at any point on the plot in the linear region. For instance, the linear gain can be found as the value of the output

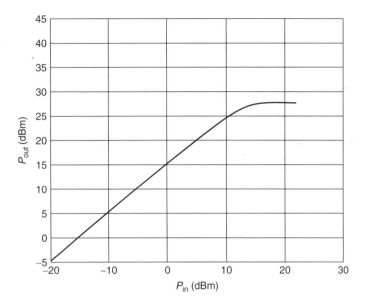

Figure 4.1   The P_{in}/P_{out} plot for a power amplifier

power in dBm when the input power is 0 dBm if this point lies in the linear region of the plot; in Figure 4.1, the linear gain value is found to be 15 dB.

In the case of an amplifier behaving as in Figure 4.1, the power level is expressed in a more physically meaningful way, referring to the performances of the amplifier. From the plot in Figure 4.2, it is easily seen that the gain decreases for increasing input power level, as already mentioned; the gain is usually shown on the same plot for quantitative evaluation. If suitable, the logarithmic scale used for the output power, interpreted as dBm, can be used also for the gain, interpreted as dB.

The gain decreases from its maximum value in the linear region down to 0 or $-\infty$ in logarithmic scale; this behaviour is referred to as gain compression. The power level can be expressed with reference to the corresponding gain compression. For instance, the power level where the gain is 1 dB less than its maximum value is commonly referred to as the 1-dB gain compression power level. The corresponding powers are as in Figure 4.3.

The corresponding power levels are similarly determined for any gain compression level. This terminology defines a power level with reference to the behaviour of the amplifier and results in a meaningful indication of the amount of distortion the amplifier is expected to introduce.

Another important quantity for the design of a power amplifier, as mentioned above, is the DC power delivered by the power supply. Amplifiers are usually biased at constant voltage, and the DC power is usually computed as the constant voltage times the average DC current:

$$P_{DC} = V_{bias\_supply} \cdot \frac{1}{T} \int_0^T I_{bias\_supply}(t) \cdot dt \qquad (4.11)$$

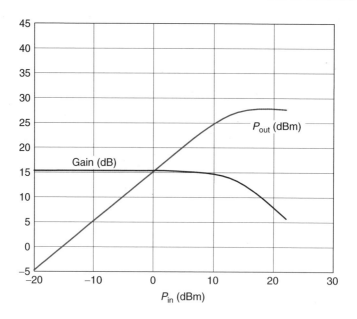

Figure 4.2 Power and gain plot

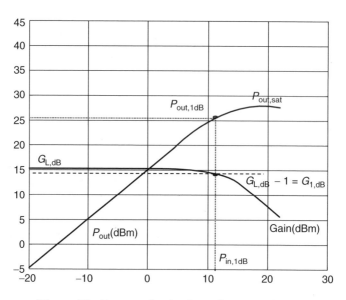

Figure 4.3 Power and gain plot and compression level

The average DC current, in general, is the bias current plus a rectified component when the amplifier is driven into significantly nonlinear operations.

The DC power is partly converted into the output signal and partly into harmonic or spurious frequencies, and the rest is dissipated inside the amplifier (Figure 4.4), where

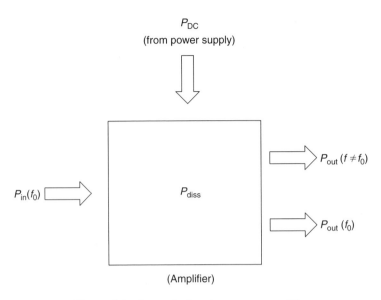

Figure 4.4 Power budget in a power amplifier

the frequency f_0 stands for the frequency band of interest. The power balance is

$$P_{DC} + P_{in}(f_0) = P_{out}(f_0) + P_{out}(f \neq f_0) + P_{diss} \qquad (4.12)$$

A quality factor for DC power consumption is the efficiency. A physically mean-ingful general expression is computed as the useful output power divided by the total input power:

$$\eta = \frac{P_{out}}{P_{in,tot}} = \frac{P_{out}(f_0)}{P_{in}(f_0) + P_{DC}} \qquad (4.13)$$

If the gain is large and components at harmonic and spurious frequencies are limited, then this is the drain or collector efficiency:

$$\eta \simeq \frac{P_{out}}{P_{DC}} \qquad (4.14)$$

The most widely adopted definition of efficiency is the so-called 'power-added efficiency', that is, the ratio between the RF power 'added' by the amplifier and the DC power required for this addition:

$$\eta_{add} = \frac{P_{out} - P_{in}}{P_{DC}} = \frac{P_{out}}{P_{DC}} \left(1 - \frac{1}{G} \right) \qquad (4.15)$$

This is not a physically correct definition since it is not a ratio between power out and power in; it even becomes negative if the gain is lower than unity. However, the weight of the input power is higher than that in the correct formula, and this is the reason for its success. The input power usually comes from a preceding driver amplifier stage, where in turn it is obtained by means of the conversion of DC to RF power with an efficiency not better than that of the power amplifier. This figure of merit, therefore, stresses the advantage of a high gain for the requirements of high output power from the preceding stages.

For a Class-A amplifier with high gain, the efficiency has a linear dependence on input power for small to medium input power levels. This is easily seen from the formula above: the DC power is approximately constant since no rectification takes place until nonlinear effects appear, and the average current from the bias supply is the bias current:

$$P_{DC} = V_{bias\_supply} \cdot I_{bias\_supply} \cong V_{bias} \cdot I_{bias} \tag{4.16}$$

The output power is proportional to the input power as long as the amplifier behaves approximately linearly:

$$P_{out} = G \cdot P_{in} \cong G_L \cdot P_{in} \tag{4.17}$$

therefore,

$$\eta \cong \frac{G_L \cdot P_{in}}{P_{DC}} \propto P_{in} \tag{4.18}$$

When the input power increases, the gain begins to decrease (gain compression); the drain or collector efficiency tends to saturate to a maximum:

$$\lim_{P_{in} \to \infty} \eta \cong \frac{P_{out,sat}}{P_{DC}} = \text{const.} \tag{4.19}$$

The power-added efficiency reaches a maximum, then starts decreasing because of the decreasing gain:

$$\lim_{P_{in} \to \infty} \eta_{add} = \lim_{G \to 0} \frac{P_{out,sat}}{P_{DC}} \left(1 - \frac{1}{G}\right) = -\infty \tag{4.20}$$

Efficiency is usually expressed as a percentage:

$$\eta_\% = \eta \cdot 100 \tag{4.21}$$

and as such it is shown in the same plot as output power and gain on a linear scale; usually, its scale is shown on the right y-axis because of the different range with respect to output power and power gain (Figure 4.5).

The dependence of efficiency on input power as shown in the figure is exponential in the low- and medium-power region because the x-axis is logarithmic while the y-axis is linear:

$$\eta = \frac{G}{P_{DC}} \cdot P_{in} \cdot 100 = \frac{G}{P_{DC}} \cdot \frac{10^{\frac{10 \cdot \log_{10}(1000 \cdot P_{in})}{10}}}{1000} \cdot 100 = \frac{G}{10 \cdot P_{DC}} \cdot 10^{\frac{P_{in,dBm}}{10}} \tag{4.22}$$

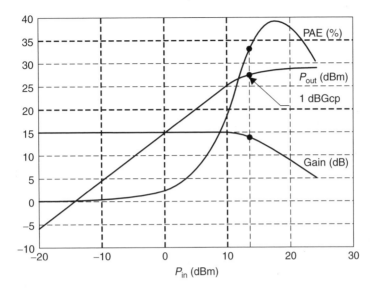

Figure 4.5   Output power, power gain and power-added efficiency

The efficiency of an amplifier is limited by the saturation of the output power because of nonlinear voltage- and current-limiting phenomena. Before, the nonlinear behaviour was so strong as to cause output power saturation; however, the distortion can be so high as to degrade the quality of the signal beyond acceptable levels. Therefore, distortion must be defined and evaluated, and usually is one of the design specifications of a power amplifier.

For a single-tone signal, a meaningful figure of merit of distortion is the harmonic content of the output signal. It is expressed as

$$HD_2 = \frac{P_{out}(2f_0)}{P_{out}(f_0)} \qquad HD_3 = \frac{P_{out}(3f_0)}{P_{out}(f_0)} \tag{4.23}$$

or correspondingly in logarithmic scale:

$$HD_{2,dBc} = 10 \cdot \log_{10}\left(\frac{P_{out}(2f_0)}{P_{out}(f_0)}\right) \qquad HD_{3,dBc} = 10 \cdot \log_{10}\left(\frac{P_{out}(3f_0)}{P_{out}(f_0)}\right) \tag{4.24}$$

These logarithmic expressions are said to be in dBc or decibel over carrier power. Obviously, the harmonic distortion depends on the operating power level; a clear effect is the distortion of the sinusoidal waveform of the output signal (Figure 4.6).

As a global figure of merit, the total harmonic distortion is also defined:

$$THD = \frac{\displaystyle\sum_{n\geq2} P_{out}(nf_0)}{P_{out}(f_0)} \qquad THD_{dBc} = 10 \cdot \log_{10}\left(\frac{\displaystyle\sum_{n\geq2} P_{out}(nf_0)}{P_{out}(f_0)}\right) \tag{4.25a}$$

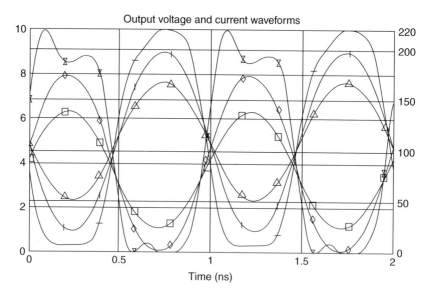

Figure 4.6 Output voltage and current waveforms for increasing input power

An alternative expression for second-order, third-order or arbitrary-order harmonic distortion is the following. It is clear from Volterra series formulations (Section 1.3.2) that for small amplitudes of a periodic signal the second-harmonic component has a quadratic dependence on input power; the third harmonic has a cubic dependence, and so on for higher-order harmonics. In a logarithmic plot as that used so far, the slope of the power of a harmonic component of arbitrary order is the order of the harmonic itself:

$$P_{out}(nf_0) \propto P_{out}^n(f_0) \Rightarrow P_{out,dBm}(nf_0) \propto n \cdot P_{out,dBm}(f_0) \tag{4.25b}$$

This is true as far as the Volterra series approach holds, that is, for mildly nonlinear behaviour. If the slope of the plots of the harmonic powers are extrapolated, they intercept the prolongation of the fundamental-frequency component power plot at the so-called nth order intercept points (Figure 4.7).

The intercept points are a measure of the power level that can be obtained with a given margin of the fundamental power to harmonic power. They are a compact figure of merit for an amplifier, while the harmonic distortion must be given at all operating power levels of interest.

Normal signals, however, are not single tone, but they are modulated; therefore they occupy a frequency band. If the signal is narrowband, it can be seen either as a carrier modulated by a relatively slow envelope or as an array of closely spaced spectral lines within the frequency band of the total signal. We have seen above (Section 1.3.2) that two tones at different frequencies produce intermodulation tones of all orders at frequencies different from those of the two signals. The most meaningful ones are the third-order intermodulation tones because they appear at frequencies near the fundamental frequency of the signal, and therefore within the band of a practical signal, where they

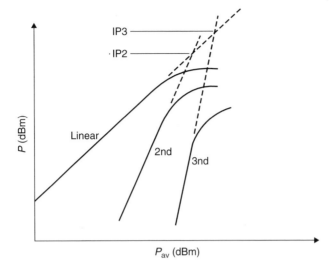

Figure 4.7 Harmonic output power plot and intercept points

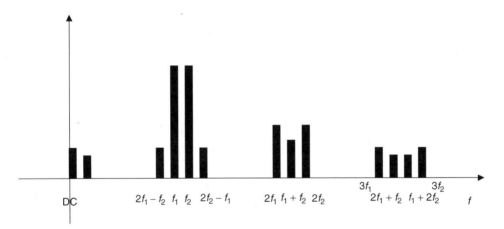

Figure 4.8 Spectrum of a two-tone signal at the output of a power amplifier, showing intermodulation lines

interfere with adjacent signal lines; this is true also for all even-order intermodulation terms, but for weak nonlinearities the lowest-order term tends to dominate. A simple way to evaluate third-order intermodulation, though still rather unrealistic, is the measure of the intermodulation generated by two closely spaced tones in the power amplifier. The frequency spectrum is as in Figure 4.8 the distortion is evaluated as

$$IMD_3 = \frac{P_{\text{out}}(2f_2 - f_1)}{P_{\text{out}}(f_2)} = \frac{P_{\text{out}}(2f_1 - f_2)}{P_{\text{out}}(f_1)} \tag{4.26}$$

The two expressions given above are identical for narrowband signals; they differ somewhat if the two tones are not very close to one another. Seen from a different point of

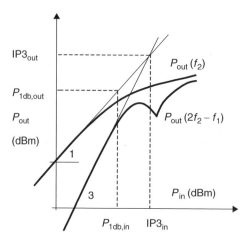

Figure 4.9 Third-order intercept point for a power amplifier

view, if the system is not frequency-independent within the signal frequency band, the two expressions differ. For practical purposes, however, given the approximations and errors involved in both simulation and measurement of these quantities, the two formulations can be considered as equivalent. The corresponding expressions in logarithmic scale are

$$IMD_{3,dBc} = 10 \cdot \log_{10}\left(\frac{P_{out}(2 \cdot f_2 - f_1)}{P_{out}(f_2)}\right) = 10 \cdot \log_{10}\left(\frac{P_{out}(2 \cdot f_1 - f_2)}{P_{out}(f_1)}\right) \quad (4.27)$$

Also, third-order intermodulation distortion can be expressed as intercept point since its dependence on input power is cubic, at least at low power level (Section 1.3.2). The intercept point is shown in Figure 4.9.

The situation in which only two tones are present within the signal band is less unrealistic than the one with a single-tone signal but is still very far from being practical. However, designers usually manage to understand the relation between this simple figure of merit and the behaviour under more complicated and realistic conditions. In case a better evaluation of the distortion properties of the power amplifier is needed, more realistic figures are given.

It is worth noting that gain compression and distortion, while apparently not directly connected, share the same origin. If we look at Volterra series expressions, we see that gain compression has its origin in the third-order nucleus and so does the third-order intermodulation. If the two tones are closely spaced and the nuclei are not very frequency-sensitive, the two terms are similar. When the gain compression is specified, therefore, distortion usually assumes predictable values; this is not absolutely true but true enough for practical design purposes, at least when distortion is not a very critical issue.

4.2 CLASSES OF OPERATION

In this paragraph, the classes of operation of power amplifiers are introduced.

Power amplifiers are usually classified by the class of operation. This is a traditional scheme, but not always illuminating; on the contrary, it may be misleading. Class-A operation means that the bias current is such that the transistor is not pinched off or cut off by the input signal anytime in the signal period, at least for moderate power levels. Class-B operation means that the transistor is pinched off or cut off for one half of the signal period. Class-AB is an intermediate situation, that is, the bias current is smaller than that of Class-A operations, but not zero; this implies that for small power levels, a Class-AB amplifier behaves as a Class-A circuit and that the fraction of the period when the device is off depends on the power level. Class-C means that the transistor is pinched off or cut off for more than one half of the signal period. In all these cases nothing is said about the load and, in particular, about the loading at the harmonic frequencies; the voltage and current waveforms are not specified either, even if they are usually assumed to be truncated sinusoids. On the other hand, Class-F operation traditionally means that the output of the transistor is loaded by a suitable load at fundamental frequency, by short circuits at even harmonics and by open circuits at odd harmonics, whatever be the bias current and the pinching-off or cutting-off time of the transistor; this ideally produces a square-wave voltage waveform for a sufficiently high power level. In fact, the transistor itself is supposed to work either at zero voltage (ohmic region or saturation) or at zero current (pinch-off or cut-off region), that is, either as a short circuit or as an open circuit, drastically reducing the power dissipated inside the device. This is similar for Class-E operations, where the transistor is supposed to work as a switch, loaded by a suitable RLC network, ensuring optimum voltage and current waveforms during switching. In Classes G and FG, a suitable combination of the loads at even and odd harmonics both at input and at output of the transistor ensures a favourable output voltage and current wave shaping for high power and efficiency; once more, the transistor can be biased in a whole range of operating points between Class-A and Class-B.

In order to clarify the situation, the two aspects of bias point and harmonic loading will be clearly distinguished in the following. In particular, all classes of operations referring to specific loads at harmonic frequencies will be treated by means of a unified theory (harmonic manipulation approach); for each of them the suitable bias conditions will be identified and described. All this description will be carried out by means of simplified transconductance models for the transistor, and by means of piecewise-linear analysis method for performance evaluation of the circuit. This approach is rather exhaustive as far as the main design goals for the amplifier are high output power and efficiency.

The case when a low distortion is the main design goal will be treated separately. This is because special arrangements must be adopted when the distortion level must be really low. Very few means are available to the designer so far for getting low distortion in the design of a power amplifier stage; much more is available for the correction of the distortion by means of external arrangements (predistortion linearisation, feedforward linearisation, etc.).

Stability of the amplifier will not be treated in this chapter, but a general stability theory under nonlinear operations is described in Chapter 8.

4.3 SIMPLIFIED CLASS-A FUNDAMENTAL-FREQUENCY DESIGN FOR HIGH EFFICIENCY

In this paragraph, a fundamental-frequency Class-A design of a power amplifier is descri-
bed, as an introduction to more sophisticated approaches involving harmonic frequencies.

4.3.1 The Methodology

Class-A amplifiers are supposed to have not only low efficiency but also low distortion.
The transistor is biased in the middle of its linear region and the load curve stays within
this region until the power level becomes high and gain compression begins. This is due
to the effect of the strong voltage- and current-limiting nonlinearities; in an FET, these
are the ohmic region or the forward conduction of the gate-channel junction and the
pinch-off or breakdown regions. For a bipolar transistor, these are the saturation and the
cut-off or breakdown regions (Figure 4.10).

In principle, very little harmonics and distortion are generated until the strong
nonlinearities of the transistor are reached by the load curve. The design of this amplifier
can therefore be treated by a quasi-linear approach. A very simple but effective approach
had been proposed some time ago [1] and was extended and improved by later works [2,
3]; it is proposed here for the clarity of the approach.

The device is modelled by a very simple equivalent circuit with a single nonlinearity, that is, the drain–source or collector–emitter-controlled current source; parasitics
and feedback elements are also neglected (see Chapter 3). In the case of an FET, the
equivalent circuit is shown in Figure 4.11.

In the case of a bipolar transistor, the input mesh changes from a series RC to a
shunt RC network. The current source is modelled as a piecewise-linear element, as in
Figure 4.12.

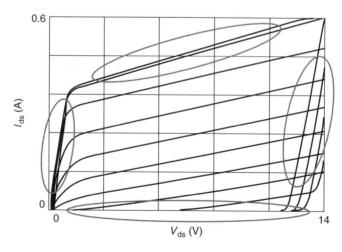

Figure 4.10 The limiting nonlinearities in the output characteristics of a transistor

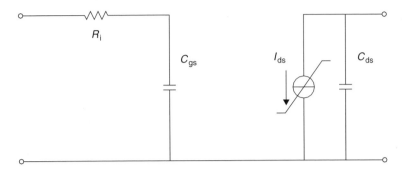

Figure 4.11   Equivalent circuit of an FET for simplified analysis

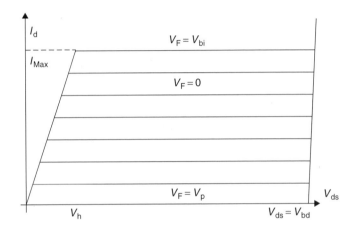

Figure 4.12   Piecewise-linear output characteristics for simplified analysis

Assumption is made that the transconductance be constant within the linear region of the device. The device is also conjugately matched at the input port (either gate or base) by the input matching network for maximum power transfer into the device.

The output matching network as seen by the output of the active device is represented for this analysis by a shunt RL network (see Figure 4.13).

The inductance resonates the drain–source or collector–emitter capacitance of the device at the fundamental frequency of operation; the resistance is the load where the active power supplied by the active device will be dissipated. The value of the resistance must be such that the design specifications are met. In practice, this is not the actual load: in a practical circuit, an external 50 Ω resistance is transformed by the output matching network into the optimum shunt RL, as required by the active device.

The performances of the active device will now be studied at the port of the nonlinear current source. In other words, the parasitic capacitance is included in the external circuitry for this study; it will be restored as an internal element afterwards (Figure 4.14).

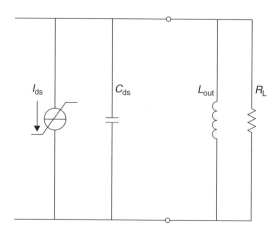

Figure 4.13   Output mesh of the power amplifier for simplified analysis

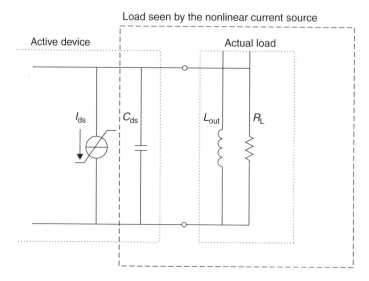

Figure 4.14   Load seen by the nonlinear current source for simplified analysis

If the capacitance and inductance resonate at the frequency of operation, the load seen by the current source is purely resistive; then, the load curve on the I/V plane is a straight line. Since the current source within its linear region is modelled as a pure transconductance, the minimum and maximum output current reached at the extremes of the line depend only on the amplitude of the input voltage signal, that is, on the input power level. The slope of the line, or correspondingly the minimum and maximum voltages reached at the extremes of the load line for a given input power, depend on the value of the resistance R_L seen by the current source. In other words, if the amplitude of the input signal is small enough not to reach the current-limiting nonlinearities, the output current waveform is dictated by the input voltage waveform; for a given input voltage, the output voltage waveform is equal to the output current waveform times the load resistance, provided that the resistance is small enough not to cause the output voltage to reach the voltage-limiting nonlinearities. Together, output current and voltage waveforms determine the extension and slope of the load line within the linear region.

Conversely, for any given load resistance and bias point, the input voltage amplitude will determine the extension of the load line, that is, the maximum and minimum values of the output current and voltage. For increasing input power levels, a level is reached when the load line touches a current- or voltage-limiting nonlinearity, depending on the bias point and slope of the load line (Figure 4.15). Any further increase in input power beyond this point will cause either the current or the voltage waveform to be clipped and distortion to appear. We will assume this limit level where the amplitude of the load line reaches the nonlinear limits of the linear region as the operating power level of the amplifier.

At this limit level, the output power is easily computed as one half the product of the sinusoidal voltage amplitude times the sinusoidal current amplitude. If the power is voltage-limited (case (a) in Figure 4.15), the output power at the beginning of compression is (Figure 4.16)

$$P_{\text{out,a}} = \frac{1}{2} \cdot \frac{\Delta V_{\text{ds}}}{2} \cdot \frac{\Delta I_{\text{d}}}{2} \cong \frac{1}{8} \cdot (V_{\text{bd}} - V_{\text{k}}) \cdot \frac{(V_{\text{bd}} - V_{\text{k}})}{R_{\text{L,a}}} = \frac{(V_{\text{bd}} - V_{\text{k}})^2}{8 \cdot R_{\text{L,a}}} = \frac{V_{\text{max}}^2}{8 \cdot R_{\text{L,a}}}$$

$$(4.28)$$

where the maximum voltage swing has been approximated by

$$V_{\text{max}} = V_{\text{bd}} - V_{\text{k}} \tag{4.29}$$

If the power is current-limited (case (b) in Figure 4.15), the output power at the beginning of compression is (Figure 4.17)

$$P_{\text{out,b}} = \frac{1}{2} \cdot \frac{\Delta V_{\text{ds}}}{2} \cdot \frac{\Delta I_{\text{d}}}{2} \cong \frac{1}{8} \cdot (I_{\text{max}} \cdot R_{\text{L,b}}) \cdot I_{\text{max}} = \frac{I_{\text{max}}^2 \cdot R_{\text{L,b}}}{8} \tag{4.30}$$

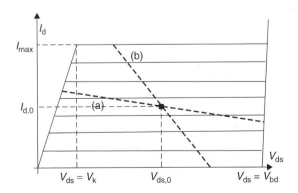

Figure 4.15   Load lines in the case of voltage-limited (a) or current-limited (b) output power

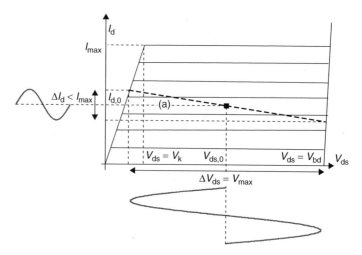

Figure 4.16 Voltage and current amplitudes in the case of voltage-limited output power

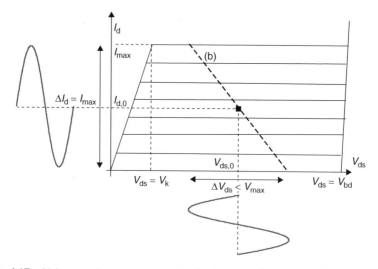

Figure 4.17 Voltage and current amplitudes in the case of current-limited output power

In both the cases, the output power is not the maximum that could be delivered by the active device because either voltage or current swing is not maximum. The maximum output power is obtained by using a load line that maximises both the voltage and current:

$$R_{L,\text{opt}} = \frac{V_{\text{max}}}{I_{\text{max}}} \tag{4.31}$$

$$P_{\text{out,max}} = \frac{1}{2} \cdot \frac{\Delta V_{ds}}{2} \cdot \frac{\Delta I_d}{2} \simeq \frac{1}{8} \cdot (I_{\text{max}} \cdot R_{L,\text{opt}}) \cdot I_{\text{max}}$$

$$= \frac{1}{8} \cdot V_{\text{max}} \frac{V_{\text{max}}}{R_{L,\text{opt}}} = \frac{1}{8} \cdot I_{\text{max}} \cdot V_{\text{max}} \tag{4.32}$$

and by a bias point such that

$$I_{d,0} = \frac{I_{max}}{2} \qquad V_{ds,0} \cong V_k + \frac{V_{bd} - V_k}{2} = V_k + \frac{V_{max}}{2} \qquad (4.33)$$

This situation is illustrated in Figure 4.18.

It is clear from Figures 4.16, 4.17 and 4.18 that

$$R_{L,b} < R_{L,opt} < R_{L,a} \qquad (4.34)$$

and also

$$P_{out,b} < P_{out,max} > P_{out,a} \qquad (4.35)$$

So far, we have computed the maximum output power delivered to a resistive load by the current source, and we also have determined the corresponding optimum resistive load; we have also computed the power corresponding to other resistances, assuming that the amplifier operates at the beginning of compression, that is, at the limit operating input power level where the current and voltage waveforms begin to be distorted. We now compute the output power delivered to the load resistance if the reactances are not resonated, that is, if the load seen by the current source is not purely resistive. Let us assume a load curve as in Figure 4.19.

This load curve corresponds to a load resistance as in the case of voltage-limited output power (case (a), dotted line in Figure 4.19) with an added shunt susceptance. The

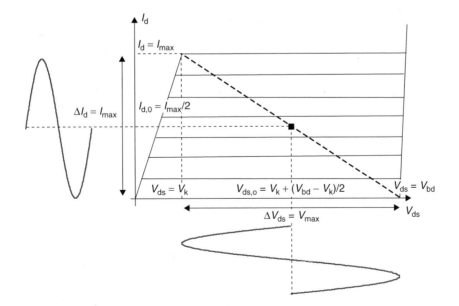

Figure 4.18 Load line ensuring maximum output voltage and current amplitudes

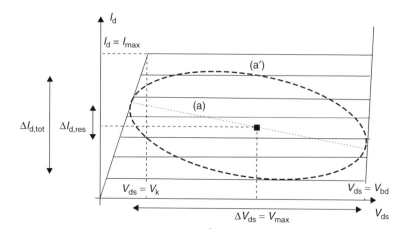

Figure 4.19   Load curve for a complex load in the case of voltage-limited output power

current is the same as before, with an extra susceptive component:

$$I_d = I_{d,res} + jI_{d,susc} = V_{max} \cdot \left(\frac{1}{R_{L,a}} + j\omega B_{L,a} \right) \qquad (4.36)$$

It is clear that only the resistive component contributes to the active power, which is therefore identical to the case when the load is purely resistive (case (a)):

$$P_{out,a'} = \frac{1}{8} \cdot \frac{V_{max}}{R_{L,a}} \cdot V_{max} = P_{out,a} \qquad (4.37)$$

Similarly, let us assume a load curve as in Figure 4.20.

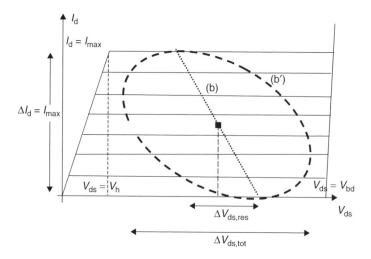

Figure 4.20   Load curve for a complex load in the case of current-limited output power

This load curve corresponds to a load resistance as in the case of current-limited output power (case (b), dotted line in Figure 4.20) with an added series reactance. The voltage is the same as before, with an extra reactive component:

$$V_{ds} = V_{ds,res} + j V_{ds,react} = I_{max} \cdot (R_{L,b} + j\omega X_{L,b}) \tag{4.38}$$

It is clear that only the resistive component contributes to the active power, which is therefore identical to the case when the load is purely resistive (case (b)):

$$P_{out,b'} = \frac{1}{8} \cdot I_{max} \cdot (I_{max} \cdot R_{L,b}) = P_{out,b} \tag{4.39}$$

We can now plot the loads described so far on the Smith Chart and tag them with their corresponding output power. We start with the optimum resistance corresponding to the maximum output power (Figure 4.21).

Then, we compute and plot the two resistances that correspond to an output power 1 dB lower than the maximum one; one will be higher (case (a)) and the other will be lower (case (b)) than the optimum one:

$$(P_{out,a}^{(1)})_{dBm} = (P_{out,max})_{dBm} - 1 \qquad R_{L,a}^{(1)} = \frac{V_{max}^2}{8 \cdot P_{out,a}^{(1)}} \tag{4.40}$$

$$(P_{out,b}^{(1)})_{dBm} = (P_{out,max})_{dBm} - 1 \qquad R_{L,b}^{(1)} = \frac{8 \cdot P_{out,b}^{(1)}}{I_{max}^2} \tag{4.41}$$

Now, the complex loads as defined above are plotted on the Smith Chart (Figure 4.23). We start from the resistances of cases (a) and (b) and add a shunt susceptance and a

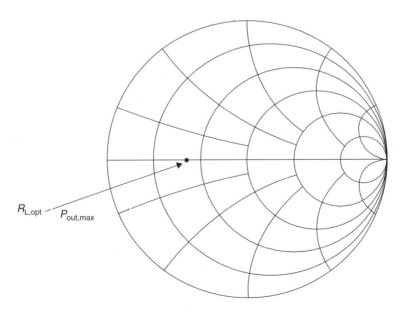

$R_{L,opt}$ $P_{out,max}$

Figure 4.21 Resistive load corresponding to maximum output power

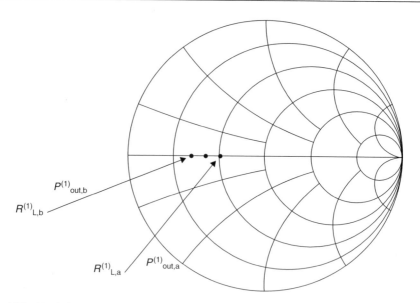

Figure 4.22 Resistive loads corresponding to voltage-limited (a) and current-limited (b) output power

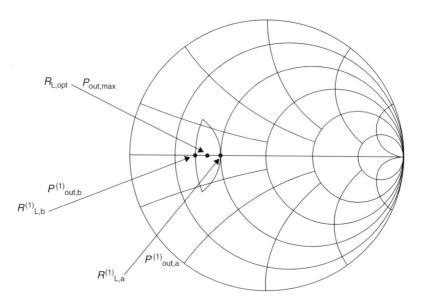

Figure 4.23 Closed constant-power contour

series reactance respectively; this causes the loads to move along the constant-conductance and constant-resistance circles respectively. The closed contour in Figure 4.23 is obtained by stopping at the intersections of the two circles.

Now the procedure can be repeated for output powers decreasing by 1-dB steps (Figure 4.24).

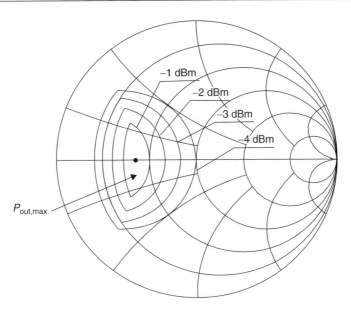

Figure 4.24   Closed constant-power contours for decreasing values of the output power

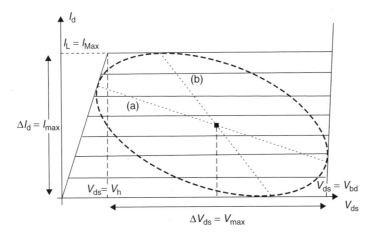

Figure 4.25   Load curve corresponding to the intersection of constant-resistance and constant-conductance curves for a given constant-power contour on the Smith Chart (Figure 4.22)

The points where the constant-conductance and constant-resistance circles intersect correspond to a susceptive or reactive load such that the load curve is as in Figure 4.25.

In this case, the current- and voltage-limiting nonlinearities are reached simultaneously. The power delivered to the load is the same for both the shunt susceptance and for the series reactance case.

The resulting closed curves on the Smith Chart are the equivalent to load-pull contours, with the assumptions of simple piecewise-linear transconductance model as described earlier. The load computed so far is, however, the one seen by the internal current source of the transistor that includes the parasitic output capacitance (see Figure 4.14). The load that must be actually presented to the output of the transistor by the output matching network is easily computed by removing the capacitance. In the example above, let us assume that we have computed a resistive optimum load $R_{L,opt}$ and that a parasitic capacitance C_{ds} is present. The actual load to be synthesised is

$$(Y_{L,opt})_{ext} = \frac{1}{R_{L,opt}} - j\omega C_{ds} \tag{4.42}$$

On the Smith Chart, this optimum admittance lies on the same constant-conductance curve as $R_{L,opt}$, shifted into the inductive half-plane by the amount of the susceptance corresponding to the parasitic capacitance (Figure 4.26).

Correspondingly, all points of the constant-power curves are moved along constant-conductance circles. The load-pull-like contours look as in Figure 4.27.

4.3.2 An Example of Application

An example is now given of a design for a power-matched amplifier for 4.5–5.5-GHz frequency band. A full nonlinear model and a commercial CAD software (HP-MDS) have been used. The device is a medium-power MESFET by GMMT (UK) and the corresponding model is a modified Materka one derived from pulsed-DC and multi-bias S-parameter measurements. The topology of the model and the DC characteristics are shown in Figure 4.28. The bias voltages are $V_{DD} = 5$ V and $V_{GG} = -0.5$ V (i.e. a Class-A condition).

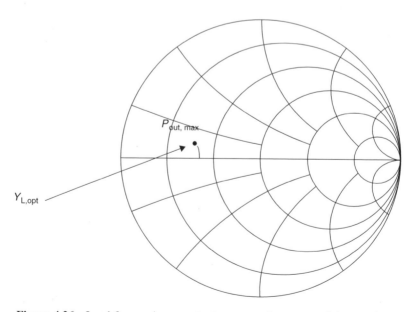

Figure 4.26 Load for maximum output power at the output of the transistor

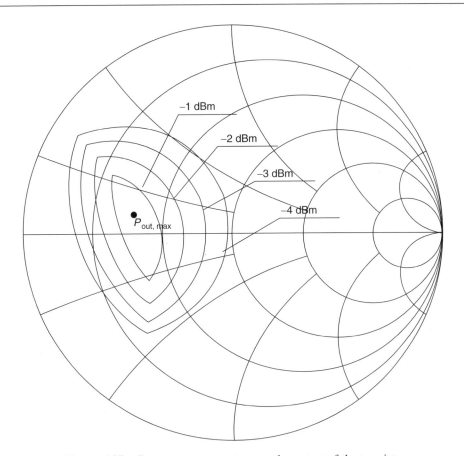

Figure 4.27 Constant-power contours at the output of the transistor

From the analysis of the device performances, that is, maximum voltage and current swing, the optimum load is assumed to be R_{opt}, while from the small-signal model the value for C_{ds} can be obtained. Assuming an $R_{opt} = 21.9\ \Omega$ and $C_{ds} = 0.31$ pF and treating the drain capacitance as an external element, the matching network has to transform the 50 Ω termination down to the optimum 21.9 Ω and absorb the drain capacitance and bondwire inductance.

In Figure 4.29, the networks used to synthesise the load-pull contours are depicted.

A possible circuit that performs the job using lumped elements is shown in Figure 4.30.

Note the advantage of keeping the drain capacitance and the bondwire inductance as an external element; the optimum load is resistive and therefore constant with frequency, so that the power match can be tracked using a single set of contours.

Finally, the input network is synthesised in order to fulfil the input conjugate match condition. In Figure 4.31, the power amplifier synthesised is depicted.

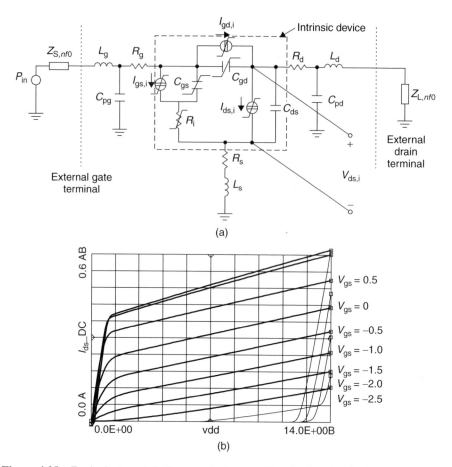

Figure 4.28   Equivalent model of power device considered (a) and DC characteristics (b)

In Figure 4.32, the S-parameters of the power amplifier (a) and its power performance (b) for three different input drive levels are depicted. As it can be noted, there is quite a good agreement between the Cripps theory and the full nonlinear simulation.

4.4 MULTI-HARMONIC DESIGN FOR HIGH POWER AND EFFICIENCY

In this paragraph, a multi-harmonic manipulation approach is described for the design of high-power, high-efficiency power amplifiers. The approach allows direct synthesis of the input and output load, although by means of piecewise-linear model and calculations.

4.4.1 Introduction

It is well established that a high efficiency can be obtained by a proper selection of bias point, say the drain or collector voltage and current DC levels, and a proper termination

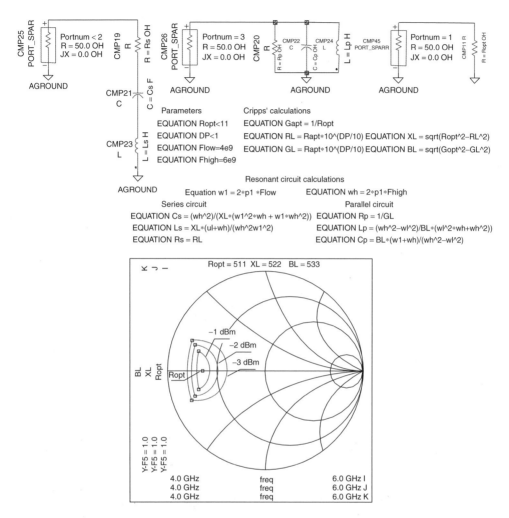

Figure 4.29 Parallel RLC circuit required to create the left side of the power contour (HP-MDS)

for all the relevant harmonics of the RF drain or collector current. This second aspect, which is obviously related to high-frequency applications, when only few harmonics can be handled, most of them being short-circuited by the device parasitics, corresponds, from a different point of view, to the best-known approach devoted to assure, at lower frequencies, a proper output voltage and/or current waveform shaping. It is easy to see, in fact, that the output voltage waveform has to have the maximum instantaneous value occurring at low (zero) output current level, while the maximum instantaneous value for the output current must correspond to very low voltages in order to minimise the overall dissipated power inside the active device.

For these reasons, starting from the pioneering work of Snider in 1967 [4], describing how it was possible in principle to improve the collector efficiency up to 100% while assuring a corresponding improvement also for the output power and the power

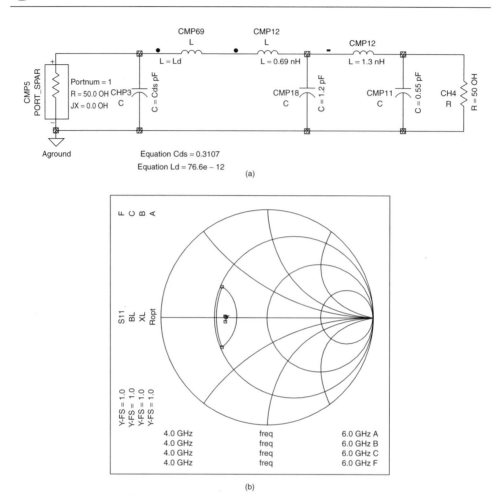

Figure 4.30 Output network (a) and its frequency behaviour (b)

gain, many 'harmonic-terminating strategies' have been proposed in the literature and presented as a natural evolution of classical and well-known A, AB, B and C biasing classes. Some of these new strategies, as the Class-F, and more recently, the inverse Class-F approaches [4–10] or the more unusual harmonic reaction amplifiers [11–13], focus on the output network terminations only in order to utilise the harmonic content generated by the current waveform clipping due to pinch-off and overdrive. Alternatively, different approaches based on device switching-mode operation, where the device itself is not considered at all since it is only an ideal switch, originated different operating classes, as Class-D (or S) [8] and Class-E [14–18], and demonstrated to be extremely effective in terms of output power and efficiency in the low-operating frequency range (up to a few GHz).

In recent years, demanding applications at ever higher operating frequencies have been forcing many designers to a critical review of most of the basic assumptions

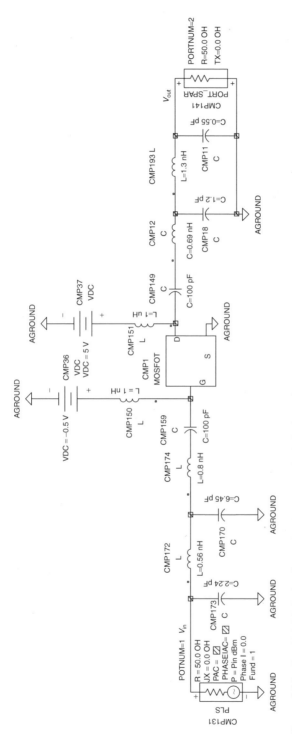

Figure 4.31 Parallel RLC circuit required to create the left side of the power contour (HP-MDS)

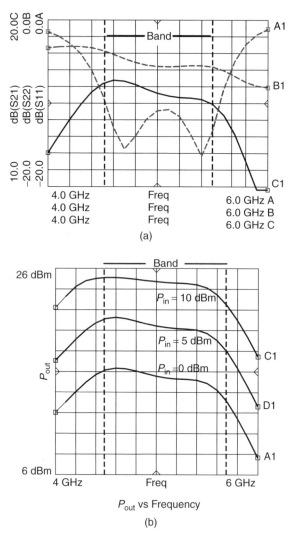

Figure 4.32 S-parameter (a) and power performance for three different input drive levels (b)

underlying the above-mentioned strategies. Design approaches based on device switching-mode operation, for instance, while maintaining the transition time at acceptable levels, demonstrated to lose a major part of their effectiveness when used to design high-efficiency microwave or millimetre-wave power amplifiers (PAs). In fact, it is easy to understand that it is practically impossible to perform the requested wave shaping for the output waveforms because of the difficulty in controlling more than a couple of voltage or current harmonic components, say the second and the third one, thus overcoming the inherent output shorting capacitive behaviour of the active device itself.

On the other hand, other aspects like the influence of input harmonic terminations on the output device performances have to be examined. With the exception of Class-F, for which particular driving waveforms have been suggested [19], or the inverse Class-F, where a possible solution is to drive the device with a proper rectangular voltage waveform instead of overdriving it with a sinusoidal signal while biasing the device itself at the pinch-off, most of the amplifiers can be assumed to be driven by a pure sinusoid at a chosen fundamental frequency, whose harmonics are carefully short-circuited at the input port. This choice represents a true limitation, as it was already demonstrated in [20–25] where the use of second-harmonic terminating control schemes, both at input and output ports, are fruitfully employed in order to significantly improve the overall transistor performances. At the same time, the lack of understanding of the real role of the harmonic control was the cause of some misinterpretation of the experimental performances of such power amplifiers. In open literature, in fact, such second-harmonic approaches (mainly experimental and with lack of physical insight) often resulted in confusing contributions, reporting even contradictory results and conclusions.

For these reasons, a different approach was proposed in the last years starting from some physical consideration on the possible mechanisms to be used in order to improve the power amplifier performances in general and the power efficiency in particular, putting into evidence the possible expected improvement when moving toward the highest frequency range. More precisely, a comprehensive theory of a multi-harmonic manipulation design strategy was underlined, moving from a weighting procedure for the second- and third-order harmonic output voltage components and giving a methodology in order to assure the proper phase and amplitude ratio between them and the fundamental one, operating at both the input and output port of the amplifier. As a result, an effective improvement of microwave PA performances is demonstrated in terms of large-signal gain, output power and power-added efficiency (PAE).

Measurements performed on sample PAs, designed using the proposed technique to operate at 5 GHz and realised in hybrid form or to operate at 20 GHz and realised in monolithic form, will be presented and discussed in the following sections. In both the cases, the results, compared with those of companion amplifiers designed under the usual tuned load approach, clearly demonstrate both the feasibility and effectiveness of the proposed methodology.

4.4.2 Basic Assumptions

Every active device that can be fruitfully used at high frequency as an amplifying element exhibits major power limitations, leading to the well-known output power saturation mechanism. In the devices based on field-effect mechanism, like the FET for instance, such physical constraints reside in the gate-source junction forward conduction and channel pinch-off (determining the maximum current swing) together with triode region and gate-drain junction breakdown (fixing the maximum voltage swing) (see Figure 4.10).

A careful technological optimisation of doping profiles and gate recess, for instance, can alleviate the effects of the above physical limitations, but a further power performance improvement has necessarily to be based on smart design methodologies that, in turn, may be approached from two different starting points, both leading to optimum performances.

On the one hand, since device efficiency is strongly dependent on the amount of power dissipated in the device itself, a possible strategy consists in its minimisation that could be obtained by a proper shaping of voltage and current waveforms. Because of the fact that P_{diss} depends on the 'product' of the two waveforms, in fact, the shaping aims at avoiding or minimising the possibly overlapping regions. Moreover, the requested waveform shaping can be realised by a proper output network design strategy, that is, properly and differently loading the harmonic content of the output current, as in the Class-F or Class inverse–F approaches [4–10], or by a careful design of the output network, both in lumped or in distributed form, while using the active device as a pure switch, as in Class-E design [14–18, 26].

On the other hand, quite a different approach may be attempted by trying to maximise the fundamental output voltage (or current) components, implying therefore higher output power and efficiency while maintaining the DC power supplied to the amplifier at the same level. This aim can be obtained, for instance, by loading the active device with a purely resistive fundamental load, that is, resonating the reactive part of the output impedance [1] while using the harmonic content of the current (or voltage) in order to flatten the voltage (or current) waveform, while approaching the device physical limitations that result in a potentially higher fundamental-frequency component and while allowing the overall output voltage to respect the above-mentioned limitations (as it will be clarified in the next paragraphs).

From a physical point of view, the two briefly underlined strategies are not so different as it can be easily derived from power balance considerations. In fact, starting from the following relation

$$P_{\text{in}} + P_{\text{DC}} = P_{\text{diss}} + P_{\text{out}} \qquad (4.43)$$

it is easy to reach the same conclusions, following one of the two roadmaps: for a given power supplied to the active device (both from the DC bias supply P_{DC} and from the RF input P_{in}), design methodology devoted to increase the device output power or to decrease the dissipated power in the active device itself seem to be equivalent, leading to the improvement of the device efficiency $\left(\eta = \dfrac{P_{\text{out}}}{P_{\text{diss}}} \right)$ while stressing the role of one of the two relevant terms through a proper 'waveform engineering' approach, which results in a careful selection of harmonic terminations.

In order to infer some useful design criteria for the input and output networks, it is helpful to make some simple considerations about the active devices, FETs for instance, used for microwave applications. As seen above, they can be effectively treated as voltage-controlled current sources [2, 3], at least while operating in their active region. As a consequence, the resulting output current waveform is considered to be imposed by the controlling input voltage and, at least to a first approximation, does not depend on the chosen output terminating impedances that actually contribute only to the shaping of output voltage waveform. Under these assumptions and assuming steady-state conditions with a fundamental frequency f, time-domain drain current and voltage can be expressed by their Fourier series expansions

$$i_D(t) = I_0 + \sum_{n=1}^{\infty} I_n \cdot \cos(n\omega t + \xi_n) \tag{4.44}$$

$$v_{DS}(t) = V_{DD} - \sum_{n=1}^{\infty} V_n \cdot \cos(n\omega t + \psi_n) \tag{4.45}$$

where
$\omega = 2\pi f$,
ξ_n is the phase of the current nth harmonic component I_n,
ψ_n is the phase of the voltage nth harmonic component V_n,

the current and voltage harmonic components being related through the load on the transistor's output port $Z_{L,n}$ (i.e. the impedances across drain-to-source device terminals at harmonic frequencies nf):

$$Z_{L,n} = \frac{V_n}{I_n} \cdot e^{j(\psi_n - \xi_n)} = Z_{L,n} \cdot e^{j\phi_n} \tag{4.46}$$

From Figure 4.33, the supplied DC power and dissipated power on the active device are

$$P_{dc} = V_{DD} \cdot I_0 \tag{4.47}$$

$$P_{diss} = \frac{1}{T} \int_0^T v_{DS}(t) \cdot i_D(t) dt = P_{dc} - P_{out,f} - \sum_{n=2}^{\infty} P_{out,nf} \tag{4.48}$$

where
$$P_{out,nf} = \tfrac{1}{2} V_n I_n \cos(\phi_n) \quad n = 1, 2, \ldots \tag{4.49}$$

represents the active power delivered from the device to the output matching network at fundamental ($P_{out,f}$) and harmonics ($P_{out,nf}$). It is to be noted that in most normal

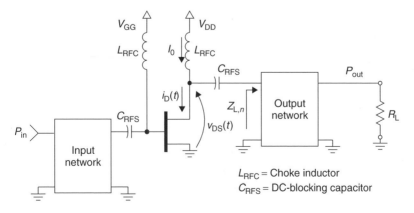

Figure 4.33   Simplified single-stage PA scheme

applications, fundamental output power alone is considered to be allowed to reach the output load R_L, filtering out harmonic components, thus leading to the following definition for drain efficiency η, which does not take into account the RF contribution P_{in}:

$$\eta = \frac{P_{out,f}}{P_{dc}} = \frac{P_{out,f}}{P_{diss} + P_{out,f} + \sum_{n=2}^{\infty} P_{out,nf}} \tag{4.50}$$

In this expression, P_{diss} and $P_{out,nf}$ take into account the output network charac-teristics: if the latter is a lossless ideal low-pass filter, then $P_{out,nf} = 0$ for $n > 1$, while P_{diss} already accounts for the power reflected by the filter towards the device; otherwise, if the output network is a frequency multiplexer, that is, if it can be seen as a one-input multi-output ports, each tuned at a different harmonic, then $P_{out,nf}$ for $n > 1$ is the power delivered on the relevant terminations at these harmonic frequencies.

From the expression above, maximum drain efficiency ($\eta = 100\%$) is obtained if

$$P_{diss} + \sum_{n=2}^{\infty} P_{out,nf} = 0 \tag{4.51}$$

that is, *if and only if* the following conditions are *simultaneously* fulfilled:

$$P_{diss} = \frac{1}{T} \int_0^T v_{DS}(t) \cdot i_D(t)dt = 0 \tag{4.52}$$

$$\sum_{n=2}^{\infty} P_{out,nf} = \frac{1}{2} \sum_{n=2}^{\infty} V_n I_n \cos(\phi_n) = 0 \tag{4.53}$$

Relevance of condition (4.53) is stressed if squared waveforms are assumed for both output current and voltage (i.e. the output network is simply resistive at any fre-quency) (Figure 4.34). In this case, while $P_{diss} = 0$ (no waveform overlapping), maximum drain efficiency is only 81.1% [27, p. 151] because of power dissipation on output ter-minations at harmonic frequencies ($P_{out,nf>0}$ for odd n).

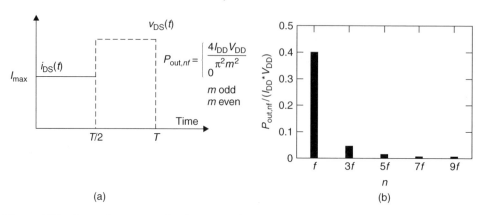

(a)   (b)

Figure 4.34   Squared current and voltage waveforms (a) and corresponding power spreading (b)

As a preliminary conclusion, condition (4.52) does not suffice to assure maximum theoretical drain efficiency, as often assumed: output power dissipated at harmonic frequencies must be simultaneously put to zero. Maximum drain efficiency can be therefore obtained if

- fundamental output power $P_{out,f}$ is maximised

or

- the *sum* of P_{diss} and $P_{out,nf}$ ($n > 1$) is minimised.

However, it is to be noted that many of the previous assumptions are valid to a first approximation only and are introduced for sake of clarity but can be easily removed in actual designs, where a full nonlinear model for the active device and a nonlinear simulator is used, without affecting the validity of the result of the presented theory.

Another very important assumption arises when considering the number of frequency components that can be effectively controlled in an actual design. On the one hand, in fact, the circuit complexity issue suggests the use of a minimum number of circuit idlers that are necessary to assure the proper termination to each harmonic. This is principally due to their physical dimensions that often result in too large a chip area occupancy and also due to the lack of availability and effectiveness of the components' models at highest frequencies, which could represent a practical limitation in their large utilization.

On the other hand, the benefits that can be obtained by controlling a larger number of harmonic components normally do not justify this increase. A reasonable and satisfactory compromise, as already anticipated, is in controlling the first two voltage harmonics (namely the second and third components), considering the other higher ones effectively shorted by the prevailing capacitive behaviour of the active device output. As a further justification of such an assumption, it is to be noted that the control, up to the fifth harmonic component, has been implemented only at the low-frequency range [13], resulting more in higher circuit complexity than in a major efficiency improvement. Therefore, the control scheme depicted in Figure 4.35 represents more than a simple theoretical solution, being a practical reasonable compromise among the various issues and constraints.

A further consideration is regarding the maximum output power condition for a given device, which, in Class-A operation, can be obtained by simultaneously maximizing voltage and current swings [1], as schematically depicted in Figure 4.36. As it is well known, in fact, the inherent nonlinear behaviour of the power amplifier, that is, the existence of hard physical limitations makes the optimum load different from the conjugate one of the output impedance while maintaining the necessity of resonating the reactive part of such an impedance.

Such a condition can be easily extended to a Class-AB operation [28], and it can be shown to be, once again, equivalent to a purely resistive loading of the controlled source, that is, to resonate, also in this case, the reactive part of the output impedance, so delivering to the external load only a pure active power.

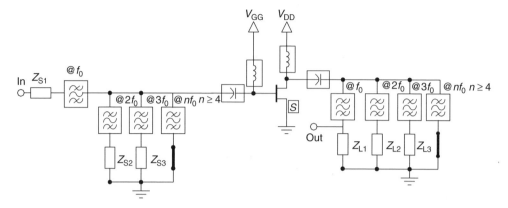

Figure 4.35   Input and output terminating scheme of a multi-harmonic manipulated PA

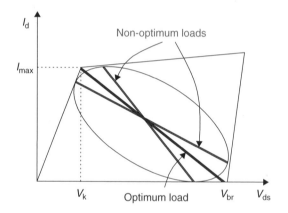

Figure 4.36   Class-A optimum and sub-optimum load curves

In order to examine this aspect, Figure 4.37 shows the extension of the optimum load concept to the Class-AB bias conditions when the tuned load approach is used. The V_{DD} value is the same and only the biasing current I_D has changed. The expected performances, in terms of output power of the Class-AB amplifier, are shown in Figure 4.38, where the output power, normalised to the one obtainable in Class-A, is given as a function of the circulation angle θ.

These results, which show in particular the existence of a maximum for the output power for a circulation angle chosen in the range 3.81 to 4.83, are obtainable if the optimum load is chosen according to the values, once again normalised to Class-A, given in Figure 4.39. Also in this case, it easy to note that the optimum load reaches equal values in Class-A and Class-B bias conditions, but assumes different values in the whole Class-AB, being lower up to 7% when operating in the above indicated range. A proper choice of the Class-AB load thus allows an improvement in the output power of the amplifier.

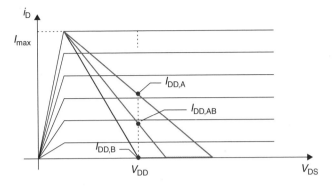

Figure 4.37   Load curves for different bias conditions

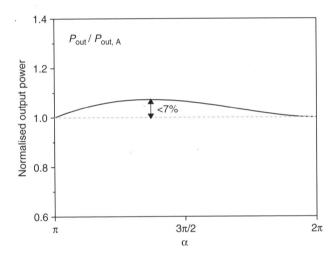

Figure 4.38   Output power normalised to the Class-A reference value as a function of the drain current conduction angle α

Finally, since a resistive termination is the optimum load for output power maximisation, the same holds for harmonic frequencies. In fact, complex terminations at harmonic frequencies generate a phase lag between the fundamental component and harmonic ones, that is, a different situation from being under a purely in-phase or out-of-phase condition, leading to an overlapping between current and voltage waveforms, thus increasing the dissipated power and decreasing the overall efficiency. This effect can be derived from eq. (4.46) if a complex load Z_{nfo} is considered (in [29], the effect of Z_{2fo} has been analysed and graphically shown).

For these reasons, in order to perform an effective control of the harmonics, while simplifying the choice of the relevant loads, a proper passive resistive termination is assured to each harmonic component after resonating the output capacitance with a proper inductive termination.

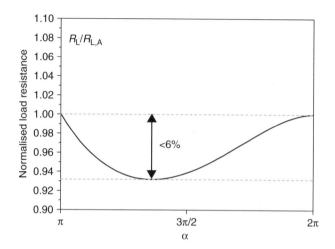

Figure 4.39 Optimum load resistance at fundamental frequency normalised to the Class-A reference value as a function of the drain current conduction angle α

4.4.3 Harmonic Tuning Approach

For low-frequency applications, assuming an infinite number of controllable harmonic terminations, two possibilities are available to fulfil condition (4.53), that is, making the active power delivered to the harmonics to vanish while assuming no overlapping between the current and the voltage waveforms according to condition (4.52).

- *Class-F* [4] or *inverse Class-F* [9, 10] strategies, in which $V_n I_n = 0$ for $n > 1$, due to the fact that the voltage (current) waveform has only *odd* harmonics, while the current (voltage) waveform has only *even* harmonics. It is to be noted that these are idealised approaches since voltage and current harmonic components, which in a real device are related by load impedances as in eq. (4.56), are separately considered. In the above approaches in fact, ideal short- or open-circuit terminations generate voltage (or current) components starting from *null* values of the corresponding current (or voltage) harmonic components. If more realistic assumptions are adopted, accounting also for the actual phase relationships between voltage and current harmonic components, both Class-C and deep Class-AB (near B) operating conditions lead to poor efficiency performances [30]. Nevertheless, the Class-F strategy, for instance, has been successfully applied in Class-AB [6, 31].

- *Class-E* strategy, in which $\phi_n = \pi/2$ for $n > 1$, because of the fact that all the harmonics apart the fundamental one have a pure reactive termination, an output capacitance C_{out} that includes also the output main parasitics, thus identically nulling the active power given to them. The active device is operated as a switch and closed-form design expressions are available [32]. In such conditions, the stage acts more as a DC/RF converter rather than as an amplifier. In this case, the power gain of the stage is not controlled and specified during the design phase; it is a specification to be fulfilled by a separately designed driver circuit using information about the input-port characteristics of the output transistor that is to be driven. Moreover, nothing is said about the input network except that the input voltage waveform has to properly drive the device to operate as a switch (i.e. deeply pinched off and saturated).

However, if the operating frequency enters the microwave region, both the approaches exhibit a degradation in performances. For instance, actual Class-F amplifiers are usually designed making use of two or three idlers only to control second and third output harmonic impedances. As frequency increases (e.g. >20 GHz), the control of both the second- and third-harmonic output impedances becomes troublesome since the active device output capacitive behaviour practically short-circuit higher components, not allowing the desired wave shaping. Moreover, for low-voltage applications, a Class-F strategy is not the best solution, since different methodologies (based on second-harmonic output impedance tuning) have demonstrated better performances [25].

On the other hand, the switching-mode operation of the active device, necessary to implement Class-E strategy, is not feasible in microwave communication systems since it requires that the power stage operates in saturated conditions, thus often increasing intermodulation distortion levels.

As a consequence, while designing high-frequency power amplifiers for communication systems, the number of the voltage harmonics that are effectively controlled is limited to the second and third ones, while the highest are assumed to be short-circuited.

With such hypotheses the drain efficiency becomes

$$\eta = \frac{P_{\text{out},f}}{P_{\text{diss}} + P_{\text{out},f} + P_{\text{out},2f} + P_{\text{out},3f}} \tag{4.54}$$

with $P_{\text{out},nf} = V_n I_n = 0$ in eq. (4.50), having V_n identically zero (short-circuited) for $n > 3$. As a consequence, the device's physical constraint $v_{\text{DS}}(t) \geq 0$ must be attained through the superposition of the few remaining harmonics (namely first, second and third). Therefore, both an overlapping between drain current and voltage waveforms ($P_{\text{diss}} > 0$) and a lower fundamental voltage component (decreasing $P_{\text{out},f}$) result, thus decreasing the achievable drain efficiency values (lower than the ideal 100%).

Under the assumptions stated above, several different solutions are proposed in literature in order to maximise η for high-frequency applications. Most of them are based on the already mentioned traditional approaches (Class-E [33], Class-F or inverse Class-F [9, 10]) and assume the same impedance values as in the ideal (i.e. infinite number of *controllable* harmonics) case. The result is that $P_{\text{out},2f}$ and $P_{\text{out},3f}$ still remain nulled and an increase on P_{diss}, due to the overlapping between the resulting voltage and current waveforms, is accepted.

Such approaches however exhibit several drawbacks. One of the latter resides in the necessity to increase the bias voltage V_{DD} in order to prevent negative drain voltage values, thus increasing the supplied DC power (otherwise, a lower saturated output power is expected), so further lowering the achievable efficiency.

On the other hand, some improvements in efficiency can be achieved by properly choosing the harmonic voltage ratios, as it was demonstrated in the high frequency

Class-F approach [6]. In this case, in fact, assuming the third to first harmonic voltage ratio (k_3 in this paper) higher (namely $k_3 = -1/6$) than in the ideal squared voltage waveform (corresponding to $k_3 = -1/3$), a slight improvement in the drain efficiency was achieved. Moreover, it is worth noting that the minimisation of the drain voltage $v_{DS}(t)$ when $i_D(t)$ reaches its maximum value (the so-called 'maximally flat condition' in [34]) is not sufficient to minimise P_{diss}. In this respect, the theoretical values of P_{diss} (normalised to $I_{max} \cdot V_{DD}$) as a function of the bias current (normalised to I_{max}) are reported in Figure 4.40, assuming the control of first and third harmonic components only with different voltage ratios k_3.

Finally, a further improvement can be obtained by increasing by a factor of $2/\sqrt{3}$ the load at fundamental frequency [30]. Nevertheless, the proposed approaches (F or inverse F) usually neglect the relationships between the voltage and current harmonic components imposed by eq. (4.46), and thus limiting the analysis to ideal (i.e. short- or open-circuit) terminations. In general, no attempt has been made to classify the various strategies and to unify them in a systematic way.

Recently, a new approach has been suggested [35]:

- *Harmonic Manipulation* (HM) based on the fulfilment of the first or second condition (Section 4.4.2, page 191), allowing non-zero values also for both $P_{out,2f}$ and $P_{out,3f}$ if a higher fundamental output power can be achieved. This methodology, which accepts the active power supplied to the harmonics to be different from zero, while diminishing the P_{diss} dissipated inside the active device, is clearly losing, in comparison with the two above-mentioned strategies, when a very high number of harmonics is involved,

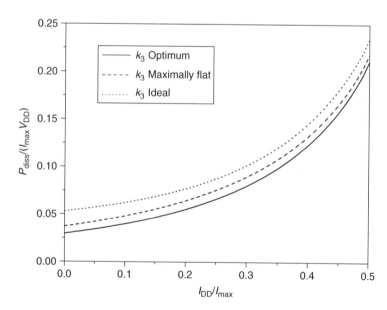

Figure 4.40   Plot of P_{diss} vs bias current I_{dc} with different voltage ratios $k_3 = V_3/V_1$. Optimum value ($k_3 = -1/6$, solid), maximally flat condition ($k_3 = -1/9$, dashed) and ideal ($k_3 = -1/3$, dotted)

but reveals to be challenging in the case under consideration. At high frequency, in fact, the practical limitation on the number of the harmonics renders the circuital solutions devoted to minimise the quantity $P_{diss} + P_{out,2f} + P_{out,3f}$ as an interesting alternative to be explored. Moreover, it is to be noted that even if this condition is equivalent to the one maximising the output power at the fundamental frequency $P_{out,f}$, from a mathematical point of view it is more convenient to utilise the latter that involves a lower number of variables to handle.

Details of the proposed HM approach will be briefly recalled in the next paragraph.

4.4.4 Mathematical Statements

On the basis of the assumptions in Section 4.4.3, expression (4.45) can be newly rewritten, utilizing second and third-harmonic components only, as follows:

$$V_{ds}(t) = V_{ds,DC} - V_{ds,fo} \cdot \cos(2\pi f_o t) - V_{ds,2fo} \cdot \cos(2 \cdot 2\pi f_o t) - V_{ds,3fo} \cdot \cos(3 \cdot 2\pi f_o t)$$

$$(4.55)$$

Normalising to the fundamental-frequency component $V_{ds,fo}$, the eq. (4.55) becomes

$$V_{ds,norm}(\vartheta) = \frac{V_{ds}(\vartheta) - V_{ds,DC}}{V_{ds,fo}} = -\cos(\vartheta) - k_2 \cdot \cos(2 \cdot \vartheta) - k_3 \cdot \cos(3 \cdot \vartheta) \quad (4.56)$$

where

$$k_2 = \frac{V_{ds,2fo}}{V_{ds,fo}}, \quad k_3 = \frac{V_{ds,3fo}}{V_{ds,fo}}, \quad \vartheta = \omega_o t \quad (4.57)$$

As it is easy to infer, the drain voltage waveform is constrained to swing within the range dictated by the device physical boundaries, that is, the drain knee voltage V_k, here assumed as a first approximation to represent a hard limit, and the drain–source breakdown voltage $V_{ds,br}$, where the gate-drain junction becomes forward biased. It is therefore necessary that

$$V_k \leq V_{ds}(\vartheta) \leq V_{ds,br} \quad (4.58)$$

It can be observed that without the contribution of harmonic components the maximum drain voltage amplitude in linear conditions is given by

$$V_{ds,fo,max} = \min[V_{ds,DC} - V_k, V_{ds,br} - V_{ds,DC}] \quad (4.59)$$

As previously mentioned, the goal of such a multi-harmonic manipulation procedure is to obtain an increase in the fundamental-frequency voltage component with respect to the case when no voltage harmonic component is allowed. This effect can be obtained by means of a proper shaping of the overall voltage waveform, constrained to swing between the same physical limitations, that is, through a proper choice and utilization of the harmonic content.

Such a statement implies that the target is to obtain $V_{ds,fo} \geq V_{ds,fo,\max}$, which is equivalent, for the physical constraints, to the inequalities:

$$V_{ds,\text{norm}}(\vartheta, k_2, k_3) \geq -1 \quad if \quad V_{ds,fo,\max} = V_{ds,DC} - V_k \tag{4.60a}$$

$$V_{ds,\text{norm}}(\vartheta, k_2, k_3) \leq -1 \quad if \quad V_{ds,fo,\max} = V_{ds,BR} - V_{ds,DC} \tag{4.60b}$$

For the sake of simplicity, only the case represented by eq. (4.60a) will be discussed, since it is the most common situation, but an equivalent analysis can be performed for the case of eq. (4.60b). From a mathematical point of view, the problem of eq. (4.60a) is equivalent to finding the values of k_2 and k_3, which allow an increase in fundamental-frequency voltage component over the not manipulated one while respecting the same physical limitations.

Such an increase can be quantitatively evaluated by means of a *voltage gain function* δ (k_2, k_3), defined by

$$\delta(k_2, k_3) \equiv \frac{V_{ds,fo}}{V_{ds,fo,\max}} = \frac{-1}{\min_{\vartheta}[V_{ds,\text{norm}}(\vartheta, k_2, k_3)]} \tag{4.61}$$

As a consequence, the resulting fundamental-frequency voltage component can be expressed as

$$V_{ds,fo}|_{\text{MHM}} = \delta(k_2, k_3) \cdot V_{ds,fo,\max} \tag{4.62}$$

The selection of optimum design points (i.e. values for k_2 and k_3 maximising the fundamental-frequency voltage component) therefore implies the study of the voltage gain function. The simplest case is related to the analysis of the problem of a harmonic manipulation based on the use of a single harmonic component. In fact, assuming $k_3 = 0$, that is, considering the third harmonic to be short-circuited, a particular kind of high-efficiency amplifier, the Class-G one, can be studied. In this case, by properly generating and properly terminating the second harmonic of the drain current, very interesting features for the power amplifier have been demonstrated and experimentally tested [25, 36, 37]. Similarly, assuming $k_2 = 0$, that is, considering the second harmonic to be short-circuited, another kind of high-efficiency amplifier, the Class-F one, has been largely studied, after the first suggestion from Snider [4]. In particular, the crucial role of the phase relationship between the fundamental and the third harmonic has been put into evidence, so explaining the necessity to bias the Class-F actual amplifier close to the pinch off (deep Class-AB) but not in Class-B, as theoretically provided, in order to profit the improvement in the amplifier performances, as forecasted by the theory of Snider [30].

More complex, but manageable following the same roadmap, is the case when both the second and third-harmonic components are used ($k_2 \neq 0$, $k_3 \neq 0$, for extension Class-FG), that is, when both the harmonics have to be generated with a proper phase relationship with respect to the fundamental one and must be terminated on a proper resistive load while resonating the output capacitance at the relevant frequencies. The mathematical treatment is quite long and, unfortunately, the results cannot be expressed in closed form. The surface of the voltage gain function, δ (k_2, k_3) in the k_2, k_3 plane is given in Figure 4.41 while its contour plot is given in Figure 4.42.

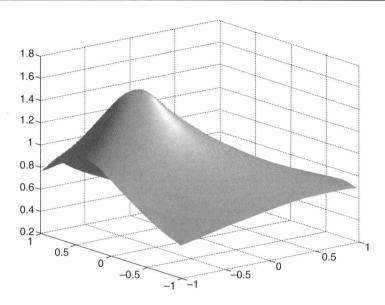

Figure 4.41   The voltage gain function $\delta\,(k_2, k_3)$ vs k_2 and k_3

It is evident that a wrong choice of the harmonics could lower the overall performances ($\delta\,(k_2, k_3) < 1$), while a proper choice can result in a significant improvement.
A clear maximum, in fact, is visible for the voltage gain function, reaching the optimum
zone for $k_2 < 0$ and $k_3 > 0$: in this case, the fundamental component is in-phase with the
third harmonic and out-of-phase with the second one. It is worthwhile to note, in particular, that in this case the proper phase relationship between the third and the fundamental
component is opposite to the one stated [30] for obtaining the Class-F behaviour.

Moreover, Figure 4.42 shows that Class-F operation corresponds to points lying
on the negative side of the vertical axis ($k_3 < 0$, $k_2 = 0$), while Class-G corresponds to
points lying on the negative side of the horizontal axis ($k_2 < 0$, $k_3 = 0$). The more classical
tuned load (TL) approach, imposing short-circuit terminations at harmonic frequencies,
is represented by the origin ($k_2 = k_3 = 0$).

Basic considerations can be carried out regarding the sign of the k_2 and k_3 harmonic coefficients. If Class-F or Class-G operation is considered, a narrow range of k_3
and k_2 can be fruitfully used for harmonic manipulation corresponding to the regions of
the respective axes in which the voltage gain function is greater than unity. In both the
cases, such condition corresponds to harmonic components out of phase (i.e. with opposite sign) with respect to the fundamental one [30, 36, 37], giving rise to a 'flattening'
of the resulting drain voltage waveform while it approaches the physical limitation of
the device (as in Figure 4.43(a) for the Class-F case). On the other hand, an in-phase
combination results in a *peaking effect* on the voltage waveform thus approaching the
physical limitation for a lower fundamental-frequency component and hence decreasing the maximum achievable fundamental-frequency voltage amplitude, as shown on
Figure 4.43(b).

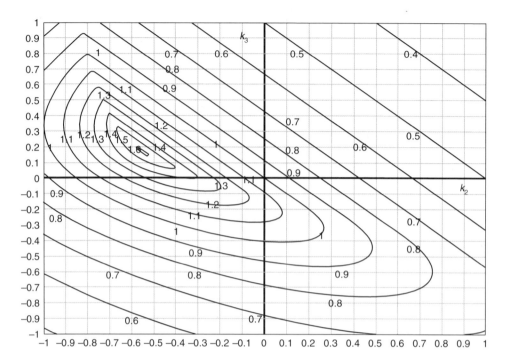

Figure 4.42 Contour plot of voltage gain function δ (k_2, k_3) in the k_2 and k_3 plane

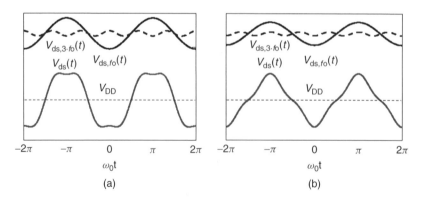

Figure 4.43 Drain voltage Class-F waveforms: (a) out-of-phase and (b) in-phase first and third components

If the waveform for the Class-G case (Figure 4.44) is considered, a further observation may be done: a flattening of the voltage waveform can be effectively obtained when the drain current is at its maximum (Figure 4.44(a), out-of-phase condition), while a peaking effect occurs in the remaining part of the cycle. On the contrary, the flattening in the voltage waveform is obtained when the drain current reaches its minimum, while the peaking occurs at its maximum if the in-phase condition stands (Figure 4.44(b)).

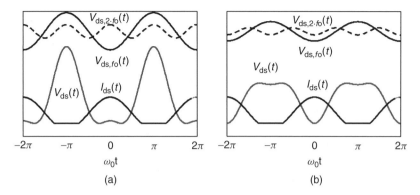

Figure 4.44 Drain voltage for Class-G design: (a) out-of-phase components and (b) in-phase components

The contribution of the second harmonic, in fact, while being out-of-phase with the fundamental one when the drain voltage reaches its minimum (maximum), results to be in-phase when it reaches its maximum (minimum) value, which in turn results to be larger.

Limiting the consideration to the out-of-phase condition and looking at the device output characteristics, it is clear that this effect leads the operating point to potentially enter the device breakdown region with evident detrimental effects on device reliability and, at the least, to a lowering of the efficiency.

To account for the peaking effect obtained when using the proper-phased second harmonic for the manipulation, a voltage overshoot function $\beta\,(k_2, k_3)$ may be introduced, defined as

$$\beta(k_2, k_3) \equiv \frac{\max_\vartheta [V_{\mathrm{ds,denorm}}(\vartheta)]}{\max_\vartheta [V_{\mathrm{ds,denorm}}(\vartheta)|_{k_2=0,k_3=0}]} = \max_\vartheta [V_{\mathrm{ds,norm}}(\vartheta)] \cdot \delta(k_2, k_3) \qquad (4.63)$$

As it can be easily inferred, $\beta\,(k_2, k_3)$ directly gives the amount of the overshoot for a given (k_2, k_3) combination and must be accounted for in order to avoid unwanted breakdown occurrence.

The contour plot for the voltage overshoot function is shown in Figure 4.45: the maximum values for such a function, $2.77 \le \beta\,(k_2, k_3) \le 3$, reside *close* to the region giving optimum values for the voltage gain function, stressing its relevance in actual designs.

Finally, another statement can be developed when examining the properties of the flattening of voltage waveform while approaching the minimum drain voltage as allowed by the relevant device physical limitation.

Figure 4.46 shows what happens to the voltage waveform for a generic choice of k_2 and k_3 in the second quadrant of k_2 and k_3 plane. As it is easy to see, the minima are not at the same level, thus resulting in a sub-optimum condition. A better choice is achievable if an 'equiripple condition' is imposed upon the voltage waveform, that is, its multiple minimum values are imposed to be equal (Figure 4.47).

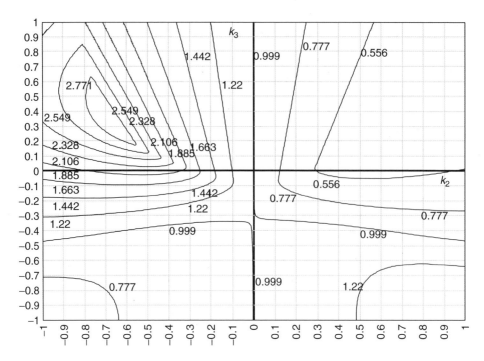

Figure 4.45   Contour plot of voltage overshoot function β (k_2, k_3) in the k_2 and k_3 plane

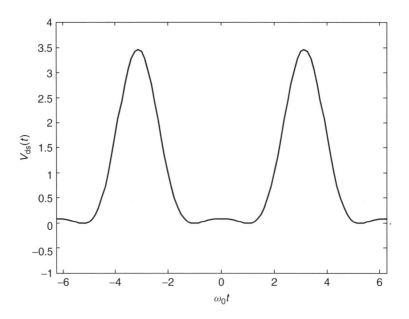

Figure 4.46   Voltage drain waveform for a generic choice of k_2 and k_3 in the second quadrant of k_2 and k_3 plane ($k_2 = -0.5, k_3 = 0.1$)

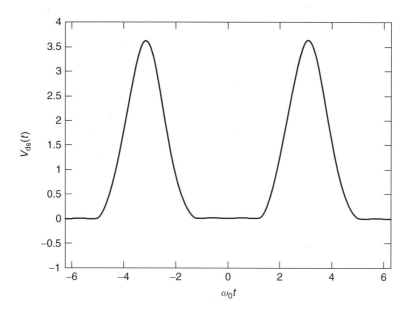

Figure 4.47 Voltage drain waveform for k_2 and k_3 fulfilling the 'equiripple condition' ($k_2 = -0.5$, $k_3 = 0.125$)

In this case, a simple equation linking the k_2 and k_3 values can be derived:

$$k_3 = \frac{k_2^2}{4 \cdot (k_2 + 1)} \tag{4.64}$$

The use of eq. (4.64) allows an explicit representation for the voltage gain function under the equiripple condition, given by

$$\delta(k_2) = \frac{4 \cdot (1 + k_2)}{5 \cdot k_2^2 + 8 \cdot k_2 + 4} \tag{4.65}$$

The plot of such a function, superimposed on the contour plot for the general voltage gain function is shown in Figure 4.48.

The maximum value for $\delta\ (k_2, k_3)$ in the equiripple condition is given by

$$\delta(k_{2,\delta\,\text{max}}, k_{3,\delta\,\text{max}}) = \frac{1 + \sqrt{5}}{2} \approx 1.62 \tag{4.66}$$

and it is obtained for the couple k_2, k_3:

$$[k_{2,\delta\,\text{max}}, k_{3,\delta\,\text{max}}] = \left[-1 + \frac{1}{\sqrt{5}}, \frac{3 \cdot \sqrt{5} - 5}{10} \right] \approx [-0.55, 0.17] \tag{4.67}$$

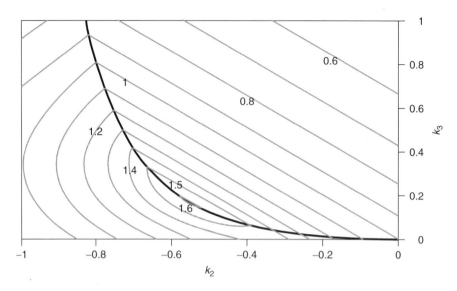

Figure 4.48 The voltage gain function δ (k_2, k_3) under the equiripple condition

Such a maximum value is coincident with the absolute maximum value obtainable for the voltage gain function δ (k_2, k_3). On the other hand, a different approach may be attempted, trying to flatten as much as possible the voltage waveform ('maximally flat' condition) as suggested in [34], that is, imposing to be null both the first and second derivatives on the waveform itself.

Such a condition is a subset of the equiripple one and the resulting value for the voltage gain function is given by

$$\delta(k_{2,\text{maximallyflat}}, k_{3,\text{maximallyflat}}) = \tfrac{3}{2} = 1.5 \tag{4.68}$$

corresponding to

$$[k_{2,\text{maximallyflat}}, k_{3,\text{maximallyflat}}] = \left[-\frac{2}{5}, \frac{1}{15}\right] = [-0.4, 0.067] \tag{4.69}$$

Once again, it means that the maximally flat condition is not the optimum choice while leading to a sub-optimum design.

This kind of result could be better understood if some physical aspects are put into the proper evidence. Weighting the harmonics, in order to assure the *maximally flat condition* for the drain voltage waveform, in fact, involves into the calculation of the power dissipated in the transistor P_{diss},

$$P_{\text{diss}} = \frac{1}{T} \int_0^T v_{\text{DS}}(t) \cdot i_{\text{D}}(t) dt \tag{4.70}$$

only the minimisation of *the function to be integrated* instead of *the integral* itself. This means that other choices, like the one previously indicated, involving the maximisation

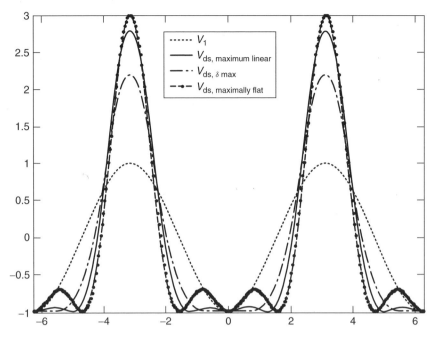

Figure 4.49 Drain voltage waveforms under different conditions: maximum linear (dotted line); maximally flat (dot-dashed line); maximum of δ (k_2, k_3) (solid line); maximum of β (k_2, k_3) (dashed line)

of the output power and consequently the minimisation of P_{diss}, results in an actual optimum choice.

Finally, for sake of comparison, Figures 4.49 and 4.50 show, as an example, the drain voltage waveforms synthesised for three different conditions: a first one corresponding to the weighting of the second and third-harmonic contribution according to the maximum value of δ (k_2, k_3), another one obtained with harmonics corresponding to the maximum value of β (k_2, k_3), and a third waveform synthesised according to the maximally flat conditions.

4.4.5 Design Statements

The voltage harmonic shaping described in the previous section must now be related to the actual increase in power performances and to the output networks' design. To this goal, let us briefly recall the rationale behind multi-harmonic manipulation.

For a given device with its physical limits, a given maximum linear swing is allowed for the drain voltage (from eq. (4.59)), whose time-domain waveform is constrained to swing between the ohmic and breakdown regions. The intrinsic drain current is imposed by the drive level of the input waveform, therefore fixing its harmonic components. The maximum output power that can be obtained under such linear operating conditions is simply given by the product of the maximum linear fundamental voltage

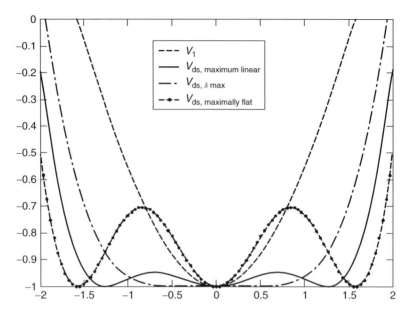

Figure 4.50 Same as Figure 4.49. Drain voltage waveforms' details

component ($V_{\mathrm{ds},fo,\max}$) times the drain current fundamental component ($I_{\mathrm{d},fo}$). Their ratio uniquely determines the load impedance at fundamental frequency (Z_{fo}) to be imposed, that is, on the basis of the discussion in Section 4.4.2, a purely resistive termination:

$$R_{\mathrm{TL,opt}} = \frac{V_{\mathrm{ds},fo,\max}}{I_{\mathrm{d},fo}} \tag{4.71}$$

In this case, harmonic terminations can be thought to be set to short-circuit ones, and the obtained design is the well-known tuned load (TL) strategy.

Starting from such a situation and supposing that the harmonic components of the drain current are not influenced by their terminations (Section 4.4.2), voltage harmonic components (second and third) can be added to the fundamental one according to their weights k_2 and k_3 computed in Section 4.4.4. The result of such a wave shaping is a new voltage waveform with the same fundamental component but with a reduced swing. The fundamental drain voltage component can be now increased by the factor δ (k_2, k_3) to reach the device limitations. In this way, for the same drive level and with the same voltage swing, a higher fundamental-frequency voltage component and therefore higher output power is obtained.

Applying multi-harmonic manipulation, the fundamental-frequency voltage component is increased by the factor δ (k_2, k_3), as indicated in eq. (4.62), here repeated for convenience:

$$V_{\mathrm{ds},fo}|_{\mathrm{MHM}} = \delta(k_2, k_3) \cdot V_{\mathrm{ds},fo,\max} \tag{4.72}$$

Therefore, the load to be imposed at fundamental frequency to obtain this goal is

$$R_{fo}|_{\mathrm{MHM}} = \delta(k_2, k_3) \cdot R_{\mathrm{TL,opt}} \tag{4.73}$$

Similarly, the harmonic terminations that have to be imposed at second- and third-order components can be computed by

$$R_{nfo}|_{\text{MHM}} = \delta(k_2, k_3) \cdot k_n \cdot \frac{I_{d,fo}}{I_{d,nfo}} \cdot R_{\text{TL,opt}} \quad n = 2, 3 \tag{4.74}$$

Fundamental frequency drain current component is, to a first approximation, unaffected by the resulting increase in the respective drain voltage component. Output performances are therefore increased by the same amount, that is,

$$\begin{aligned} a) & \quad P_{\text{out,MHM}} = P_{\text{out,TL}} \cdot \delta(k_2, k_3) \\ b) & \quad G_{\text{out,MHM}} = G_{\text{out,TL}} \cdot \delta(k_2, k_3) \\ c) & \quad \eta_{\text{d,MHM}} = \eta_{\text{d,TL}} \cdot \delta(k_2, k_3) \end{aligned} \tag{4.75}$$

Equation (4.73) in particular gives the optimum fundamental-frequency termination, and in its simplicity reveals a potential source of error while performing PA design.

In fact, a widely used procedure to investigate the power performances of a given device is to measure its load-pull contours. Load-pull systems are nowadays becoming extremely sophisticated, providing the possibility to perform load-/source-pull measurements not only at fundamental but also at harmonics. The usual procedure, in the case of harmonic load pull, consists in finding the optimum fundamental-frequency termination for fixed values of harmonic loads. Once such value is determined, it is held fixed and the harmonic loads are varied until an optimum value for them is found. On the basis of the theory outlined in the previous section, such a combination of loads is not the optimum one since the fundamental-frequency load without (or for a fixed) harmonic tuning is not the same that can be obtained by properly varying the harmonic loads.

A correct load-pull procedure should vary harmonic load together with the fundamental one to find the global optimum combination. [38] On the other hand, eq. (4.73) may be used in order to find a step-by-step procedure starting from the tuned load case.

4.4.6 Harmonic Generation Mechanisms and Drain Current Waveforms

In this section, the problem of the proper current harmonic generation will be addressed. In fact, since passive terminations only have to be employed, the properly phased voltage harmonic components must result from eq. (4.44), that is, starting from the output drain current harmonic components while choosing suitable terminations. Different approaches can be explored in order to obtain the proper phase relationships among the drain current harmonic components, and will be briefly examined in the following.

A first possibility consists in the use of the output clipping phenomena, that is, in the generation of current harmonic components by means of hard device nonlinearities as the pinch-off and the input gate-source junction forward conduction. Since this phenomena is related to the input drive level and to the selected bias point, it implies a proper selection of the active device operating conditions.

If a simple sinusoidal drive is used as input signal, the resulting drain current is simply a truncated sinusoid, whose conduction angle (ϑ_c), defined in Figure 4.51(a),

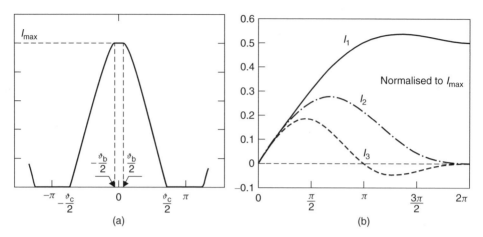

Figure 4.51 Ideal output drain current: (a) truncated sinusoid; (b) relevant harmonic content

completely determines the resulting amplitude and phase relationships among current harmonics and reveals some important properties.

By a simple Fourier transformation, the drain current harmonic components can be computed as plotted in Figure 4.51(b), where only the first three harmonics are reported for the sake of brevity. As it is possible to note, for conduction angles ranging from Class-A to Class-B, the second-harmonic component I_2 is always in-phase, while the third one I_3 remains always out-of-phase with respect to the fundamental component I_1 (i.e. having the same and opposite sign respectively).

As a consequence, the direct application of the multi-harmonic manipulation procedure described earlier, that is, with purely resistive harmonic loads, is not allowed at all. Only a Class-F design is therefore directly applicable [30]. Moreover, it is worth noting that the choice of a Class-C bias conditions becomes deleterious, resulting in an uncorrected phase relationship, while for a Class-B bias conditions, as suggested in [4], only a mathematical solution corresponding to the 'opening' of the odd harmonics seems to be affordable in order to assure the forecasted benefits, their amplitudes being identically zero.

Moreover, even if a second nonlinear phenomenon (i.e. the input diode forward conduction (ϑ_b)) is encountered, the behaviour of the harmonics versus the circulation angle ϑ seems to be modified only a bit, as shown in Figure 4.52. In fact, only at the highest circulation angles and for a heavy diode conduction ϑ_b, the second-harmonic component changes it sign, thus allowing the Class-G [25, 29] operation (trapezoidal waveform).

As a first remark, if simple resistive harmonic terminations appear to be not useful, complex ones can be experienced, also at the fundamental frequency, partially reducing the improvement obtained through the multi-harmonic manipulation due to the amount of reactive power involved at the fundamental frequency itself. In this case, a simple, suitable design criteria is obtained choosing the harmonic terminations as dictated by the high-frequency Class-E approach, while paying at least a higher overshot factor $\beta \approx 3.65$ [33].

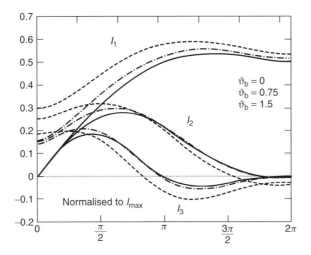

Figure 4.52 Drain current relevant harmonic content for different input diode forward conduction angle ϑ_b

A second opportunity consists in using the effect of device input nonlinearities. A Volterra analysis of the input circuit [36, 39] shows, in fact, that the main contribution to the harmonic generating mechanism at the device input is given by the nonlinear input capacitor C_{gs}, thus confirming the numerical and experimental results in [20, 21].

If reasonable drive levels are considered, without the necessity of entering the turn-on zone of the input diode, thus improving at least the reliability of the device, the control voltage V_{gs} exhibits a major second-harmonic content leading to an asymmetrical gate-source voltage waveform, as depicted in Figure 4.53. In order to avoid this effect, often considered a detrimental one, the input harmonic terminations are frequently set to short-circuit values [20, 22] or compensated by means of a counteracting nonlinearity [21]. Nevertheless, major improvements of power performances are obtained if such a second-harmonic input voltage component is used to implement the technique described beforehand. In fact, the input signal nonlinear distortion implies the generation of a second-harmonic gate voltage component that is *out of phase* with respect to the fundamental one and therefore usable for the generation of output current components with the same phase relationship. Moreover, also a third-harmonic component 'in-phase' with the fundamental one is generated, suitable for a Class-FG multi-harmonic manipulation.

Up to now, while nonlinear *output* clipping phenomena determine a 'wrong' phase relationship among current harmonics, *input* nonlinearities effectively act in a reverse direction, generating second- and third-harmonic components with the proper phasing. These two counteracting effects cooperate in a very complex way in real devices. On the other hand, it is clear that the input nonlinearities dominate at moderate drive levels, while output clipping phenomena should prevail for higher levels. Such a behaviour strongly depends on biasing conditions since the latter fix the drive level at which physical limitations are incurred: roughly speaking, the closer is the bias point to the Class-A reference, the higher will be the drive level at which the counteracting output harmonic generation prevails [40].

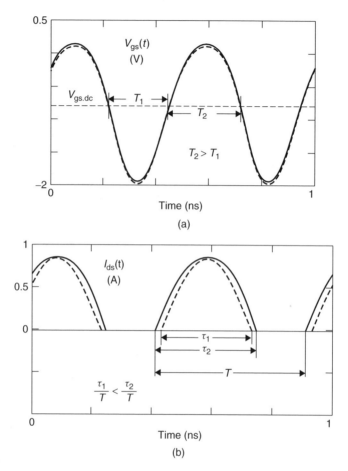

Figure 4.53 (a) V_{gs} voltage waveform as computed by the truncated Volterra expansion (dashed line) and by a full nonlinear HB method (solid line); (b) output current (solid line), evidencing the increase in the duty cycle over the undistorted condition (dashed line)

Another aspect that must be considered is related to the amplitudes of the ratios among the voltage harmonic components, that is, the values of k_2 and k_3. In fact, even if the phase relationships are correct, the values of k_2 and k_3 are related to the drain current harmonic components and to the harmonic load resistances by eq. (4.74). While the amplitude of the harmonic components increase with the input drive signal, the harmonic load resistances are upper limited by the output device resistance value R_{ds}. Such a behaviour is demonstrated for a typical power stage in Figure 4.54, where the relative amplitudes of second- and third-harmonic drain voltage components with respect to the fundamental one (k_2, k_3) are plotted as a function of the input power for a fixed bias point and loading (both input and output). As it is easy to note, because of the actual device and the circuital solution adopted, both the harmonic generation mechanism and the R_{ds} values are not able to produce the wished voltage harmonics. The obtainable values for k_2 and k_3, in fact, result to be lower than the optimum ones.

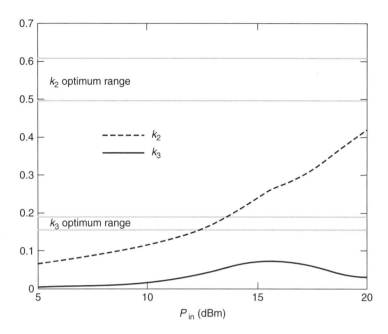

Figure 4.54 Class-FG design: normalised amplitude of drain voltage second (dashed line) and third (solid line) harmonics

The above-mentioned upper limitation for the load terminations of the harmonics could limit the effectiveness of the proposed methodology. For this reason, an approach that gives the possibility to fix independently the requested amplitudes of the starting current harmonics seems to be interesting. Since, as already mentioned, the drain voltage waveform is built from the drain current harmonic components, resulting from eq. (4.44), in fact, proper drain current harmonic components could be generated in order to obtain the proper phase and amplitude relationships. As noted earlier, in order to reach the latter goal, the input nonlinearities can be fruitfully employed, but this is not the unique possibility: a pre-shaped waveform may be fed to the input of the power stage containing, already, the correct phasing between its harmonics.

Even if it is possible to analyse the best input drive waveform for each harmonic strategy, as it is presented in [19] for a Class-F amplifier design, this methodology could be practically unfeasible because of its difficult implementation and also because of being too sensitive to the chosen active device input model. Because of these reasons, more practical approaches based on the analysis of realistic and easy-to-implement cases alone can suggest to the designer how to solve the problem of an effective application of the harmonic manipulation.

For instance, a class of eligible waveform is obviously a rectangular waveform in general and a square one in particular, the latter being easily obtainable using a Class-F amplifier [30] as the driver stage. In this case, the analysis may start directly from the drain current waveforms, that in the following will be assumed as a rectangular waveform, as a truncated sinusoid (as the reference case before examined) and finally

Table 4.1 Drain current circulation angles allowing Class-FG approach

Current waveform model	Class-FG
Truncated sinusoid	Never possible
Quadratic	$6.06 < \vartheta_c < 2\pi$
Rectangular	$4.18 < \vartheta_c < 2\pi$

as a quadratic waveform, to take into account a more realistic active device pinch off nonlinear behaviour, as suggested in [41]. Through a Fourier analysis on the three-current waveforms, the corresponding relevant harmonics are easily derived and the regions where purely resistive output loading allows an effective harmonic manipulation can be evidenced, as reported in Table 4.1.

The drain current conduction angle ϑ_c has to be considered for all the three cases. It represents the portion of the period when the drain current assumes non-zero values, corresponding to the duty cycle for the rectangular waveform.

Using a piecewise-linear simplified model for the active device, for the regions of Table 4.1, the expected improvements in terms of output power and drain efficiency can be evaluated through eq. (4.75). The theoretical output power (normalised to the performances of a standard Class-A amplifier design) and the drain efficiency for a tuned load (TL) and Class-FG amplifiers are depicted in Figure 4.55 and Figure 4.56 respectively, while more detailed results, including the Class-F and the Class-G solutions, can be found in [28].

It is to be noted that the theoretical purely resistive multi-harmonic manipulation seems to be useful only for a narrow range of the drain current circulation angle ϑ_c when limited to the output port only. Moreover, the efficiency improvements could be not satisfactory: for a high-efficiency design, for instance, it appears to be more appropriate to choose values of ϑ_c closer to Class-B bias condition, obtaining higher efficiency values. As a consequence, the optimum design is a trade-off among all the above-mentioned parameters.

In summary, many different solutions seem to be available using output and/or input manipulation in order to obtain significant improvements over the classical tuned load amplifier solution. Obviously, a combined action, activated both at input and output ports of the amplifier, could represent the best solution depending on the acceptable growth in circuit complexity. Moreover, especially for the simplified analysis based on the three different driving signals listed above, the reported results represent only a first-order approximation, while the effects of the input and output nonlinearities are not accounted for. In any case, a more accurate analysis based on a full nonlinear model of an actual device, including all the sources of nonlinear behaviour, demonstrates the validity of the main conclusions with only minor modifications, mainly on the ranges listed in Table 4.1.

4.4.7 Sample Realisations and Measured Performances

In order to demonstrate the effectiveness of the proposed harmonic manipulation strategy for high-efficiency design, two sets of power amplifiers have been designed and realised.

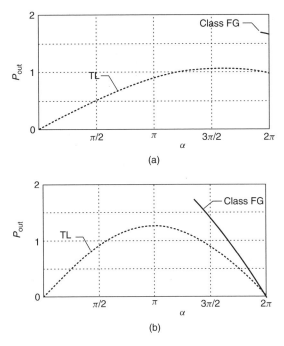

(a)

(b)

Figure 4.55 Normalised output for a tuned load PA (dotted) and Class-FG PA (solid): (a) quadratic and (b) square drain current waveform

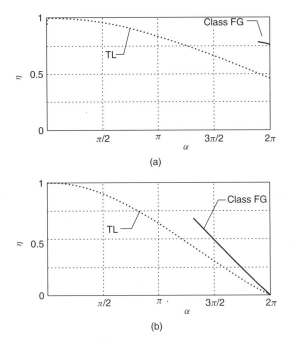

(a)

(b)

Figure 4.56 Drain efficiency according to a simplified model for a tuned load PA (dotted) and Class-FG PA (solid): (a) quadratic and (b) square drain current waveforms

More precisely, two amplifiers were designed and realised in hybrid form in order to operate at $f_0 = 5$ GHz (fundamental frequency), namely a tuned load reference stage, and a Class-FG amplifier utilising second- and third-harmonic voltage tuning. Another pair of amplifiers, one as a reference stage and one working in Class-G, were designed and realised in monolithic form to operate at 20 GHz.

For the first set of two amplifiers, the device used has been a medium-power MESFET by Alenia Marconi Systems with a Class-AB bias condition ($I_q = I_{dss}/3 \approx 80$ mA; $V_q = 5$ V) and 1-mm gate periphery. In particular, the bias point was chosen to be the same for both the amplifiers and precisely for the one that optimised the tuned load performances. For the sake of comparison, in fact, it was preferred not to choose the bias conditions that the simulations demonstrated to be the best for the Class-FG amplifier.

The device has been modelled in-house by a full nonlinear model, whose topology is depicted in Figure 4.57, after a characterisation procedure based on the extraction of multi-bias S-parameters and on pulsed-DC measurements [42].

For the design of the two amplifiers, the choice of the fundamental-frequency output termination has been optimised by means of the technique in [43] and an input matching network has been synthesised to get maximum input power transfer at large-signal (conjugate large-signal input match) and to generate, if necessary, the drain current harmonic components with the appropriate phase relationships, thus implementing the considerations performed in the previous section.

In particular, for the tuned load amplifier design, the input capacitor C_{gs} whose terminal voltage directly controls the output current source, is nearly short-circuited for the

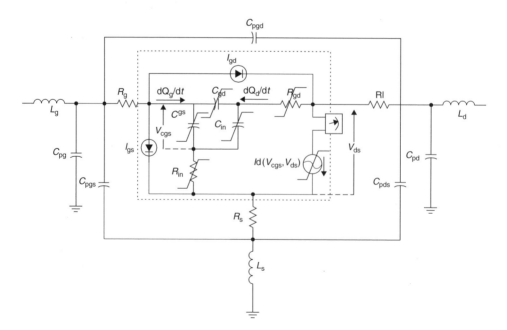

Figure 4.57 Device nonlinear model

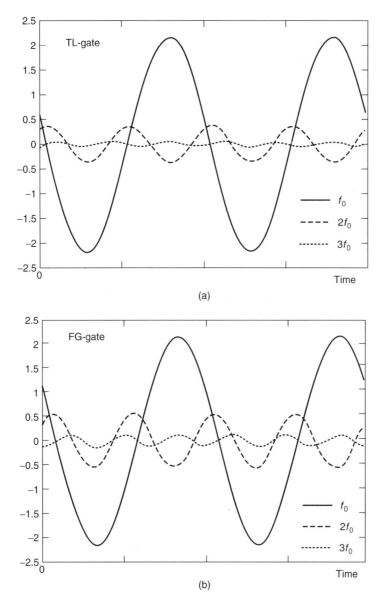

Figure 4.58 Simulated intrinsic gate voltage components at 1-dB compression point for TL- and Class-FG-designed amplifiers

harmonic components at $2f_0$ and $3f_0$, while for the Class-FG amplifier, it is properly terminated to allow the multi-harmonic manipulation, that is, to increase the input harmonic content, as depicted in Figure 4.58. The two solutions have been obtained loading the input circuit by an almost open (TL approach) or short-circuit (Class-FG approach) external terminations at $2f_0$ and $3f_0$. Such loads are transformed by the parasitic network in the proper short- (TL approach) or open-circuit (Class-FG approach) loading across C_{gs}.

In the same way, the output matching networks are designed following different criteria for each amplifier.

For the TL amplifier, the output network actually shorts the harmonic components of the intrinsic drain current at $2f_0$ and $3f_0$, so obtaining an almost-sinusoidal voltage output waveform. On the contrary, for the Class-FG amplifier, the second and third drain current harmonics are resistively terminated in order to shape the drain voltage waveform as described in Section 4.4.4.

With reference to Figure 4.57 and accounting for the consideration in the previous sections, the values to be synthesised at the intrinsic drain terminals at $f_0, 2f_0$ and $3f_0$ are summarised in Table 4.2 for both the TL and FG amplifiers.

It is to underline that, for the latter case, the proper phase relationship between the second- and third-harmonic voltage components and the fundamental one has been obtained by means of both the input and output harmonic terminations. This is because of the fact that, in this case, an output harmonic manipulation alone is not sufficient, while the bare input nonlinearities cannot assure the proper drain current phase relationships. Both the ports have therefore been properly loaded at the relevant frequencies. The Class-FG design has been performed synthesising a purely resistive load at the fundamental frequency and two complex ones at the two higher harmonics, so assuring the proper drain voltage components phase relationships, and has been shown in Figure 4.59.

The values of the optimum terminations at the extrinsic device terminals are summarised in Table 4.3.

In order to synthesise the external loads in Table 4.3, the distributed approach schematically drawn in Figure 4.60 (TL design) and Figure 4.61 (Class-FG design) has been followed for the input and output network respectively.

To explain in detail, the TL output network has been realised by means of two stubs ($\lambda/8$ open-circuit stub and $\lambda/6$ short-circuit stub) controlling second- and third-harmonic terminations respectively. The fundamental load has been synthesised by a standard LC cell. The Class-FG output network is simpler since it is obtained starting with a $\lambda/12$ short-circuit stub controlling third-harmonic component and an LC cell to control fundamental and second-harmonic impedances. Biasing voltages have been applied through the RF signal connectors.

Simulated results are reported in Figure 4.62 and Figure 4.63 where the drain voltage waveforms and the corresponding I/V load curves, computed at -1-dB gain compression point, are indicated as obtained by a full nonlinear simulator (HP-MDS). It is to be noted that the use of a second-harmonic component for the Class-FG PA

Table 4.2 Intrinsic drain termination for the realised tuned load and Class-FG amplifiers

Frequency (GHz)	Tuned load	Class-FG
5	$26.5 + j0.1$	$49.7 + j3.7$
10	$2.4 + j0.1$	$22.7 + j131.1$
15	$2.2 - j0.1$	$8.7 + j11.8$

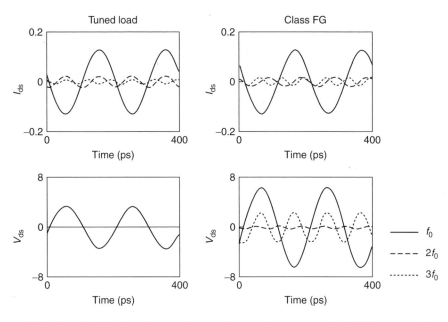

Figure 4.59 Simulated drain voltage and current harmonic components at 1-dB compression point for TL and Class-FG designed amplifiers

Table 4.3 External input and output impedances for the realised tuned load and Class-FG amplifiers

Frequency	Tuned load		Class-FG	
(GHz)	Input	Output	Input	Output
5	$14.4 + 25.9j$	$22.3 + 6.1j$	$14.5 + 25.8j$	$32.6 + 20.5j$
10	$42.9 - 223.4j$	$1.6 - 4.6j$	$0.5 + 14.8j$	$3.4 + 68.1j$
15	$23.9 - 130.4j$	$1.3 - 7.2j$	$2.0 + 26.8j$	$42.9 - 469.2j$

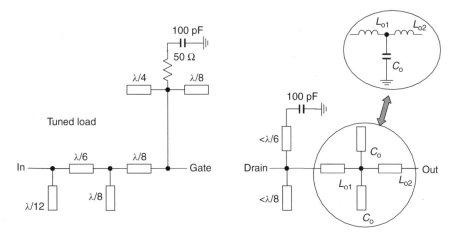

Figure 4.60 Tuned load design network criteria

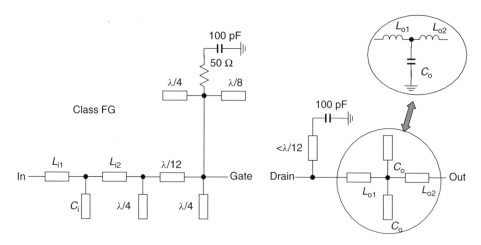

Figure 4.61 Class-FG design network criteria

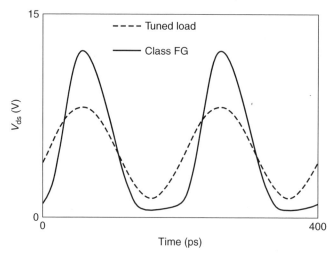

Figure 4.62 Simulated drain voltage waveforms at 1-dB compression point for TL- and Class-FG-designed amplifiers

produces, as expected, a peaking effect on the drain voltage waveform, whose value can be predicted by eq. (4.63) and must be accounted for in order to avoid device breakdown.

In Figure 4.64, the layouts of the two PAs are reported, as realised in hybrid form on alumina substrates. Plots of measured output power and power-added efficiency as functions of the input drive at 5 GHz are shown in Figure 4.65 and Figure 4.66 respectively. As expected, the use of the multi-harmonic manipulation significantly improves the PA's performances. In particular, for an input drive level of 18.3 dBm, a maximum power-added efficiency was obtained for a Class-FG power amplifier, with measured output power and power-added efficiency levels of 25.6 dBm and 60% respectively, corresponding to a drain efficiency of 73.7%.

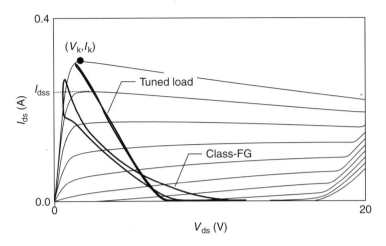

Figure 4.63   Simulated I/V load curves at 1-dB compression point for TL- and Class-FG-designed amplifiers

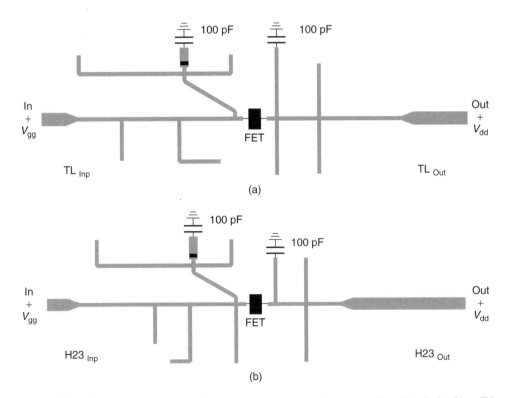

Figure 4.64   5-GHz power amplifiers layouts and mounting schemes: (a) Tuned load; (b) Class-FG

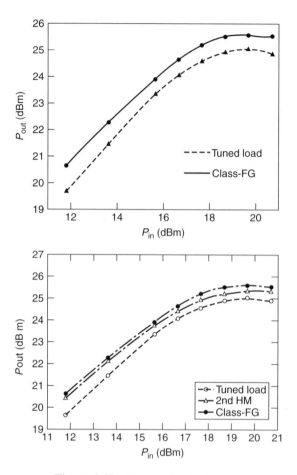

Figure 4.65   Measured output power

The remarkable improvements in output power and power-added efficiency are synthesised in Table 4.4, where a measured improvement factor of 1.43 for Class-FG with respect to TL amplifier is reported. This figure is not far from the theoretically expected value (1.56).

A final statement has to be discussed in order to clarify an ambiguous problem. Harmonic tuning strategies are often referred as detrimental approaches if the linearity of the stage has to be addressed, because of the fact that higher-order harmonics are allowed to circulate in a nonlinear system, so potentially increasing the effects of the nonlinearities. Among the others, one of these effects must be considered carefully for its crucial impact on the power amplifier performances in terms of linearity: the third-order intermodulation product (IMD), since it is often an in-band signal, is difficult to be filtered out. So, the question whether a high-efficiency design based on the harmonic manipulation technique is compatible or not with high linearity performances can be addressed passing through the evaluation of the role that the input and output harmonic terminations have on the overall amplifier features *including IMD*.

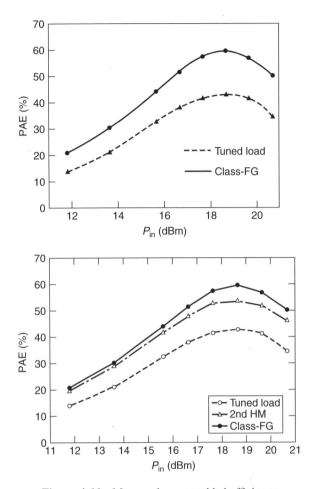

Figure 4.66 Measured power-added efficiency

Table 4.4 Performances of tuned load vs Class-FG PA

	P_{out} measured (dBm)	PAE measured (%)	PAE expected (%)
TL	25.0	42	39
FG	25.6	60	61

In earlier published works [44, 45], the remarkable effect of the *output* termination at the fundamental frequency on *IMD* has been clearly demonstrated, while the corresponding input one appeared to be much less crucial. In a similar way, the impact of a proper second-harmonic injection at the input port on the lowering of *IMD* levels has been, mainly experimentally, demonstrated [36, 46].

A more complete analysis on the effect of the various harmonic terminations, both at the input and output ports, chosen in order to implement different classes of high-efficiency power amplifier has been recently reported [47]. Moreover, the crucial role of the second-harmonic component, generated by the input nonlinearities, has been put into evidence by means of a Volterra series approach [48]. In particular, a proper choice of input terminations assures the requested phase relationship among the first three harmonic components of the input voltage driving signal, which presents the harmonics reported in Figure 4.67, including the corresponding ones for the TL amplifier, reported for sake of comparison.

A first qualitative interpretation is possible, taking into consideration the effects that the multi-harmonic manipulation has on the overall amplifier performances. The presence of the third harmonic in fact improves significantly the output power and the power-added efficiency (see Figures 4.65 and 4.66) in comparison with both the tuned load and the Class-G amplifiers, while worsening the corresponding IMD (35 of the latter one, but still improving the IMD of the former one (Figure 4.68)). Figure 4.68 in particular shows the relevant measured results, plotted versus the output back-off (OBO) for the three amplifiers, obtained by injecting two equal amplitude signals 50 MHz apart, confirming the above considerations.

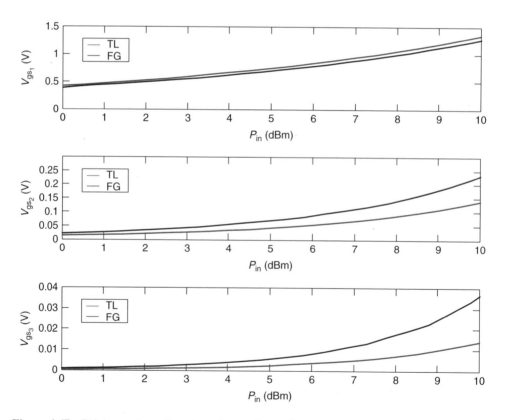

Figure 4.67 Driving voltage first three harmonics vs input power for TL amplifier (shorting input harmonic terminations) and for FG amplifier(using proper second- and third-harmonic input terminations)

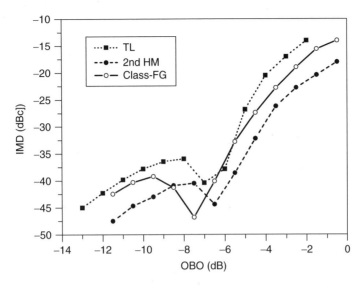

Figure 4.68   IMD measurements vs output back-off (OBO)

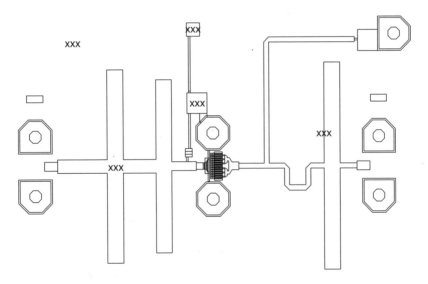

Figure 4.69   Layout of the harmonic manipulated amplifier

In conclusion, for the Class-FG approach, the generation of an input second and third harmonics [47] actually decreases the level of IM distortion, while improving significantly both P_{out} (Figure 4.65) and PAE (Figure 4.66), thus suggesting that a multi-harmonic manipulation strategy may allow an increase in the power amplifier's overall performances through a trade off with linearity.

A final consideration regards the presence of a sweet spot, that is, a null IMD value that is basically unaffected by the multi-harmonic manipulation strategy. It seems to depend on the selected bias point that is the same for the two amplifiers, but further investigations are necessary.

The second set of two amplifiers was realised in monolithic form, utilizing a 0.6-mm power P-HEMT device from Fujitsu. Also in this realisation, one of the amplifiers, used as reference one and named *Amp B*, was designed in order to have the second harmonic (40 GHz) short-circuited at both the input and the output ports. On the contrary, the harmonic manipulated version, named *Amp A* (see Figure 4.69), was designed in order to have the second harmonic not shorted at the input port and *resonated* at the output one [49]. No effort was made to try to control the third harmonic (60 GHz).

The different performances of the two amplifiers are clearly evidenced in Figures 4.70 and 4.71. In particular, the experimental results show that Amp A delivers a power 1.5 dB higher than Amp B, exhibiting at the same time a net increase of power-added efficiency of 10% (Figure 4.64). On the other hand, the frequency behaviour in Figure 4.71 clearly shows the effect of the phase of the second harmonic on power performances, in the case of second-harmonic manipulation as compared with the tuned load case. In particular, in the case of Amp A, at 20 GHz, a peak in power and efficiency can be noted. At the band edges, the efficiency exhibits a strong degradation. Still, the beneficial effect of the proper-phased second harmonic covers a 1-GHz span (5% bandwidth).

Finally, an IM3 test, shown in Figure 4.72, verified the impact of the use of such an approach on the linearity performances of the two amplifiers. The results clearly indicate that no significant differences occurred in the two cases as a function of back-off. Moreover, recalling the fact that Amp A has 1.5 dB higher power output compression, this translates to higher linearity for a given input drive level or allows to operate at lower output back-off, that is, with a higher output power for a fixed level of linearity.

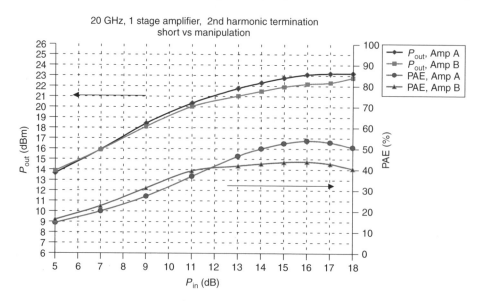

Figure 4.70   20-GHz single-stage amplifier test results: second-harmonic manipulation vs tuned load

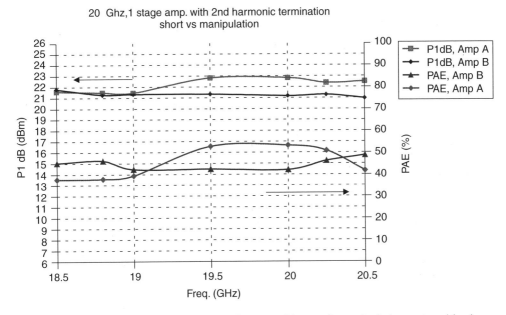

Figure 4.71 Single-stage amplifier test results: second-harmonic manipulation vs tuned load as a function of frequency

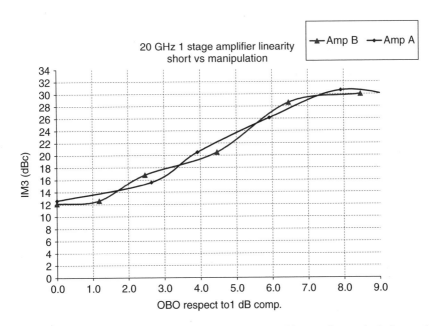

Figure 4.72 20 GHz, IM3 test: comparison between second-harmonic manipulation and tuned load

4.5 BIBLIOGRAPHY

[1] S.C. Cripps, 'A Theory for the Prediction of GaAs FET Load-Pull Power Contours', *IEEE MTT-S Symp. Dig.*, 1983, pp. 221–223.

[2] L.J. Kushner, 'Output performances of idealised microwave power amplifiers', *Microwave J.*, 103–110, October 1989.

[3] L.J. Kushner, 'Estimating power amplifier large signal gain', *Microwave J.*, 87–102, June 1990.

[4] D.M. Snider, 'A theoretical analysis and experimental confirmation of the optimally loaded and overdriven RF power amplifiers', *IEEE Trans. Electron Devices*, **ED-14**(6), 851–857, 1967.

[5] V.J. Tyler, 'A new high-efficiency high power amplifier', *Marconi Rev.*, **21**(130), 96–109, 1958.

[6] C. Duvanaud, S. Dietsche, G. Pataut, J. Obregon, 'High efficient Class F GaAs FET amplifiers operating with very low bias voltages for use in mobile telephones at 1.75 GHz', *IEEE Microwave Guided Wave Lett.*, **3**(8), 268–270, 1993.

[7] F.H. Raab, 'Introduction to Class-F power amplifiers', *RF Des.*, **19**(5), 79–84, 1996.

[8] H.L. Krauss, C.W. Bostian, F.H. Raab, *Solid State Radio Engineering*, John Wiley & Sons, New York (NY), 1980.

[9] A. Inoue, T. Heima, A. Ohta, R. Hattori, Y. Mitsui, 'Analysis of Class-F and inverse Class-F amplifiers', *IEEE MTT-S Symp. Dig.*, 2000, pp. 775–778.

[10] C.J. Wei, P. DiCarlo, Y.A. Tkachenko, R. McMorrow, D. Bartle, 'Analysis and experimental waveform study on inverse Class-F mode of microwave power fets', *IEEE MTT-S Symp. Dig.*, 2000, pp 525–528.

[11] T. Nojima, S. Nishiki, 'Harmonic reaction amplifier – a novel high-efficiency and high-power microwave amplifier', *IEEE MTT-S Symp. Dig.*, 1987, pp. 963–966.

[12] S. Nishiki, T. Nojima, 'High efficiency microwave harmonic reaction amplifier', *IEEE MTT-S Symp. Dig.*, 1988, pp. 1007–1010.

[13] S. Toyoda, 'High efficiency amplifiers', *IEEE MTT-S Symp. Dig.*, San Diego (CA), 1994, pp. 253–256.

[14] N.O. Sokal, A.D. Sokal, 'Class E – a new class of high-efficiency tuned single-ended switching power amplifiers', *IEEE J. Solid-State Circuits*, **SC-10**(3), 168–176, 1975.

[15] F.H. Raab, 'Idealised operation of the Class E tuned power amplifier', *IEEE Trans. Circuits Syst.*, **CAS-24**(12), 725–735, 1977.

[16] T. Sowlati, C.A.T. Salama, J. Sitch, G. Rabjohn, D. Smith, 'Low voltage, high efficiency GaAs Class E power amplifiers for wireless transmitters', *IEEE J. Solid-State Circuits*, **30**(10), 1074–1079, 1995.

[17] B. Molnar, 'Basic limitations on waveforms achievable in single-ended switching mode tuned (Class-E) power amplifiers', *IEEE J. Solid-State Circuits*, **SC-19**(2), 144–146, 1984.

[18] N.O. Sokal, 'Class E high-efficiency power amplifiers, from HF to microwave', *IEEE MTT-S Symp. Dig.*, Baltimore (MD), 1998, pp. 1109–1112.

[19] A.N. Rudiakova, V.G. Krizhanovski, 'Driving waveforms for Class-F power amplifiers', *IEEE MTT-S Symp. Dig.*, Boston (MA), June 2000.

[20] P.M. White, 'Effect of input harmonic terminations on high efficiency Class-B and Class-F operation of PHEMT devices', *IEEE MTT-S Symp. Dig.*, 1998, pp. 1611–1614.

[21] K. Jeon, Y. Kwon, S. Hong, 'Input harmonics control using nonlinear capacitor in GaAs FET power amplifier', *IEEE MTT-S Symp. Dig.*, 1997, pp. 817–820.

[22] S. Watanabe, S. Takanuka, K. Takagi, H. Kuroda, Y. Oda, 'Simulation and experimental results of source harmonic tuning on linearity of power GaAs FET under Class AB operation', *IEEE MTT-S Symp. Dig.*, 1996, pp. 1771–1777.

[23] M. Maeda, H. Masato, H. Takeara, M. Nakamura, S. Morimoto, H. Fujimoto, 'Source second-harmonic control for high efficiency power amplifiers', *IEEE Trans. Microwave Theory Tech.*, **MTT-43**(12), 2952–2958, 1995.

[24] S.R. Mazumder, A. Azizi, F.E. Gardiol, 'Improvement of a Class-C transistor power amplifier by second-harmonic tuning', *IEEE Trans. Microwave Theory Tech.*, **MTT-27**(5), 430–433, 1979.

[25] P. Colantonio, F. Giannini, G. Leuzzi, E. Limiti, 'High efficiency low-voltage power amplifier design by second harmonic manipulation', *Int. J. RF Microwave Comput.-Aided Eng.*, **10**(1), 19–32, 2000.

[26] T.B. Mader, Z.B. Popovic, 'The transmission-line high-efficiency Class-E amplifier', *IEEE Microwave Guided Wave Lett.*, **5**(9), 290–292, 1995.

[27] S.C. Cripps, *RF Power Amplifiers for Wireless Communications*, Artech House, Boston-London, 1999, Chap. 8.

[28] P. Colantonio, F. Giannini, G. Leuzzi, E. Limiti, 'Very high efficiency power amplifiers: the harmonic manipulation approach', *Microwaves, Radar and Wireless Communications, MIKON 2000*, Wroclaw (Poland), May 2000, pp. 33–46.

[29] P. Colantonio, F. Giannini, G. Leuzzi, E. Limiti, 'Class G approach for high efficiency PA design', *Int. J. RF Microwave Comput.-Aided Eng.*, **10**(6), 366–378, 2000.

[30] P. Colantonio, F. Giannini, G. Leuzzi, E. Limiti, 'On the Class-F power amplifier design', *Int. J. RF Microwave Comput.-Aided Eng.*, **9**(2), 129–149, 1999.

[31] P. Colantonio, F. Giannini, E. Limiti, G. Saggio, 'Experimental performances of 5 GHz harmonic-manipulated high efficiency microwave power amplifiers', *Electron. Lett.*, **36**(9), 800–801, 2000.

[32] N.O. Sokal, 'Class-E switching-mode high-efficiency tuned RF/microwave power amplifier: improved design equations', *IEEE MTT-S Symp. Dig.*, 2000, pp. 779–782.

[33] T.B. Mader, E.W. Bryerton, M. Markovic, M. Forman, Z. Popovic, 'Switched-mode high-efficiency microwave power amplifiers in a free-space power-combiner array', *IEEE Trans. Microwave Theory Tech.*, **MTT-46**(10), 1391–1398, 1998.

[34] F.H. Raab, 'Class-F power amplifiers with maximally flat waveforms', *IEEE Trans. Microwave Theory Tech.*, **MTT-45**(11), 2007–2012, 1997.

[35] P. Colantonio, F. Giannini, G. Leuzzi, E. Limiti, 'Multi harmonic manipulation for highly efficient microwave power amplifiers', *Int. J. RF Microwave Comput.-Aided Eng.*, **11**(6), 366–384, 2001.

[36] P. Colantonio, F. Giannini, G. Leuzzi, E. Limiti, 'Input/Output optimum 2nd harmonic terminations in low-voltage high-efficiency power amplifiers', *Proc. of the 10th MICROCOLL*, Budapest (Hungary), Mar. 1999, pp. 401–406.

[37] P. Colantonio, F. Giannini, G. Leuzzi, E. Limiti, 'Improving performances of low-voltage power amplifiers by second-harmonic manipulation', *Proc. of the GAAS '98*, Amsterdam (The Netherlands), Oct. 1998.

[38] P. Colantonio, A. Ferrero, F. Giannini, E. Limiti, V. Teppati. 'Harmonic load/source pull strategies for high efficiency PAs design', to be published on IEEE MTT.

[39] J.J. Bussgang, L. Herman, J.W. Graham, 'Analysis of nonlinear systems with multiple inputs', *IEEE Proc.*, **62**(8), 1088–1119, 1974.

[40] P. Colantonio, F. Giannini, G. Leuzzi, E. Limiti, 'Harmonic tuned PAs design criteria', *IEEE MTT-S Int. Microwave Symp. Dig.*, Vol. **3**, Seattle (WA), June 2002, pp. 1639–1642.

[41] T.M. Scott, 'Tuned power amplifiers', *IEEE Trans. Circuit Theory*, **CT-11**, 385–389, 1964.

[42] G. Leuzzi, A. Serino, F. Giannini, S. Ciorciolini, 'Novel nonlinear equivalent-circuit extraction scheme for microwave field-effect transistors', *Proc. of the 25th European Microwave Conf.*, Bologna (Italy), Sept. 1995, pp. 548–552.

[43] P. Colantonio, F. Giannini, G. Leuzzi, E. Limiti, 'Direct-synthesis design technique for non-linear microwave circuits', *IEEE Trans. Microwave Theory Tech.*, **MTT-43**(12), 2851–2855, 1995.

[44] R.A. Minasian, 'Intermodulation distortion analysis of MESFET amplifiers using the Volterra series representation', *IEEE Trans. Microwave Theory Tech.*, **MTT-28**(1), 1–8, 1980.

[45] R.S. Tucher, C. Rauscher, 'Modelling the third-order intermodulation-distortion properties of GaAs FET', *Electron. Lett.*, **3**, 508–510, 1977.

[46] M.R. Moazzam, C.S. Aitchison, 'A low third order intermodulation amplifier with harmonic feedback circuitry', *IEEE MTT-S Int. Microwave Symp. Dig.*, 1996, pp. 827–830.

[47] P. Colantonio, F. Giannini, G. Leuzzi, E. Limiti, 'IMD performances of harmonic-tuned microwave power amplifiers', *Proc. of the European Gallium Arsenide Applications Symposium*, Paris (France), Oct. 2000, pp. 132–135.

[48] P. Colantonio, F. Giannini, G. Leuzzi, E. Limiti, 'High-efficiency low-IM microwave PA design', *IEEE MTT-S Int. Microwave Symp. Dig.*, Vol. **1**, Phoenix (AZ), May 2001, pp. 511–514.

[49] A. Betti-Berutto, T. Satoh, C. Khandavalli, F. Giannini, E. Limiti, 'Power amplifier second harmonic manipulation: mmWave application and test results', *Proc. of the European Gallium Arsenide Applications Symposium*, Munich (Germany), Oct. 1999, pp. 281–285.

5

Oscillators

5.1 INTRODUCTION

In this introduction, a short description of the oscillatory circuits more commonly used in microwave circuits is given, and a brief recapitulation of the main methods available for 'unstable' circuit design is provided.

Oscillators can, in principle, be considered as linear circuits, since an instability giving rise to an oscillatory behaviour, for instance sinusoidal, is a linear phenomenon. In fact, most oscillators are designed by means of linear concepts and tools, and their performances are satisfactory, at least for basic applications. However, many oscillatory performances have an intrinsically nonlinear nature, and they are becoming increasingly important in microwave applications. First, the amplitude of the oscillation cannot be predicted by linear considerations only, and also the frequency of oscillation is often not accurately predicted; however, simple empirical considerations can yield a reasonable estimation of the power being produced by the oscillator, and the use of high-Q resonators can force the frequency to be very close to the desired value. Nonetheless, a fully nonlinear method can give a better and more accurate evaluation of the actual performances of the oscillator, ensuring a first-pass design. Still more important, there are phenomena that can be described only by means of purely nonlinear considerations. The circuits that exploit such features are becoming increasingly important in microwave systems: for instance, injection locking of an oscillator, which is fundamental in an oscillator array; phase-noise reduction for accurate phase modulation/demodulation; subharmonic generation for phase-locked loops; chaos prediction for chaotic communication or for chaos avoidance. In this chapter, the linear conditions for stability and oscillation are first recalled; then, methods for large-signal behaviour prediction are briefly summarised. The most common and practical fully nonlinear analysis methods that are becoming increasingly important for accurate oscillator design are then reviewed. Methods for noise evaluation, mostly of nonlinear nature in oscillators, are also briefly discussed together with the guidelines for low phase-noise oscillator design. Stability in nonlinear regime for the design of general microwave circuits free of spurious oscillations of nonlinear origin or for the design of intentionally unstable nonlinear circuits, as frequency dividers and chaotic oscillators, and an overview of frequency locking in microwave oscillators are treated in Chapter 8.

Nonlinear Microwave Circuit Design F. Giannini and G. Leuzzi
© 2004 John Wiley & Sons, Ltd ISBN: 0-470-84701-8

5.2 LINEAR STABILITY AND OSCILLATION CONDITIONS

In this paragraph, the stability or instability of linear circuits are described as a preliminary step for both nonlinear oscillation design and for nonlinear stability determination.

The behaviour of a linear autonomous network, that is, a network without external signals, is represented by a homogeneous system of linear equations. The standard case in electronic circuits, however, involves a nonlinear network including solid-state nonlinear components (diodes, transistors) biased by one or more power supplies establishing an operating DC point. The operating or quiescent point is usually found by approximate graphical methods (load line), by approximate nonlinear analysis making use of simple models for the nonlinear device(s) or by accurate numerical nonlinear network analysis, usually by means of a CAD program. Once the quiescent point is determined, the circuit is linearised and linear parameters are evaluated, as for instance the hybrid model, the Giacoletto model, or whatever equivalent (see Chapter 3), or by black-box data as scattering parameters, usually found by direct measurement. As long as any RF signal establishing itself in the circuit remains small, that is, as long as its amplitude does not exceed the range for which the linearisation holds, this is accurate enough for the analysis and design of the RF behaviour of the circuit. In this paragraph, this hypothesis is assumed to hold and purely linear considerations are made.

The unknowns of the homogeneous (Kirchhoff's) system of equations are voltages, currents, waves or a mixture of these, depending on the type of equations selected. In all cases, a trivial or degenerate solution is always possible when all voltages and currents are zero. This is the solution at all frequencies when the circuit is stable or at all frequencies except one or more than one when the circuit oscillates. At each of these frequencies, the determinant of the system of equations is zero and a non-trivial or non-degenerate solution exists. This or these solutions represent the oscillation(s) in the circuit. Let us assume a nodal analysis of the circuit (KCL), and therefore an admittance-matrix representation of the circuit:

$$\vec{I} = \ddot{Y} \cdot \vec{V} = \vec{0} \qquad (5.1)$$

where $\ddot{Y}$ is an $n \times n$ complex matrix and $\vec{V}, \vec{I}$ and $\vec{0}$ are $n \times 1$ complex vectors of the unknown voltages, of the node currents and the zero vector respectively, for a circuit with n nodes. The admittance matrix $\ddot{Y}$ is a function of the values of the elements of the circuit, both linear (passive elements and parasitic elements of the active device) and linearised (intrinsic elements of the active device); it is also a function of the (angular) frequency ω. The condition for the existence of an oscillation at a generic frequency ω_0 therefore is the scalar equation:

$$\det(\ddot{Y}) = 0 \qquad (5.2)$$

The left-hand side of the equation can be seen as a function of the frequency, which is the unknown of eq. (5.2), for fixed values of the elements of the circuit: this is the case in which an existing circuit is analysed for determination of its stability:

$$\det(\ddot{Y}) = F(\omega_0) = 0 \qquad (5.3)$$

If the equation has no solution for any frequency ω, then the circuit is stable; otherwise, if one or more solutions exist, the circuit will oscillate at all the frequencies solution of the equation. Nothing can be said on the amplitudes of the oscillations or on the existence of other spurious frequencies generated by their interactions.

Otherwise, the left-hand side of eq. (5.2) can be a function of the value(s) of one or more circuit elements, while the frequency has a fixed value ω_0:

$$\det(\overset{\leftrightarrow}{Y}) = G(R, L, C, \ldots) = 0 \tag{5.4}$$

The set of values of the circuit elements that satisfy the equation, if any exists, is the solution to the problem of the design of an oscillator at a given frequency ω_0. In order for the solution to be satisfactory from a practical point of view, it must be verified that eq. (5.3) with the designed values of the circuit elements has no solution for any other frequency ω.

This is complete from a mathematical point of view, but is not practical from the designer's point of view. Therefore, simpler approaches are developed. First of all, the network can be divided into two subnetworks at an arbitrary port (Figure 5.1). The two subnetworks are represented in a nodal approach by two scalar complex admittances.

Systems (5.1) and (5.2) become scalar equations:

$$I_L + I_R = (Y_L + Y_R) \cdot V = 0 \tag{5.5}$$
$$Y_L + Y_R = 0 \tag{5.6}$$

Equation (5.6) is also known as the Kurokawa oscillation condition [1]. It can be shown (e.g. [2–4]) that if eq. (5.6) is satisfied, the whole network oscillates, unless there are unconnected parts of the network. A simple illustration is given here for a cascaded network with two nodes; the two-port network in the middle typically stands for a biased active device (a transistor), while the two one-port networks are the input- and output-matching networks. For the circuit shown in Figure 5.2, eq. (5.1) reads as

$$\left(\begin{bmatrix} Y_{11} & Y_{12} \\ Y_{21} & Y_{22} \end{bmatrix} + \begin{bmatrix} Y_{\text{source}} & 0 \\ 0 & Y_{\text{load}} \end{bmatrix} \right) \cdot \begin{bmatrix} V_1 \\ V_2 \end{bmatrix} = \begin{bmatrix} 0 \\ 0 \end{bmatrix} \tag{5.7}$$

and eq. (5.2) reads as

$$(Y_{11} + Y_{\text{source}}) \cdot (Y_{22} + Y_{\text{load}}) - Y_{12}Y_{21} = 0 \tag{5.8}$$

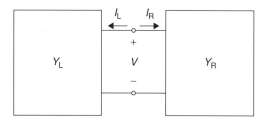

Figure 5.1 A port connecting the two subnetworks of an autonomous linear circuit

Figure 5.2   A cascaded two-node network

Equation (5.8) can be rearranged in two different ways:

$$Y_{\text{source}} = -Y_{11} + \frac{Y_{12}Y_{21}}{Y_{22} - Y_{\text{load}}} = -Y_{\text{in}} \qquad (5.9)$$

$$Y_{\text{load}} = -Y_{22} + \frac{Y_{12}Y_{21}}{Y_{11} - Y_{\text{source}}} = -Y_{\text{out}} \qquad (5.10)$$

Equations (5.9) and (5.10) correspond to the arrangements shown in Figure 5.3.

Equations (5.9) and (5.10) are equivalent, showing that the oscillation condition can be imposed equivalently at the input or at the output port of the active two-port network; we remark that no assumption has been made on the networks.

Let us now come back to eq. (5.6): this complex equation can be split into two real ones:

$$Y_{\text{L}}^{\text{r}} + Y_{\text{R}}^{\text{r}} = 0 \qquad (5.11\text{a})$$

$$Y_{\text{L}}^{\text{j}} + Y_{\text{R}}^{\text{j}} = 0 \qquad (5.11\text{b})$$

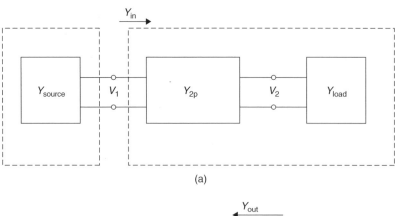

(a)

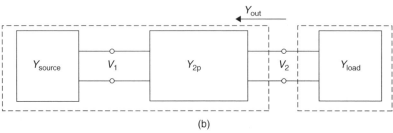

(b)

Figure 5.3   The two-node cascaded network reduced to a one-node network in two different ways

where Y^r and Y^j are the real and imaginary parts respectively of the complex admittance parameter $Y = Y^r + jY^j$. System (5.11) can be interpreted in the following way: one of the two subnetworks must exhibit a negative conductance, whose absolute value must equal the positive conductance of the other subnetwork; moreover, the two susceptances must resonate. In practice, the subnetwork including the active device provides the negative conductance, while the other subnetwork must be designed in order that eq. (5.11) be satisfied.

Equivalently, if Kirchhoff's voltage law impedance parameters are used, the circuit can be represented as in Figure 5.4, and eqs. (5.2) and (5.11) become

$$\det(\overleftrightarrow{Z}) = 0 \tag{5.12}$$

$$Z_L^r + Z_R^r = 0 \tag{5.13a}$$

$$Z_L^j + Z_R^j = 0 \tag{5.13b}$$

Similar considerations as above can be repeated by replacing conductance and susceptance with resistance and reactance respectively.

In microwave circuits, waves and scattering parameters are normally used instead of voltages, currents and impedance parameters. Equivalently, the network shown in Figure 5.3 is modified to the network as shown in Figure 5.5, and eqs. (5.1), (5.2), (5.9), (5.10) and (5.11) become

$$\vec{b} = \overleftrightarrow{\Gamma} \cdot \vec{a} = \vec{a} \tag{5.14}$$

$$\det(\overleftrightarrow{\Gamma} - \overleftrightarrow{1}) = 0 \tag{5.15}$$

$$\Gamma_{source} = \left(S_{11} + \frac{S_{12}S_{21}}{1 - S_{22}\Gamma_{load}} \right)^{-1} = \frac{1}{\Gamma_{in}} \tag{5.16}$$

$$\Gamma_{load} = \left(S_{22} + \frac{S_{12}S_{21}}{1 - S_{11}\Gamma_{source}} \right)^{-1} = \frac{1}{\Gamma_{out}} \tag{5.17}$$

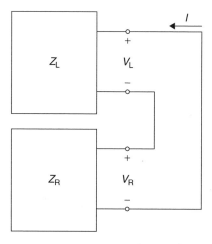

Figure 5.4   A series connection of the two subnetworks of an autonomous linear circuit

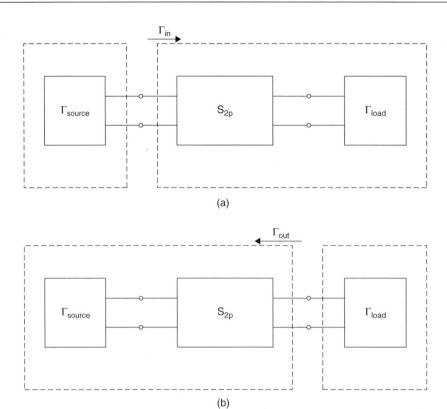

Figure 5.5 The two-node cascaded network reduced to a one-node network, using scattering parameters

$$|\Gamma_L| \cdot |\Gamma_R| = 1 \qquad\qquad (5.18a)$$

$$\angle\Gamma_L + \angle\Gamma_R = 0 \qquad\qquad (5.18b)$$

System (5.18) is equivalent to what is commonly known as the Barkhausen oscillation condition. Equation (5.18a) implies that the wave reflected by one of the two subnetworks (the one including the active device, e.g. the right one) must have an amplitude greater than that of the incident wave

$$|\Gamma_R| > 1 \qquad\qquad (5.19)$$

while the other subnetwork must attenuate the incident wave so that gain and loss of the two subnetworks compensate:

$$|\Gamma_L| = \frac{1}{|\Gamma_R|} \qquad\qquad (5.20)$$

Equation (5.18b) states that phase delays of the two subnetworks must compensate, yielding zero total phase delay.

This approach is well known and very simple; however, this is not the situation the designer actually has to look for. Typically, an oscillator must be designed in such a way that a growing instability is present in the circuit, so that the noise always present in the circuit can increase to a fairly large amplitude at the oscillation frequency only. Therefore, all signals have a time dependence of the form

$$v(t) = v_0 \cdot e^{(\alpha+j\omega)t} \text{ or } a(t) = a_0 \cdot e^{(\alpha+j\omega)t} \tag{5.21}$$

Admittance, impedance or scattering parameters in eqs. (5.2), (5.3) and (5.14) respectively must be computed as functions of the complex Laplace parameter $s = \alpha + j\omega$ instead of the standard (angular) frequency ω. In the analysis case, eq. (5.3) and its equivalent condition for an impedance or wave representation are

$$F_Y(\alpha_0 + j\omega_0) = 0 \tag{5.22a}$$

$$F_Z(\alpha_0 + j\omega_0) = 0 \tag{5.22b}$$

$$F_\Gamma(\alpha_0 + j\omega_0) = 0 \tag{5.22c}$$

The circuit is an oscillator if one or more solutions exist for one or more values of ω_0 with also $\alpha_0 > 0$. Conversely, an oscillator must be designed from eq. (5.4), or its equivalent condition for an impedance or wave representation, so that

$$G_Y(R, L, C, \ldots) = 0 \tag{5.23a}$$

$$G_Z(R, L, C, \ldots) = 0 \tag{5.23b}$$

$$G_\Gamma(R, L, C, \ldots) = 0 \tag{5.23c}$$

are satisfied for the desired value of the Laplace parameter $\omega = \omega_0$ and $\alpha = \alpha_0 > 0$.

A practical problem when using eq. (5.22) or eq. (5.23) instead of eq. (5.3) and eq. (5.4) arises from the fact that CAD programs do not usually compute network parameters in the Laplace domain. Equivalent conditions must therefore be available requiring only standard frequency-domain expressions. Typically, an oscillator includes a resonator that forces the circuit to oscillate near its resonant frequency, more or less independently of the amplitude of voltages and currents in the circuits. Therefore, eq. (5.11b) or eq. (5.18b) mainly involving the frequency-dependent elements can typically be computed in the frequency domain, that is with $\alpha = 0$, to a good degree of approximation. Contrariwise, eq. (5.11a) or eq. (5.18a) mainly involving the negative- and positive-resistance terms typically are sensitive to voltage and current amplitudes in the circuit. From what has been said above, it seems to be a reasonable assumption that the circuit be designed in such a way that the total conductance or resistance be negative or that the total reflection be greater than one:

$$Y_L^r + Y_R^r < 0 \tag{5.24a}$$

$$Z_L^r + Z_R^r < 0 \tag{5.24b}$$

$$|\Gamma_L| \cdot |\Gamma_R| > 1 \tag{5.24c}$$

While conditions (5.24a) and (5.24b) are correct, condition (5.24c) is not generally true; a simple example is sufficient to clarify the point. Let us consider a simple parallel resonant circuit as in Figure 5.6.

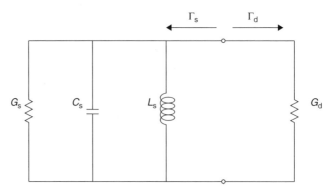

Figure 5.6   A parallel resonant circuit

Kirchhoff's current law at the only node, written in the form of eq. (5.22a), is

$$G_s + G_d + sC + \frac{1}{sL} = 0 \tag{5.25}$$

whence

$$s_0 = -\frac{G_{tot}}{2C} \pm \sqrt{\left(\frac{G_{tot}}{2C}\right)^2 - \frac{1}{LC}} \tag{5.26}$$

where

$$G_{tot} = G_s + G_d \tag{5.27}$$

For a growing oscillation, we must have

$$\alpha_0 > 0 \Rightarrow G_{tot} = G_s + G_d < 0 \tag{5.28}$$

If the left subcircuit is a biased active device behaving as a negative conductance and the right subcircuit is a passive network, so that

$$G_s < 0 \quad G_d > 0 \tag{5.29}$$

we must have

$$|G_s| > G_d \quad \text{or} \quad -G_s > G_d \tag{5.30}$$

This condition gives a growing instability, thus confirming the validity of eq. (5.24a). In particular, if the quality factor (Q) of the circuit is high, that is, if

$$G_{tot} \ll 2\sqrt{\frac{C}{L}} \tag{5.31}$$

the complex Laplace parameter can be approximated by

$$s_0 = \alpha_0 \pm j\omega_0 \cong -\frac{G_{tot}}{2C} \pm j\frac{1}{\sqrt{LC}} \tag{5.32}$$

and the oscillation frequency is

$$\omega_0 = \frac{1}{\sqrt{LC}} \tag{5.33}$$

independent of the resistive elements in the circuit.

Let us now check whether eq. (5.24c) is also valid. If condition (5.31) holds, the reflection coefficients of the left and right subcircuits in Figure 5.6 can be approximated by

$$\Gamma_s \cong \frac{G_0 - G_s}{G_0 + G_s} \quad \Gamma_d \cong \frac{G_0 - G_d}{G_0 + G_d} \tag{5.34}$$

where $G_0 = \dfrac{1}{Z_0} = 20$ mS. It is by no means true that if eq. (5.30) holds, then eq. (5.24c) is satisfied. Let us show this by assigning actual values to the resistive elements of the circuit. For instance, we can take

$$G_s = -30 \text{ mS} \quad G_d = 25 \text{ mS} \tag{5.35}$$

which gives growing instability since eq. (5.30) is satisfied. For the reflection coefficients, we have

$$\Gamma_s = -5 \quad \Gamma_d = \tfrac{1}{9} \quad |\Gamma_s| \cdot |\Gamma_d| = \tfrac{5}{9} < 1 \tag{5.36}$$

and eq. (5.24c) is not satisfied. If we take

$$G_s = -15 \text{ mS} \quad G_d = 10 \text{ mS} \tag{5.37}$$

eq. (5.30) is again satisfied, and the circuit is unstable. For the reflection coefficients, we have

$$\Gamma_s = 7 \quad \Gamma_d = \tfrac{1}{3} \quad |\Gamma_s| \cdot |\Gamma_d| = \tfrac{7}{3} > 1 \tag{5.38}$$

Equation (5.24c) is now satisfied. It is therefore clear that eq. (5.22b) is not correct.

The above considerations can be repeated for a series resonant circuit as in Figure 5.7. We get

$$s_0 = -\frac{R_{\text{tot}}}{2L} \pm \sqrt{\left(\frac{R_{\text{tot}}}{2L}\right)^2 - \frac{1}{LC}} \tag{5.39}$$

where

$$R_{\text{tot}} = R_s + R_d \tag{5.40}$$

For a growing oscillation, we must have

$$\alpha_0 > 0 \Rightarrow R_{\text{tot}} = R_s + R_d < 0 \tag{5.41}$$

If

$$R_s < 0 \quad R_d > 0 \tag{5.42}$$

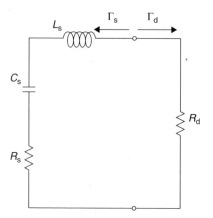

Figure 5.7 A series resonant circuit

we must have

$$|R_s| > R_d \quad \text{or} \quad -R_s > R_d \tag{5.43}$$

This condition gives a growing instability. In particular, if the Q of the circuit is high, that is, if

$$R_{tot} \ll 2\sqrt{\frac{L}{C}} \tag{5.44}$$

the complex Laplace parameter can be approximated by

$$s_0 = \alpha_0 \pm j\omega_0 \cong -\frac{R_{tot}}{2L} \pm j\frac{1}{\sqrt{LC}} \tag{5.45}$$

and the oscillation frequency is again

$$\omega_0 = \frac{1}{\sqrt{LC}} \tag{5.46}$$

independent of the resistive elements in the circuit. Equation (5.34) becomes

$$\Gamma_s \cong \frac{R_s - R_0}{R_s + R_0} \quad \Gamma_d \cong \frac{R_d - R_0}{R_d + R_0} \tag{5.47}$$

It can be shown that the origin of the ambiguity in the instability criterion for the amplitudes of the reflection coefficients (5.24c) lies in the range of values that the circuit conductances or resistances assume with respect to the normalising conductance or resistance respectively. A practical arrangement for working with reflection coefficients in the frequency domain with a design criterion similar to eq. (5.24a) or eq. (5.24b) is the following. The instability criterion is computed at the port where the external resistive load is connected to the oscillator (Figure 5.8).

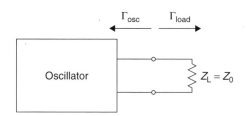

Figure 5.8   An oscillator partitioned at the output port

In this case, no ambiguity is present. Since the load is real, for stable oscillations the phase of Γ_{osc} must be zero and its amplitude must be infinite so that

$$\Gamma_{osc} \cdot \Gamma_{load} = 1 \tag{5.48}$$

This situation corresponds to $Z_{osc} = -Z_0 = -50 \ \Omega$. For growing instability, the oscillator can be approximated as a series or parallel resonator in the vicinity of the resonant frequency ω_0 where $\angle\Gamma_{osc}(\omega_0) = 0$. In the case that the resonance is a series one, then

$$R_{osc}(\omega_0) < -50 \ \Omega \Rightarrow 1 < \Gamma_{osc}(\omega_0) < \infty \tag{5.49}$$

If the resonance is a parallel one, then from eq. (5.34),

$$G_{osc}(\omega_0) < -20 \ \text{mS} \Rightarrow -\infty < \Gamma_{osc}(\omega_0) < -1 \tag{5.50}$$

The type of resonance is easily evaluated on the Smith Chart if the reflection coefficient is plotted as a function of frequency around ω_0: in a parallel resonance, the impedance or admittance of the oscillator changes from inductive below ω_0 to capacitive above ω_0; the reverse is true for a series resonance. Therefore, four situations are possible: two stable ones and two unstable ones giving rise to a growing oscillation. They are depicted in Figure 5.9 for the sake of illustration.

A more rigorous and general formulation is as follows [5, 6]. Let us come back to the general equation system in eq. (5.1); its determinant in the Laplace domain has been introduced in eq. (5.22). For typical oscillators, a solution $s_0 = \alpha_0 + j\omega_0$ of eq. (5.22) in the complex Laplace plane is located in the vicinity of the frequency ω_1 where the phase of the function $F(j\omega)$ becomes zero (Figures 5.10, 5.11 and 5.12); from an electrical point of view, this corresponds to resonating the circuit reactances computed in periodic regime $s = j\omega$.

We can therefore write the solution of eq. (5.22) as

$$s_0 = \alpha_0 + j\omega_0 = \alpha_0 + j(\omega_1 + \delta\omega) \tag{5.51}$$

where both α_0 and $\delta\omega$ are small compared to ω_1. Expanding in Taylor series $F(s + j\omega)$ around $s = j\omega_1$ we get (see Appendix A.10)

$$F(s + j\omega) \cong F(j\omega_1) - \left.\frac{\partial F(j\omega)}{\partial\omega}\right|_{\omega=\omega_1} (\partial\omega - j\alpha_0) + \cdots = 0 \tag{5.52}$$

Figure 5.9 The four resonance types of an oscillator at its output port: parallel unstable, parallel stable, series unstable and series stable

Equation (5.52) is solved for α_0, yielding

$$\alpha_0 = -F(j\omega_1) \cdot \frac{\text{Im}\left[\left.\dfrac{\partial F(j\omega)}{\partial \omega}\right|_{\omega=\omega_1}\right]}{\left|\left.\dfrac{\partial F(j\omega)}{\partial \omega}\right|_{\omega=\omega_1}\right|^2} \tag{5.53}$$

Four cases are possible, depending on the sign of $F(j\omega_1)$ and $\text{Im}\left[\left.\dfrac{\partial F(j\omega)}{\partial \omega}\right|_{\omega=\omega_1}\right]$; are listed in Table 5.1.

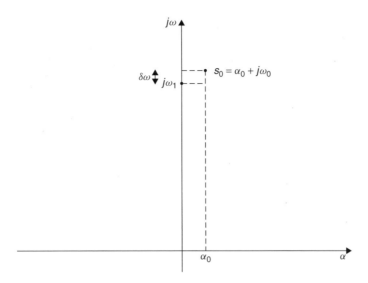

Figure 5.10   The zero of the function $F(s) = F(\alpha + j\omega)$ in the complex Laplace plane

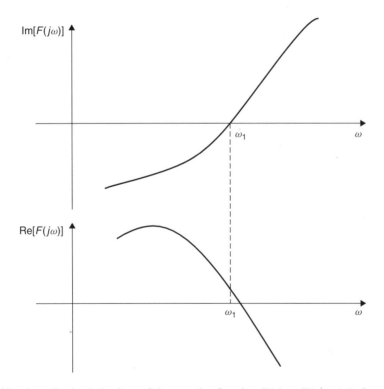

Figure 5.11   A qualitative behaviour of the complex function $F(s) = F(\alpha + j\omega)$ along the imaginary axis of the complex Laplace plane

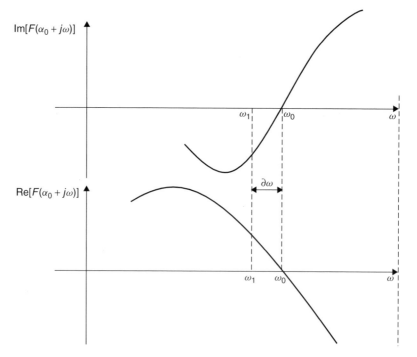

Figure 5.12 A qualitative behaviour of the complex function $F(s) = F(\alpha + j\omega)$ along the $\alpha = \alpha_0$ line of the complex Laplace plane

Table 5.1 Stability test for an oscillator

$F(j\omega_1)$	$\mathrm{Im}\left[\dfrac{\partial F(j\omega)}{\partial \omega}\bigg|_{\omega=\omega_1}\right]$	Stability
Positive	Positive	Stable
Positive	Negative	Unstable
Negative	Negative	Stable
Negative	Positive	Unstable

From an electrical point of view, the four cases correspond to the series and parallel resonances described above.

Let us illustrate this result with our example parallel resonant oscillating circuit in Figure 5.6. The determinant of Kirchhoff's equation system in our case becomes (eq. (5.25)):

$$F(s) = G_s + G_d + sC + \frac{1}{sL} = G_{\text{tot}} + sC + \frac{1}{sL} = 0 \qquad (5.54)$$

The determinant computed for $s = j\omega$ is

$$F(j\omega) = G_{\text{tot}} + j\omega C + \frac{1}{j\omega L} = G_{\text{tot}} + j\left(\omega C - \frac{1}{\omega L}\right) \qquad (5.55)$$

Its imaginary part becomes zero for

$$\omega_1 = \frac{1}{\sqrt{LC}} \tag{5.56}$$

From the above,

$$\frac{\partial F}{\partial \omega} = -\frac{2\alpha\omega}{(\alpha^2 - \omega^2)^2 + 4\alpha^2\,\omega^2} \cdot \frac{1}{L} + j\left(C - \frac{\alpha^2 - \omega^2}{(\alpha^2 - \omega^2)^2 + 4\alpha^2\,\omega^2} \cdot \frac{1}{L}\right) \tag{5.57}$$

$$\left.\frac{\partial F}{\partial \omega}\right|_{s=j\omega_1} = j\left(C + \frac{1}{\omega_1^2} \cdot \frac{1}{L}\right) = j2C \tag{5.58}$$

Therefore,

$$\alpha_0 = -\frac{G_{\text{tot}}}{2C} \tag{5.59}$$

Referring to Table 5.1, we have

$$\text{Im}\left[\left.\frac{\partial F(j\omega)}{\partial \omega}\right|_{\omega=\omega_1}\right] = 2C > 0 \tag{5.60}$$

always; therefore, if $G_{\text{tot}} > 0$, then $F(j\omega_1) > 0$, and the circuit is stable. Otherwise, when $G_{\text{tot}} < 0$, then $F(j\omega_1) < 0$, and the circuit is unstable, as found earlier. The real part of the Laplace constant is also evaluated from eq. (5.53), with the same result as above.

5.3 FROM LINEAR TO NONLINEAR: QUASI-LARGE-SIGNAL OSCILLATION AND STABILITY CONDITIONS

In this paragraph, the linear stability and oscillation conditions so far described are modified to take into account the nonlinearity, that is, the dependence of the oscillator parameters on the amplitude of the signal.

The stability and oscillation conditions given so far are valid only in linear regime. However, the behaviour of actual oscillating circuits always involves the nonlinear characteristics of the active device. A rigorous study requires the use of full-nonlinear analysis methods that will be described in Section 5.4. However, many conclusions of the previous paragraph are extended to the nonlinear regime by means of simple considerations requiring a general knowledge of the dependence of circuit parameters on the amplitude of the signal.

First of all, let us extend the stability and oscillation considerations described in the previous paragraph to a circuit with parameters varying with the amplitude of the signal. Let us first consider the reflection coefficients (e.g. those shown in Figure 5.5) dependent on the amplitude of the signal within the oscillator. Intuition and experience suggest that a growing instability will not grow forever but will saturate at a certain amplitude, because of the limitations of the active device. Therefore, let us assume that the steady-state periodic regime has been attained when the signal within the oscillator has reached the equilibrium amplitude A_0 with oscillation frequency ω_0. Equations (5.11), (5.13) and (5.18) are now rewritten, in complex form, and explicitly indicate the network parameter dependence on signal amplitude:

$$Y_s + Y_d = Y_{tot}(A_0, \omega_0) = 0 \qquad (5.61a)$$

$$Z_s + Z_d = Z_{tot}(A_0, \omega_0) = 0 \qquad (5.61b)$$

$$\Gamma_s \cdot \Gamma_d = \Gamma_{tot}(A_0, \omega_0) = 1 \qquad (5.61c)$$

We remark that a feedback oscillator can be studied in the same way as a negative-resistance oscillator by writing

$$A \cdot \beta = T(A_0, \omega_0) = 1 \qquad (5.62)$$

It is now interesting to derive a formal stability criterion [1, 4, 7]. Let us apply a small perturbation δA to the amplitude of the oscillating signal; if the oscillation is stable, the amplitude will come back to the same value as it was before the perturbation. Since the perturbation is small, the perturbed admittance in eq. (5.61a) can be expanded in Taylor series to the first order, and eq. (5.61a) becomes

$$Y_{tot}(A_0, j\omega_0) + \left. \frac{\partial Y_{tot}(A, s)}{\partial A} \right|_{\substack{A = A_0 \\ s = j\omega_0}} \cdot \delta A + \left. \frac{\partial Y_{tot}(A, s)}{\partial s} \right|_{\substack{A = A_0 \\ s = j\omega_0}} \cdot \delta s + \cdots = 0 \qquad (5.63)$$

where $s_0 = j\omega_0$ and $\delta s = \delta\alpha + j\delta\omega$. The perturbed complex Laplace parameter becomes

$$s_0' = s_0 + \delta s = \delta\alpha + j(\omega_0 + \delta\omega) \qquad (5.64)$$

Since $Y_{tot}(A_0, \omega_0) = 0$, we have

$$\left. \frac{\partial Y_{tot}(A, s)}{\partial A} \right|_{\substack{A = A_0 \\ s = j\omega_0}} \cdot \delta A + \left. \frac{\partial Y_{tot}(A, s)}{\partial s} \right|_{\substack{A = A_0 \\ s = j\omega_0}} \cdot \delta s \cong 0 \qquad (5.65)$$

If the real part of the perturbed Laplace parameter is negative, the amplitude of the perturbation will decrease to zero and the oscillation will come back to the previous state. Equation (5.65) becomes

$$\left. \frac{\partial Y_{tot}(A, s)}{\partial A} \right|_{\substack{A = A_0 \\ s = j\omega_0}} \cdot \delta A - j \left. \frac{\partial Y_{tot}(A, s)}{\partial \omega} \right|_{\substack{A = A_0 \\ s = j\omega_0}} \cdot \delta s \cong 0 \qquad (5.66)$$

whence

$$\delta s = \delta\alpha + j\delta\omega \cong -j \frac{\dfrac{\partial Y_{tot}}{\partial A}}{\dfrac{\partial Y_{tot}}{\partial \omega}} \cdot \delta A = -j \frac{\dfrac{\partial Y_{tot}}{\partial A} \cdot \dfrac{\partial Y_{tot}^*}{\partial \omega}}{\left| \dfrac{\partial Y_{tot}}{\partial \omega} \right|^2} \cdot \delta A \qquad (5.67)$$

where the partial derivatives are computed in the unperturbed oscillation state, that is for $s = j\omega_0$, and the star denotes complex conjugation. By dividing eq. (5.67) into real and imaginary parts we get

$$\delta\alpha \cong \frac{\dfrac{\partial Y_r}{\partial\omega}\dfrac{\partial Y_i}{\partial A} - \dfrac{\partial Y_r}{\partial A}\dfrac{\partial Y_i}{\partial\omega}}{\left|\dfrac{\partial Y_{tot}}{\partial\omega}\right|^2} \cdot \delta A \tag{5.68a}$$

$$\delta\omega \cong -\frac{\dfrac{\partial Y_r}{\partial A}\dfrac{\partial Y_r}{\partial\omega} + \dfrac{\partial Y_i}{\partial A}\dfrac{\partial Y_i}{\partial\omega}}{\left|\dfrac{\partial Y_{tot}}{\partial\omega}\right|^2} \cdot \delta A \tag{5.68b}$$

where

$$\frac{\partial Y_{tot}}{\partial\omega} = \frac{\partial Y_r}{\partial\omega} + j\frac{\partial Y_i}{\partial\omega} \quad \frac{\partial Y_{tot}}{\partial s} = \frac{\partial Y_r}{\partial s} + j\frac{\partial Y_i}{\partial s} \tag{5.69}$$

Therefore, the stability condition is

$$\delta\alpha < 0 \Rightarrow \frac{\partial Y_r}{\partial A}\frac{\partial Y_i}{\partial\omega} - \frac{\partial Y_r}{\partial\omega}\frac{\partial Y_i}{\partial A} > 0 \tag{5.70}$$

If eq. (5.70) is satisfied, the oscillation is stable. Similarly,

$$\frac{\partial Y_r}{\partial A}\frac{\partial Y_r}{\partial\omega} + \frac{\partial Y_i}{\partial A}\frac{\partial Y_i}{\partial\omega} = 0 \Rightarrow \delta\omega = 0 \tag{5.71}$$

If the eq. (5.71) is satisfied, the frequency of oscillation is stable and will not change for a small perturbation of the amplitude of the oscillating signal. The smaller the expression at the left-hand side of the first of eq. (5.71), the smaller the sensitivity of the oscillation frequency to an amplitude perturbation.

Equations (5.70) and (5.71) can be equivalently rewritten in terms of the impedance and reflection coefficient representation of the network, with equivalent results.

We can also rewrite eq. (5.70) in the following form:

$$0° < \text{Arg}\left[\left.\frac{\partial T}{\partial\omega}\right|_{\substack{A = A_0 \\ s = j\omega_0}}\right] - \text{Arg}\left[\left.\frac{\partial T}{\partial A}\right|_{\substack{A = A_0 \\ s = j\omega_0}}\right] < 180° \tag{5.72}$$

where the function T is any of the network functions in eq. (5.61) or eq. (5.62). Geometrically, eq. (5.72) can be interpreted as follows: the oscillatory state is stable if the angle between the derivative of the function $T(A, \omega)$ with respect to frequency and the derivative with respect to amplitude is greater that $0°$ and less than $180°$ in the complex plane of the function $T(A, \omega)$, when taken counterclockwise.

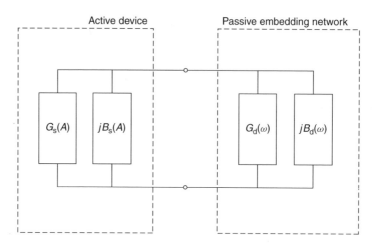

Figure 5.13   A parallel resonant circuit partitioned into an active and a passive subcircuit

Let us illustrate the rule for a parallel resonant circuit, as partitioned in Figure 5.1, and repeated in Figure 5.13, where one subcircuit includes the nonlinear amplitude-dependent active device and the other the linear amplitude-independent passive embedding network. We can assume that the reactive part of the active device is frequency-independent when the embedding network has a very strong frequency dependence, so that the former can be neglected in the narrowband near the resonant frequency; this is usually the case for single-frequency oscillators that include a high-Q resonator in the passive network for frequency stabilisation.

The admittances are

$$Y_s(A) = G_s(A) + j B_s(A) \quad Y_d(\omega) = G_d(\omega) + j B_d(\omega) \quad Y_{tot}(A, \omega) = Y_s(A) + Y_d(A) \tag{5.73}$$

The oscillation condition requires that

$$Y_{tot}(A_0, \omega_0) = Y_s(A_0) + Y_d(\omega_0) = 0 \tag{5.74}$$

In real and imaginary parts,

$$G_{tot}(A_0, \omega_0) = G_s(A_0) + G_d(\omega_0) = 0 \quad B_{tot}(A_0, \omega_0) = B_s(A_0) + B_d(\omega_0) = 0 \tag{5.75}$$

This formula can be represented graphically on the complex plane of the admittance (Figure 5.14). Two curves are traced on the plane, the former being the admittance locus of the device as a function of the amplitude of the signal and the latter being the negated admittance locus of the embedding network as a function of the frequency [1].

Equation (5.75) tells us that the amplitude and frequency of the oscillation are found at the intersection of the two curves. The stability condition of eq. (5.70) now reads as

$$\frac{\partial G_s}{\partial A} \frac{\partial B_d}{\partial \omega} - \frac{\partial G_d}{\partial \omega} \frac{\partial B_s}{\partial A} > 0 \tag{5.76}$$

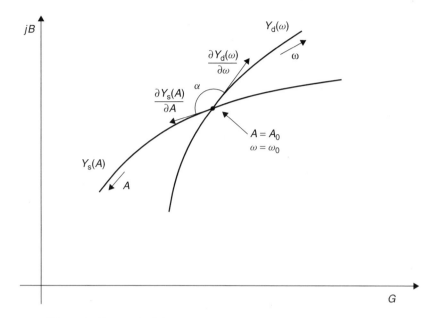

Figure 5.14 Loci of the device and embedding network admittances

and it can be interpreted graphically that the angle α between the vector of the derivative of the two curves with respect to the curve variable have a value between $0°$ and $180°$, with maximum stability when $\alpha = 90°$.

The same considerations can be repeated if the two subnetworks are represented by impedance parameters or reflection coefficients, leading to similar results. The results for the impedance parameter representation are easily deduced by the use of duality. For the reflection coefficient, the oscillation condition reads

$$\Gamma_d(\omega_0) = \frac{1}{\Gamma_s(A_0)} \tag{5.77}$$

For the derivation of the stability condition, we can remark that the reflection coefficient is related to admittance and impedance parameters by conformal transformations; therefore, angles are preserved, and so the requirement that (Figure 5.15)

$0° < \beta < 180°$ and $\beta = 90°$ for maximum stability.

This can be obtained by analytical calculations also [8]. We first write the reflection coefficients as

$$\Gamma_s(A) = \rho \cdot e^{j\vartheta} \quad \Gamma_d(\omega) = \eta \cdot e^{j\xi} \tag{5.78}$$

The stability condition eq. (5.70) reads as

$$\eta \cdot \frac{d\rho}{dA} \cdot \frac{d\xi}{d\omega} - \rho \cdot \frac{d\vartheta}{dA} \cdot \frac{d\eta}{d\omega} > 0 \tag{5.79}$$

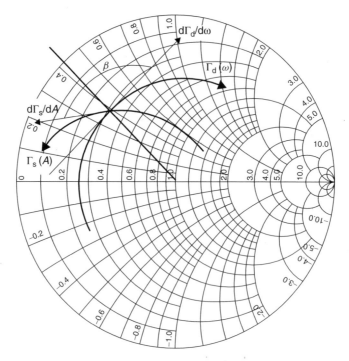

Figure 5.15 Loci of the device and embedding network reflection coefficients for stability criterion definition

If we define

$$\frac{1}{\Gamma_s(A)} = \zeta \cdot e^{j\psi} = \frac{1}{\rho} \cdot e^{-j\vartheta} \tag{5.80}$$

the stability condition becomes

$$\frac{d\zeta}{dA} \cdot \frac{d\xi}{d\omega} - \frac{d\eta}{d\omega} \cdot \frac{d\psi}{dA} < 0 \tag{5.81}$$

Let us illustrate again the rule with the oscillator as partitioned in Figure 5.8. In small-signal linear regime, the circuit has already been shown to present a growing instability when the reflection coefficient of the oscillator at the output port Γ_{osc} has a parallel resonance for $\omega = \omega_0$ with $-\infty < \Gamma_{osc} < -1$ (Figure 5.9a). This corresponds to a small-signal negative conductance:

$$G_{osc,ss} > -20 \text{ mS} \tag{5.82}$$

As the signal grows in amplitude, the negative conductance provided by the oscillator increases (algebraically) as the power amplifying capability of the active device tends to saturate. A qualitative behaviour of the output conductance of the oscillator as a function of the signal amplitude is shown in Figure 5.16.

Correspondingly, the curve traced by the reflection coefficient of the oscillator at the output port as a function of frequency shifts outwards from the centre of the Smith

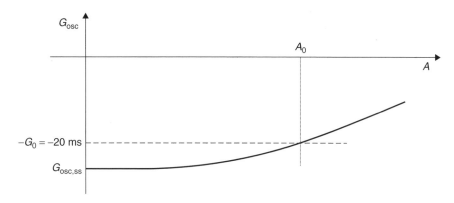

Figure 5.16 Qualitative behaviour of the output negative conductance of the oscillator as a function of the amplitude of the oscillating signal

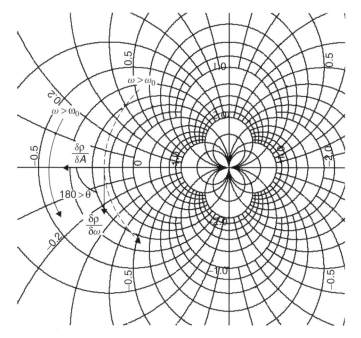

Figure 5.17 The reflection coefficient of the oscillator at the output port for increasing signal amplitude

Chart (Figure 5.17). When the conductance reaches the value $G_{osc} = -20$ mS, the curve passes through the point $\Gamma = -\infty$ and oscillation becomes stable with an amplitude A_0. If the reactances of the circuit are approximately independent of the signal amplitude, that is, if the (passive) resonator has a high quality factor, then the frequency for which $\Gamma_{osc} = -\infty$ is still approximately ω_0, as in small-signal conditions. As can be seen in Figure 5.17, the angle between the derivative of the reflection coefficient with respect to amplitude and that taken with respect to frequency is not far from 90°; this ensures a growing instability all along the path from small signal to steady-state oscillation, and satisfies the stability condition for oscillation once the steady state has been reached.

Let us illustrate the rule with the simple parallel resonant circuit in Figure 5.6, and study it in the time domain [1, 9]; let us also assume that only the total conductance depends on the amplitude of the voltage, while the capacitance and inductance are constant. This is approximately true for a circuit where the negative resistance is provided by an active device: in fact, the reactances in the active device are amplitude dependent; however, if the LC resonator has a high (loaded) quality factor, then the assumption is reasonably accurate. Kirchhoff's equation in the time domain reads as

$$C \cdot \frac{dv(t)}{dt} + G_{tot}(A) \cdot v(t) + \frac{1}{L} \cdot \int v(t) \cdot dt = 0 \qquad (5.83)$$

The oscillator will reach equilibrium when the amplitude is such that

$$G_{tot}(A_0) = 0 \qquad (5.84)$$

Let us assume that the nonlinearities are not too stiff and that the oscillating voltage at equilibrium is almost sinusoidal:

$$v(t) \cong A_0 \cdot \cos(\omega_0 t + \varphi_0) \quad \omega_0 = \frac{1}{\sqrt{LC}} \qquad (5.85)$$

Let us now perturb both amplitude and phase of the oscillating signal

$$v(t) = (A_0 + \delta A(t)) \cdot \cos(\omega_0 t + (\varphi_0 + \delta\varphi(t))) \qquad (5.86)$$

and see which is the evolution in time of the small perturbations. The derivative and integral with respect to time of the perturbed voltage, approximated to the first order in the perturbations, are

$$\frac{dv(t)}{dt} = \frac{dA(t)}{dt}\bigg|_{A=A_0} \cdot \cos(\omega_0 t + \varphi_0) - A_0 \cdot \left(\omega_0 + \frac{d\varphi}{dt}\bigg|_{\varphi=\varphi_0}\right) \cdot \sin(\omega_0 t + \varphi_0) \quad (5.87)$$

$$\int v(t) \cdot dt = \frac{1}{\omega_0^2} \cdot \frac{dA(t)}{dt}\bigg|_{A=A_0} \cdot \cos(\omega_0 t + \varphi_0) + \frac{A_0}{\omega_0} \cdot \sin(\omega_0 t + \varphi_0) - \frac{A_0}{\omega_0^2} \cdot \frac{d\varphi}{dt}\bigg|_{\varphi=\varphi_0}$$
$$\cdot \sin(\omega_0 t + \varphi_0) \qquad (5.88)$$

We also have

$$G_{tot}(A_0 + \delta A(t)) \cong G_{tot}(A_0) + \frac{dG_{tot}}{dA}\bigg|_{A=A_0} \cdot \delta A(t) = \frac{dG_{tot}}{dA}\bigg|_{A=A_0} \cdot \delta A(t) \qquad (5.89)$$

By replacing in eq. (5.87) and eq. (5.88), and separating the terms with sinus and cosinus time dependence, we get

$$\left(C + \frac{1}{\omega_0^2 L}\right) \cdot \frac{d(\delta A)}{dt} + \frac{dG_{tot}}{dA} \cdot \delta A = 0 \qquad (5.90)$$

$$\left(-\omega_0 C + \frac{1}{\omega_0 L}\right) - \left(C + \frac{1}{\omega_0^2 L}\right) \cdot \frac{d\varphi}{dt} = 0 \tag{5.91}$$

Since we are considering the oscillation at equilibrium, eq. (5.85) holds, and we get

$$2C \cdot \frac{d(\delta A)}{dt} + \frac{dG_{tot}}{dA} \cdot \delta A = 0 \tag{5.92}$$

$$2C \cdot \frac{d\varphi}{dt} = 0 \tag{5.93}$$

From eq. (5.92),

$$\delta A(t) = \delta A(t_0) \cdot e^{-\frac{dG_{tot}}{dA} \cdot \frac{t}{2C}} \tag{5.94}$$

The perturbation will vanish exponentially with time, and the amplitude of the oscillation will come back to the equilibrium value if

$$\left. \frac{dG_{tot}(A)}{dA} \right|_{A=A_0} > 0 \tag{5.95}$$

The oscillation will therefore be stable. The amplitude-dependent total conductance usually results from the sum of an amplitude-independent passive positive conductance (the load) and an amplitude-dependent active negative conductance (the biased active device). Typical conductances are as in Figure 5.16. Therefore, the total conductance has a positive derivative with respect to the amplitude of the oscillating signal, and the oscillation is stable. Otherwise, if the negative resistance becomes more negative as the signal grows, the oscillation is not stable and a small perturbation will grow exponentially in time until a stable equilibrium point is reached. An example is shown below for an amplifier with gain expansion.

From eq. (5.93), we see that a perturbation of the phase will not grow or vanish in time. This is natural in a time-invariant autonomous circuit: the time origin of the oscillation has no physical meaning, and the circuit will not react to a shift in time of the oscillating signal.

The result is a particular case of what has been found above. In particular, eq. (5.70) becomes

$$\frac{\partial Y_r}{\partial A} = \frac{dG_{tot}}{dA} \qquad \frac{\partial Y_i}{\partial \omega} = C > 0 \qquad \frac{\partial Y_r}{\partial \omega} = \frac{dG_{tot}}{d\omega} \cong 0$$

$$\frac{\partial Y_i}{\partial A} = \frac{d}{dA}\left(\omega C - \frac{1}{\omega L}\right) \cong 0 \tag{5.96}$$

$$\frac{\partial Y_r}{\partial A}\frac{\partial Y_i}{\partial \omega} - \frac{\partial Y_r}{\partial \omega}\frac{\partial Y_i}{\partial A} \cong C \cdot \frac{dG_{tot}}{dA} > 0 \tag{5.97}$$

The oscillation is stable. Equation (5.71) becomes

$$\frac{\partial Y_r}{\partial A} = \frac{dG_{tot}}{dA} > 0 \qquad \frac{\partial Y_r}{\partial \omega} = \frac{dG_{tot}}{d\omega} \cong 0 \tag{5.98}$$

$$\frac{\partial Y_i}{\partial A} \cong 0 \qquad \frac{\partial Y_i}{\partial \omega} = \frac{d}{d\omega}\left(\omega C - \frac{1}{\omega L}\right) = C + \frac{1}{\omega^2 L} = 2C > 0 \quad (5.99)$$

$$\frac{\partial Y_r}{\partial A}\frac{\partial Y_r}{\partial \omega} + \frac{\partial Y_i}{\partial A}\frac{\partial Y_i}{\partial \omega} = 0 \Rightarrow \delta\omega = 0 \tag{5.100}$$

The amplitude and the frequency of the oscillation are stable for small perturbations of the amplitude of the signal. Let us now plot the admittance on the Smith Chart:

$$Y_i(\omega) = \omega C - \frac{1}{\omega L} \qquad Y_r(A) = -G_s(A) + G_0 \tag{5.101}$$

We get the situation illustrated in Figure 5.17.

5.4 DESIGN METHODS

In this paragraph, design methods making use of simple small- and large-signal concepts are described for the design of oscillators. Guidelines for high-efficiency design are given, based on the above considerations.

So far, oscillation and stability conditions have been defined for a designer to correctly judge whether the circuit will oscillate or not. Now, general guidelines for the design of oscillating circuits are described. With this goal, let us define some typical oscillator topologies.

A typical linear design strategy for microwave oscillators [3] is based on a potentially unstable active device at the design frequency of oscillation. In case the device is stable at that frequency, or if it is desirable to enhance its potential instability at the design frequency while stabilising it at all other frequencies, a shunt or series-feedback network (Z_f) is added. Then, the load at the input of the active device (Z_s) is chosen so that a negative resistance is seen at the output of the device at the design frequency (Γ_{do}, with $|\Gamma_{do}| > 1$); if the input stability circle for the active device including the feedback network is drawn, Γ_s must lie in the unstable part of the Smith Chart. Finally, the external load is connected to the output of the active device (through Z_d) in such a way that the oscillation condition described in the previous paragraph is satisfied at the load port. Typically, the three networks Z_f, Z_s and Z_d are reactive; Z_f or Z_s usually include a high-Q resonator for frequency stability and low phase noise or a voltage-controlled reactance for frequency control (VCO). Schematic topologies for both shunt and series cases are shown in Figure 5.18 for a common-source FET device.

This approach reduces the stability analysis of the oscillator to that of a negative-resistance one-port network connected to an external 50 Ω load. However, this is not necessarily the best approach for high-efficiency or low-noise design. Let us rearrange the circuits in Figure 5.18 as in Figure 5.19, where the external load Z_0 has been included in the impedance Z_1, that therefore is no more purely reactive.

More generally, the resistive load can be included in any of the three elements of the feedback network that assumes a Π configuration for the shunt feedback and a T configuration for the series feedback [5, 7, 10, 11]. For these basic configurations, the

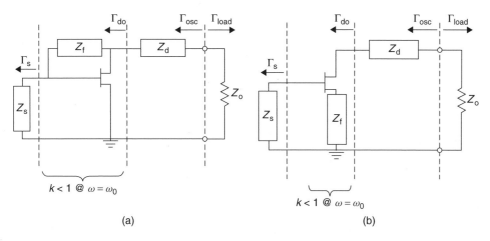

Figure 5.18   Shunt- and series-feedback oscillator topologies

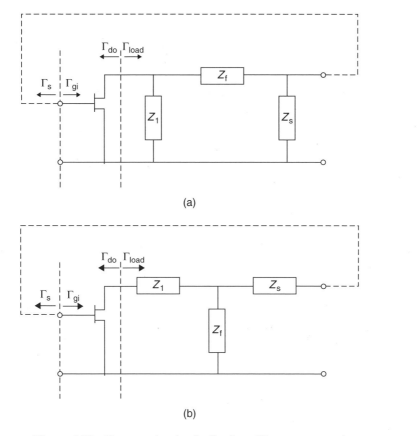

Figure 5.19   Shunt- and series-feedback oscillators rearranged

degrees of freedom of the designer are the four real numbers corresponding to the values of the two reactances (Z_f, and Z_s in our case) and to the real and imaginary parts of the complex load (Z_l in our case).

By this arrangement, it is apparent that the active device amplifies the power entering its input port; then, the feedback network feeds it back again to the input of the active device for the oscillation to be sustained, while delivering part of the power to the resistive load included in the network (Figure 5.20). Since only the external resistive load included in the feedback network dissipates power, while its other elements are reactive, the power amplified by the active device is delivered to the external load except the fraction fed back to the input of the active device.

In fact, more complex embedding networks are possible with more degrees of freedom; however, it is easily seen that only four real numbers are needed for the optimum oscillation determination. If more parameters are available, they can be used for the optimisation of other specifications, for example, noise, bandwidth, stability of oscillation, and so on. Let us demonstrate this point.

In Figure 5.21, the active device with the embedding network (including the external load) is shown. Let us assume that the optimum values for voltages and currents at the input and output ports of the active device have been found. It is still assumed that the signal is quasi-sinusoidal, and therefore only fundamental-frequency voltages and currents are taken into account. It is not relevant how the optimum values for input and output voltages and currents have been found: it can, for instance, be assumed that admittance parameters have been used, measured under large-signal conditions at the design oscillation frequency. In this case, we can write

$$I_{gs} = Y_{a,i} \cdot V_{gs} + Y_{a,r} \cdot V_{ds}$$
$$I_{ds} = Y_{a,f} \cdot V_{gs} + Y_{a,o} \cdot V_{ds} \tag{5.102}$$

or in matrix form as

$$\vec{I}_a = \overset{\leftrightarrow}{Y}_a \cdot \vec{V}_a \tag{5.103}$$

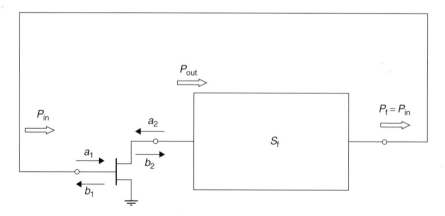

Figure 5.20   The oscillator as an amplifier with a feedback network

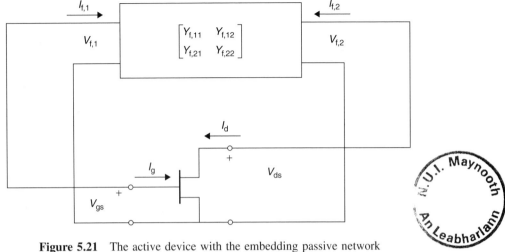

Figure 5.21 The active device with the embedding passive network

where the subscript a refers to the active device, and the relations hold only for the specified frequency and amplitude of the signals:

$$Y_{a,ij} = Y_{a,ij}(V_{gs,0}, V_{ds,0}, \omega_0) \quad i, j = 1, 2 \tag{5.104}$$

A possible choice for the optimum voltages and currents is such that the active device is simultaneously conjugately matched at input and output ports; in other words, $I_{gs,0}$, $V_{gs,0}$, $I_{ds,0}$ and $V_{ds,0}$ must be such that

$$\begin{cases} \dfrac{I_{gs,0}}{V_{gs,0}} = -(Y_{a,in})^* = -\left(Y_{a,11} - \dfrac{Y_{a,12} \cdot Y_{a,21}}{Y_{a,22} - \dfrac{I_{ds,0}}{V_{ds,0}}} \right)^* \\[4ex] \dfrac{I_{ds,0}}{V_{ds,0}} = -(Y_{a,out})^* = -\left(Y_{a,22} - \dfrac{Y_{a,12} \cdot Y_{a,21}}{Y_{a,11} - \dfrac{I_{gs,0}}{V_{gs,0}}} \right)^* \end{cases} \tag{5.105}$$

In fact, these relations only determine the ratio between currents and voltages at input and output ports of the active device; their absolute value, or rather the operating power, is fixed by other considerations, as for example gain saturation of the active device. A typical transistor has a gain saturation characteristic as in Figure 5.22.

The input operating power $P_{in,0}$ at which the admittance parameters are evaluated can be chosen as that maximising the added power $P_{out} - P_{in}$ [7, 10, 11]; this choice sets the absolute value of the voltages and currents at the input and output ports of the transistor at which the admittance parameters are measured and at which the design is performed.

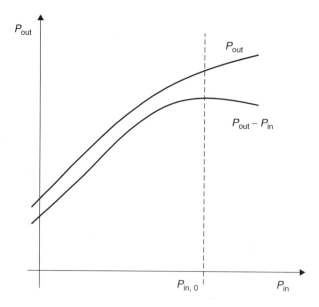

Figure 5.22 Typical gain saturation characteristic of a transistor

For the feedback network,

$$I_{f,1} = Y_{f,11} \cdot V_{f,1} + Y_{f,12} \cdot V_{f,2}$$
$$I_{f,2} = Y_{f,21} \cdot V_{f,1} + Y_{f,22} \cdot V_{f,2}$$

(5.106)

or in matrix form:

$$\vec{I}_f = \ddot{Y}_f \cdot \vec{V}_f$$

(5.107)

The oscillation condition at the chosen power level states that

$$\begin{matrix} I_{gs,0} = -I_{f,1} \\ I_{ds,0} = -I_{f,2} \end{matrix} \quad \text{with} \quad \begin{matrix} V_{gs,0} = V_{f,1} \\ V_{ds,0} = V_{f,2} \end{matrix}$$

(5.108)

Therefore, the feedback network parameters must satisfy

$$-I_{gs,0} = Y_{f,11} \cdot V_{gs,0} + Y_{f,12} \cdot V_{ds,0}$$
$$-I_{ds,0} = Y_{f,21} \cdot V_{gs,0} + Y_{f,22} \cdot V_{ds,0}$$

(5.109)

These are four real equations that can be used to set four independent parameters out of the six (if reciprocal) or eight (if non-reciprocal) real parameters of the feedback network at the design frequency ω_0. Referring to the T or Π configurations shown in Figure 5.23, the usual choice is for two purely reactive elements and a complex element, including the external resistive load. The values of the elements are easily computed [e.g. 5, 7, 11–15], and the corresponding formulae are shown in Table 5.2 and Table 5.3 (from [11]). If the feedback network has more parameters, additional degrees of freedom are available to the designer for the accomplishment of additional design specifications.

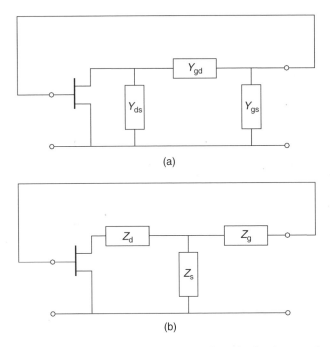

Figure 5.23 T and Π configurations of the feedback network

Table 5.2 Explicit expressions of the elements of the T feedback network in Figure 5.19

Case	Z_g	Z_s	Z_d
Load in Z_d	$j\dfrac{\mathrm{Re}[V_g \cdot (I_g^* + I_d^*)]}{\mathrm{Im}[I_g^* \cdot I_d]}$	$j\dfrac{\mathrm{Re}[V_g \cdot I_g^*]}{\mathrm{Im}[I_g \cdot I_d^*]}$	$\dfrac{V_d}{I_d} - Z_s \cdot \left(1 + \dfrac{I_g}{I_d}\right)$
Load in Z_s	$j\dfrac{\mathrm{Re}[I_d \cdot (V_d^* - V_g^*)]}{\mathrm{Im}[I_g \cdot I_d^*]}$	$\dfrac{V_g - Z_g \cdot I_g}{I_g + I_d}$	$j\dfrac{\mathrm{Re}[I_g \cdot (V_g^* - V_d^*)]}{\mathrm{Im}[I_d \cdot I_g^*]}$
Load in Z_g	$\dfrac{V_g}{I_g} - Z_s \cdot \left(1 + \dfrac{I_d}{I_g}\right)$	$j\dfrac{\mathrm{Re}[V_d \cdot I_d^*]}{\mathrm{Im}[I_d \cdot I_g^*]}$	$j\dfrac{\mathrm{Re}[V_d \cdot (I_d^* + I_g^*)]}{\mathrm{Im}[I_d^* \cdot I_g]}$

Table 5.3 Explicit expressions of the elements of the Π feedback network in Figure 5.19

Case	Y_{gs}	Y_{gd}	Y_{ds}
Load in Y_{ds}	$j\dfrac{\mathrm{Re}[I_g \cdot (V_d^* - V_g^*)]}{\mathrm{Im}[V_g^* \cdot V_d]}$	$j\dfrac{\mathrm{Re}[I_g \cdot V_g^*]}{\mathrm{Im}[V_g^* \cdot V_d]}$	$\dfrac{I_d}{V_d} - Y_{gd} \cdot \left(\dfrac{V_g}{V_d} - 1\right)$
Load in Y_{gd}	$j\dfrac{\mathrm{Re}[V_d \cdot (I_d^* + I_g^*)]}{\mathrm{Im}[V_g^* \cdot V_d]}$	$\dfrac{Y_{gs} \cdot V_g - I_g}{V_d - V_g}$	$j\dfrac{\mathrm{Re}[V_g \cdot (I_g^* + I_d^*)]}{\mathrm{Im}[V_d^* \cdot V_g]}$
Load in Y_{gs}	$\dfrac{I_g}{V_g} - Y_{gd} \cdot \left(\dfrac{V_d}{V_g} - 1\right)$	$j\dfrac{\mathrm{Re}[I_d \cdot V_d^*]}{\mathrm{Im}[V_d^* \cdot V_g]}$	$j\dfrac{\mathrm{Re}[I_d \cdot (V_g^* - V_d^*)]}{\mathrm{Im}[V_d^* \cdot V_g]}$

Similar considerations can be done if large-signal S-parameters are used instead of admittance parameters. The corresponding formulae are easily computed [7, 10, 16].

A more advanced approach for high-efficiency oscillator design requires that the active device be analysed under large-signal drive for high-power and high-efficiency amplification. Its bias point (not explicitly indicated in the figure), its embedding impedances (Γ_s and Γ_{load}) and the optimum power levels P_{in} and P_{out} are designed following the criteria described in Chapter 4. The design can be performed by fundamental-frequency quasi-linear design criteria or full-nonlinear computer-aided optimisation. For example [11], a nonlinear model can be used to derive simplified explicit expressions for optimum power and input and output loads under large-signal drive; then, the feedback network is synthesised as shown above.

If a highly nonlinear design of the power amplifier is performed, harmonic loading must also be taken into account for optimum waveform shaping [5, 17, 18]. As a consequence, the passive feedback network must provide the designed loading and feedback at all the harmonics included in the analysis. The same considerations and formulae as for fundamental frequency can be used for each harmonic independently since the feedback network is linear. In principle, there is no limitation to the number of harmonics that can be considered; however, only second and third harmonics are usually included in the design, as already pointed out when studying power amplifiers. With this approach, Class-E and Class-F operations have been demonstrated, with efficiencies as high as 67% at 1.6 GHz with 24 dBm output power [18].

A more numeric approach can however be taken [14, 15]. If an accurate full-nonlinear model is available for the active device and a nonlinear optimisation procedure is viable, then the voltage and current phasors at the input and output ports of the active device at fundamental and harmonic frequencies are numerically optimised. The optimum combination of current and voltage phasors is found relative to the design goals, which can be maximum output power, maximum efficiency, input and output match, and so on. Once the optimum phasors values are determined, the embedding network, usually in the form of a feedback network, are explicitly or numerically computed, as seen above. This approach is very general but relies on nonlinear optimisation, which may be prone to inefficient optimisation (local minima, difficult determination of the objective function to be minimised), and does not allow for clear trade-off among the achievement of different specifications by the designer.

There is, however, still an aspect to be taken care of for a correct oscillator design, that is, whether the onset of the oscillation actually takes place and whether the oscillation grows up until the desired power level is reached for permanent oscillation. The power amplifier is designed in order to have optimum power performance at the specified power level; the feedback is also designed so as to ensure that the Barkhausen oscillation condition is satisfied at that power level:

$$A(P_0) \cdot \beta = 1 \qquad (5.110)$$

where the gain of the nonlinear active device is dependent on the power level, while the linear feedback network is independent of it. For the oscillation to start, it is required that a growing instability be present from the small-signal regime to the steady-state large-signal

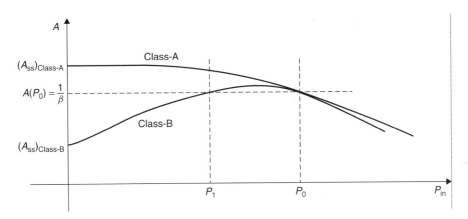

Figure 5.24 Power gain as a function of input power level in typical power amplifiers

regime of eq. (5.110). Therefore, condition (5.72) ensuring growing oscillations must be verified for all signal amplitudes, that is, for all power levels, from small signal to steady-state operating power.

From a physical point of view, an oscillation in a feedback amplifier will grow if the power gain of the active network is greater than the attenuation of the passive feedback network. Therefore, if the small-signal power gain of the power amplifier is greater than the large-signal one, then the oscillation will start, and will grow until the steady-state oscillation amplitude has been reached. This is the typical case of a Class-A power amplifier, whose gain decreases as long as the amplifier is driven into compression. However, if a low-current bias point is selected for high efficiency (Class-AB or B), the gain may actually be smaller in small-signal conditions because of the small transconductance (for FETs) or current gain (for BJTs) at the low-current operating point (Figure 5.24) [5, 18]. In the latter case, oscillation will not start at all in small-signal regime, and the oscillator will not work unless the oscillation is somehow forced with sufficient amplitude.

The case of Class-B, exhibiting gain expansion, lends itself to the illustration of the rules for determination of oscillation stability. It can be seen from Figure 5.24 that there are two equilibrium points where $A(P) \cdot \beta = 1$; however, the derivative with respect to the amplitude of the signal is positive at the first point, that is therefore unstable: if the circuit is forced to oscillate at that amplitude, any small perturbation will cause the signal amplitude to grow until the stable oscillation point is reached.

5.5 NONLINEAR ANALYSIS METHODS FOR OSCILLATORS

In this paragraph, the special formulations that the nonlinear analysis methods described in Chapter 1 assume for autonomous circuits are described, together with the main problems connected with their use for circuit analysis and design.

In Chapter 1, several numerical approaches have been briefly described, namely time-domain direct numerical integration, harmonic and spectral balance, Volterra series expansion, descriptive function, and so on. However, these methods require modifications when applied to autonomous circuits; the main algorithm modifications and their consequences on applications are described in the following.

5.5.1 The Probe Approach

The two main peculiarities of autonomous circuits with respect to non-autonomous ones are, on the one hand, the existence of a trivial or degenerate solution for Kirchhoff's system of equations, wherein all electrical variables (voltages and currents) are equal to zero; on the other hand, the value of the frequency is in general not *a priori* known for oscillator analysis or must be imposed for oscillator design.

The case of autonomous linear circuits has been described in Section 5.2. For them, the degenerate solution is avoided by imposing that the determinant of Kirchhoff's system of equations be equal to zero; this condition yields the oscillation frequency for oscillator analysis or can be used to impose it for oscillator design. This is a general approach, mainly analysis-oriented; for design purposes, the circuit is better divided into two parts connected by a single port. The same condition that the solution be non-degenerate at that port yields an easier tool for active circuit design.

In fact, this second approach can be seen from a different point of view. Let us consider the network in Figure 5.1 in a different way, as shown in Figure 5.25.

Equations (5.5) and (5.6) still read as

$$I_{probe} = I_L + I_R = (Y_L + Y_R) \cdot V_{probe} = 0 \quad Y_L + Y_R = 0 \tag{5.111}$$

If the frequency of the probing voltage source V_{probe} is swept, the oscillation frequency is found as the frequency at which the probing current I_{probe} becomes zero.

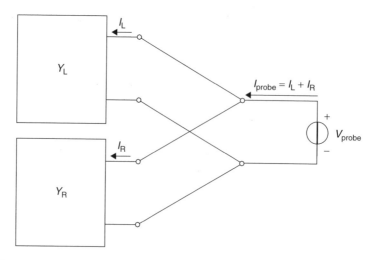

Figure 5.25 Total admittance calculation for an oscillator divided into two arbitrary shunt subnetworks

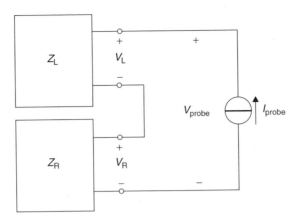

Figure 5.26 Total impedance calculation for an oscillator divided into two arbitrary series subnetworks

If phasors in frequency domain are used, both real and imaginary parts of the current phasor must be equal to zero, or alternatively, the amplitude of the current phasor must be zero. The amplitude of the probing voltage V_{probe} is arbitrary since the circuit is linear. Equivalently, if impedances are used within a Kirchhoff's voltage law (KVL) formulation, the circuit in Figure 5.4 can be redrawn as in Figure 5.26.

Equation (5.111) becomes

$$V_{probe} = V_L + V_R = (Z_L + Z_R) \cdot I_{probe} = 0 \quad Z_L + Z_R = 0 \quad (5.112)$$

Again, if the frequency of the probing constant-value current source I_{probe} is swept, the oscillation frequency is found as the frequency at which the probing voltage V_{probe} becomes zero.

In particular, both formulations can be applied to the oscillator partitioned as in Figure 5.8, as shown in Figure 5.27.

Also, in the case of a nonlinear circuit, the frequency is *a priori* unknown, and a degenerate solution with all voltages and currents equal to zero always exists. The consequences on the various types of nonlinear analysis are not the same: they are briefly described in the following.

5.5.2 Nonlinear Methods

Time-domain direct numerical integration methods are in principle very suitable for the analysis of autonomous circuits. If the circuit is unstable, the solution evolves in time from the initial state (usually zero voltages and currents) to the steady-state oscillatory regime, showing the actual transient behaviour; the onset of the oscillation can therefore be checked. It may be necessary to add a wide-band short pulse at the initial time instant of the analysis in order to simulate the spontaneous noise always present in a circuit, which

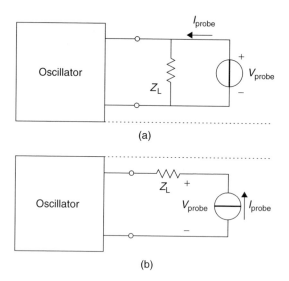

Figure 5.27   Voltage and current probes at the load port for an oscillator

triggers the oscillation. There may be however some numerical problems, especially if the oscillator is based on a high-Q resonator, and if the frequency must be accurately determined. The numerical integration scheme discretises the time variable in sufficiently small steps for the integration to converge (see Chapter 1). If frequency is to be predicted accurately, the step must be sufficiently smaller than the oscillation period. For the oscillation frequency estimation, the time-domain solution is usually Fourier-transformed when it has reached the steady-state oscillation: the time step must therefore be small enough to avoid aliasing problems in the transform. However, high-Q circuits require a long time to bring the oscillating reactances to the steady state, and the integration time can become extremely long. Moreover, the treatment of distributed elements in the circuit, either by equivalent lumped-element representation or by way of a convolution approach, can introduce additional inaccuracies; the number of past samples required by an accurate convolution approach can also substantially increase the memory requirements. These problems notwithstanding, the time-domain approach can prove to be very valuable especially when the onset of the oscillation or the stability of the oscillatory state must be checked in a direct way.

Harmonic and spectral balance analyses require modifications in order to be of any use for autonomous circuit analysis. The balance (Kirchhoff's) system of equations is written (see Chapter 1) as

$$\bar{I}_L(\bar{V}) + \bar{I}_{NL}(\bar{V}) = \bar{0} \tag{5.113}$$

where the linear currents include Norton equivalent current sources only at $\omega = 0$ (bias power supply of the active device(s)). Equation (5.113) corresponds to eq. (5.1) of the linear case and can be used for oscillator analysis, corresponding to eq. (5.3) for the linear case, or for the synthesis or tuning of the oscillator, corresponding to eq. (5.4) for the linear case. Both the alternatives are discussed below.

In the case of oscillator analysis, there is an additional unknown in the system with respect to the forced (non-autonomous) case, that is, the frequency that is not *a priori* known in the analysis of an oscillator. However, since there is no external source, the time reference can be chosen arbitrarily: this is usually done by setting to zero the phase of a frequency component with a (presumably) non-zero amplitude. A standard choice is the fundamental-frequency voltage component at the output port of the active device. In this way, the number of unknowns is again equal to the number of equations, and the system can be numerically solved by means of Newton's method. The vector of unknowns therefore becomes

$$
\vec{V} = \begin{bmatrix} V_0 \\ V_1^r \\ V_1^i \\ V_2^r \\ V_2^i \\ \cdot \\ \cdot \\ V_N^r \\ V_N^i \end{bmatrix} \Rightarrow \vec{V}_\omega = \begin{bmatrix} V_0 \\ V_1^r \\ \omega \\ V_2^r \\ V_2^i \\ \cdot \\ \cdot \\ V_N^r \\ V_N^i \end{bmatrix} \tag{5.114}
$$

This approach is very general; however, system (5.114) always has a degenerate solution for $\vec{V}_\omega = \vec{0}$, as said above. Since no equivalent of condition (5.2) for the linear case exists for a nonlinear system of equations, the degenerate solution cannot, in principle, be eliminated unless additional conditions are added. Two main approaches have been proposed that are described hereafter.

A first approach [19] consists of dividing the currents in eq. (5.113) by the amplitude of the voltage in the circuit; in case voltages and currents both tend to zero, eq. (5.113) is not satisfied, because the left-hand side tends to a finite non-zero limit. The amplitude of the voltage can be defined, for example, as the quadratic sum of the amplitudes of all voltage phasors in the circuit, so as to avoid the improper choice of a voltage with zero or negligible amplitude. System (5.113) can thus be rewritten as

$$
\frac{\overline{I}_L(\overline{V}_\omega) + \overline{I}_{NL}(\overline{V}_\omega)}{\sqrt{\sum_n |V_n|^2}} = \overline{0} \tag{5.115}
$$

This approach effectively removes the degenerate solution.

Another approach [20] makes use of a modification of the Kurokawa condition. For a linear circuit, the oscillation condition at one port for a single frequency ω is

$$
Y_L(\omega) + Y_R(\omega) = Y_{tot}(\omega) = 0 \tag{5.116}
$$

as described above; the degenerate solution is avoided by this formulation. In a nonlinear circuit with a single port connecting the linear and nonlinear subcircuits, all harmonics are present at the connecting port, and the condition can be generalised as [21]

$$Y_{\mathrm{L}}(n\omega) + Y_{\mathrm{NL}}(n\omega) = Y_{\mathrm{tot}}(n\omega) = 0 \quad n = 1, \ldots N \tag{5.117}$$

where the admittance of the nonlinear subcircuit at the nth harmonic, that is at frequency $n\omega$, is defined as

$$Y_{\mathrm{NL}}(n\omega) = \frac{I_n}{V_n} \tag{5.118}$$

The equation system (5.117) can be extended to a nonlinear circuit with an arbitrary number of connecting ports between the linear and nonlinear subcircuits. It can, in principle, replace the system of eq. (5.113), since its solution is also a solution of eq. (5.113), but the degenerate solution is not present. However, a harmonic or spectral balance system usually includes a number of harmonics high enough for the highest ones to have a very small amplitude so as to minimise the truncation error in the Fourier series expansion. Thus, the corresponding admittance in eq. (5.118) can suffer from numerical instability. Moreover, some voltage components V_n could be zero with a non-zero corresponding current I_n, and the formulation of the admittances as in eq. (5.118) would force eq. (5.117) away from the correct solution. An alternative formulation includes the generalised Kurokawa condition (5.117) only at the fundamental frequency ($n = 1$) of a physically meaningful port, for example the output port of the active device. This condition can replace the equivalent KCL equation at that port and harmonic, leaving the number of equations and unknowns unchanged; or it can be added to the KCL equations, yielding an augmented set of equations. In the latter case, a minimisation procedure is required to find the values of the unknowns that satisfy the equation system [20].

We remark that the complex condition (5.117) for a single frequency can be split into real and imaginary parts

$$\mathrm{Re}[Y_1] = 0 \quad \mathrm{Im}[Y_1] = 0 \tag{5.119}$$

and that inclusion of both the conditions is advisable: the real part is usually sensitive to voltage amplitude and the imaginary part to frequency. Since both voltages and frequency are present in the vector of unknowns, both the equations are needed to avoid the degenerate solution.

Both approaches remove the degenerate solution; however, convergence of the solution is still problematic. Practical oscillators usually include elements that are very sensitive to frequency, such as high-Q resonators; therefore, the harmonic or spectral balance error can be very large even when the frequency is not very different form the actual one, and convergence to the actual frequency value can be very slow or impossible. Moreover, a continuation method such as the source-stepping procedure cannot be applied (see Chapter 1) because no RF sources are present in an autonomous circuit. Several procedures have been proposed to overcome the problem.

Different continuation methods include an initial reduction of the number of harmonics in the Fourier series expansions, which reduces the dimension of the system, and therefore eases the numerical solution. Once a solution is found, this is used as an initial guess for the vector of unknowns in the analysis with the number of harmonics increased by one. This is repeated until higher harmonics have a negligible amplitude [20]. Another continuation method starts from a circuit including simplified nonlinearities of the active device, which must therefore be expressed in parametric form in the nonlinear model. Again, once a solution is found to this simplified analysis, it is used as an initial guess for the vector of unknowns in the analysis with more realistic nonlinearities, until the analysis with the correct model is used. As a limit case, the solution of the small-signal circuit can be used for first-guess frequency estimation, especially if the oscillator includes a highly selective resonator; in this case, the conditions described in Section 5.2 must be satisfied for verification of the onset of oscillations.

Another approach randomly selects a number of values of the frequency and voltage components within a 'reasonable' range. The sets of values are repeatedly used as initial guesses for the analysis until one of them leads the algorithm to convergence. A more deterministic scheme fixes the amplitude of a voltage frequency component, for example the fundamental-frequency component at the output port of the active device, at a 'reasonable' value. This component acts as a forcing term of the equation system, which is considerably simpler that the autonomous case. Even an approximate solution can be useful as an initial guess of the actual oscillator analysis [19].

Let us now come to the case of oscillator synthesis or tuning, corresponding to eqs. (5.1) and (5.4) for the linear case. Frequency is now fixed to the design value $\omega = \omega_0$. The unknowns of system of eq. (5.113) are the voltage harmonics, where the phase of a reference harmonic is set to zero, as said above, and the values of some elements of the circuit that determine the frequency of oscillation. If these elements are reduced to a single tuning parameter, the number of unknowns equals the number of equations, and the system is numerically solved by means of Newton's method. The vector of unknowns now becomes

$$
\vec{V} = \begin{bmatrix} V_0 \\ V_1^r \\ V_1^i \\ V_2^r \\ V_2^i \\ \cdot \\ \cdot \\ V_N^r \\ V_N^i \end{bmatrix} \Rightarrow \vec{V}_T = \begin{bmatrix} V_0 \\ V_1^r \\ T \\ V_2^r \\ V_2^i \\ \cdot \\ \cdot \\ V_N^r \\ V_N^i \end{bmatrix} \tag{5.120}
$$

where T is the value of the tuning parameter; typically, this is the value of a capacitance or inductance, or a parameter of a resonator, or the value of a DC control voltage, for example the reverse bias voltage of a varactor diode.

In the case of more than one tuning parameter, optimisation must be performed in order to find the set of values of the unknowns that satisfy Kirchhoff's equation. Still, the same approaches described for the analysis case must be used for avoiding the degenerate solution and for correctly initialising the unknowns for convergence to the non-degenerate solution [22].

The harmonic balance system of equations thus defined can be solved repeatedly for several values of the frequency in a given range: a tuning curve is found, giving the dependence of the oscillation frequency on the value of the tuning parameter. Whenever the system does not converge, then the requested oscillation frequency cannot be forced by the tuning parameter only, provided that numerical problems have been avoided. This approach is suitable for VCO design or optimisation.

It must be pointed out that the calculation of the tuning curve of an oscillator can be used for oscillator analysis also. In this approach, the frequency is swept within a suitable range around an initial guess, say ±25%; the initial guess may come from a linear analysis or be taken as the resonant frequency of a high-Q resonator, when present. Then, a tuning parameter is chosen, for example, the value of a capacitance or the length of a line, which can shift the oscillation frequency. For each frequency in the sweep, the corresponding value of the tuning parameter is found. The free oscillation frequency is the one corresponding to the original value of the tuning parameter. The advantage of this approach is that each fixed-frequency analysis does not require the recalculation of the admittance (or equivalent) matrix of the linear part of the circuit, and it is therefore quite fast; this is not true if the frequency is an unknown, whose value is changed at each iteration of the Newton's algorithm for the solution of the harmonic balance problem. Moreover, some physical insight is gained on the sensitivity of the circuit to the value of the tuning parameter, if needed. A qualitative example of the plot is shown in Figure 5.28, where the tuning parameter is a capacitance with actual value $C = C_{osc}$, the initial guess for the oscillation frequency is $\omega^{(0)}$ and the oscillation frequency ω_{osc} corresponds to the actual value C_{osc} of the capacitance in the circuit.

The described modifications to the harmonic or spectral balance algorithms for the analysis of oscillators can be extended to multi-tone autonomous circuits, where some external sources excite an oscillating circuit, producing a multi-tone spectrum [23]. This approach is particularly interesting for the analysis of self-oscillating mixers and will be described in more detail in Chapter 7.

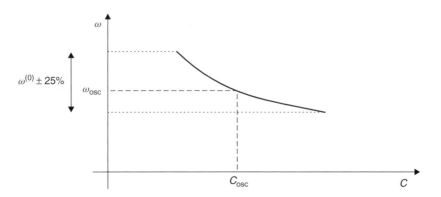

Figure 5.28   Tuning curve of the oscillation frequency vs the tuning capacitance

In the formulation described so far, the harmonic or spectral balance approach has been applied to the nonlinear equivalent of eq. (5.1), that is, a system of autonomous equations (5.113), solved as such. From a different point of view, a modification to that depicted in Figures 5.25 and 5.26 and expressed in eqs. (5.111) and (5.112) for the linear case, the analysis of an autonomous circuit can be reduced to repeated analyses of an equivalent non-autonomous circuit forced by a probing voltage or current at a single port. Frequency and amplitude of the probing signal are swept within a suitable range; the correct solution is found when the control quantity (the probing current or voltage respectively) is zero, indicating that the removal of the probing signal does not perturb the circuit. The scheme is similar to that depicted in Figure 5.27 and is shown in Figure 5.29.

The probing signal (either voltage or current) is applied at one frequency, usually the fundamental; a filter 'masks' the presence of the probe at all other frequency components. The probe injects a signal at a probing frequency with a probing amplitude, both *a priori* unknown, and a phase arbitrarily set to zero; therefore, it introduces two additional real variables. The complex equation, requiring that the (complex) control quantity (either current or voltage) be zero, is added to the system of equations (5.65):

$$I_{\text{probe}}(V_{\text{probe}}, \omega) = 0 \quad \text{or} \quad V_{\text{probe}}(I_{\text{probe}}, \omega) = 0 \qquad (5.121)$$

Therefore, the problem is well posed, and can be solved by means of iterative methods such as Newton's method. However, this formulation is also prone to convergence to the degenerate solution wherein the frequency and all phasors amplitudes are equal to zero. Therefore, a 'good' initial guess for the unknown frequency and phasors amplitudes must be available.

A possible continuation scheme for improving the convergence of the algorithm starts with the determination of the frequency of oscillation startup by means of a linear

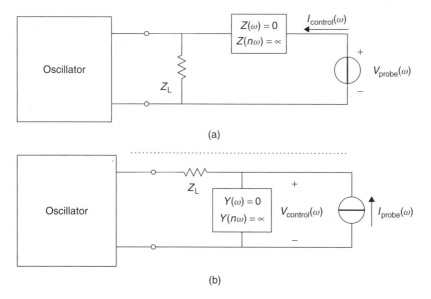

(a)

(b)

Figure 5.29   Voltage and current probes at the load port for a nonlinear oscillator

analysis of the oscillator circuit linearised at the bias point, without the probe. Then, the probe is introduced, injecting a probing signal with a frequency equal to the linear approximation and with a small amplitude. This situation corresponds to the initial oscillation startup, when the oscillation is still growing in time because the power generation at RF by the negative resistance or conductance of the active device prevails on the power dissipation on the positive resistance or conductance of the passive elements, including the load. The control quantity is not zero, as the oscillator delivers power to the probing generator. Then, the amplitude of the probing signal is stepped up until the amplitude of the oscillation brings the active device into saturation, reducing thus the value of the power-generating negative resistance or conductance (see Section 5.3). The value of the control quantity will approach zero for a given value of the amplitude of the probing signal; it will not probably be exactly zero because the oscillation frequency usually shifts in large-signal regime with respect to the small-signal linear calculation. However, this is a good starting point for the self-consistent simultaneous solution of eq. (5.113) and (5.121), avoiding the degenerate solution. It must be pointed out that a quasi-linear determination of the start-up frequency can alternatively be obtained by performing repeated non-autonomous analyses with a small amplitude of the probing signal and fixed frequency; the frequency value is swept within a suitable range, as said above. The frequency at which the control quantity has zero imaginary part and negative real part is a suitable candidate for oscillation startup and a good first guess of the large-signal oscillation frequency.

This method is easily modified to automatically avoid the degenerate solution in the same way as described in eq. (5.118) above. Instead of eq. (5.121), the Kurokawa condition can be written for the probing port:

$$Y_{\text{probe}}(V_{\text{probe}}, \omega) = \frac{I_{\text{probe}}(V_{\text{probe}}, \omega)}{V_{\text{probe}}} = 0 \quad \text{or} \quad Z_{\text{probe}}(I_{\text{probe}}, \omega) = \frac{V_{\text{probe}}(I_{\text{probe}}, \omega)}{I_{\text{probe}}} = 0$$

(5.122)

Convergence may still be problematic, and suitable procedures for accurate first-guess determination are still needed for improving convergence; however, as said, convergence to the degenerate solution is avoided.

Similar to the above formulation, the probe approach also lends itself to oscillator synthesis or tuning. The frequency is now fixed to the design value, and the values of one or more circuit elements are left free to vary in order that the design requirement be met; if a single element is chosen as a tuning parameter, the number of equations equals the number of unknowns. The system of equations is formed by eq. (5.113) and eq. (5.121), and its solution requires care in order to avoid the degenerate solution. A tuning curve can be computed if the analysis is repeated for several frequencies in a suitable range; this approach can also be used for oscillator analysis, as said above.

Volterra series formulation is also a viable approach for oscillator analysis and tuning; so far, a 'probing' approach has been demonstrated that is analogous to that implemented with harmonic or spectral balance [24]. The basic arrangement is that shown in Figure 5.29: the circuit is forced by a probing voltage or current at a single port. Frequency and amplitude of the probing signal are *a priori* unknown: their correct values, corresponding to the fundamental oscillation frequency and voltage or current amplitude at the probing node or branch of the oscillator, give zero-control current or voltage at the probing branch or node of the oscillator, indicating that the removal of the probing signal does not perturb the oscillating circuit. In this formulation, the control current or voltage is computed by standard application of the Volterra series algorithm to the one-port nonlinear circuit being probed. Instead of zero-control current or voltage, a zero admittance or impedance of the one-port circuit at fundamental frequency can be sought; admittance and impedance are computed by simply dividing the control current or voltage by the probing voltage or current. The correct amplitude and frequency of the probing signal are the unknowns of a complex equation; their values can be found by very simple iterative methods with excellent convergence properties. A more general formulation has also been proposed [25], relying on the Volterra series expression of a mildly nonlinear oscillator formulated as a single-loop nonlinear feedback system.

This approach shares the same advantages and drawbacks of the Volterra series analysis described in Chapter 1: it is limited to weak nonlinearities, but it is very fast and reliable. Increasing the order of the Volterra series may prove cumbersome, even though automated methods have been implemented for high-order nuclei calculations based on general procedures [26]. An application to feedback amplifiers based on the same principle has also been demonstrated [27].

5.6 NOISE

In this paragraph, the noise characteristics of oscillator are briefly described, together with some methods for noise prediction.

One of the most important characteristics of an oscillator is the spectral purity of the oscillating signal; this is affected by noise, which causes the spectral line of the oscillating signal to widen, causing degradation of the oscillator performances (Figure 5.30). The noise is generated by the internal noise source of the active device and by resistive elements within the embedding network. While the latter are easily avoided in a careful design, the former cannot be avoided, and their effect must be minimised.

Phase noise in an oscillator is so called because it randomly changes the phase of the oscillating signal and therefore its instantaneous frequency, causing the widening of the spectral line. Noise perturbs the noiseless oscillatory state with two different mechanisms: for very low frequency noise, that is, for noise components at a small frequency offset from the (noiseless) oscillation frequency, the small noise generators modulate the oscillating signal quasi-statically. Formally, the noise sources are added to the noiseless Kirchhoff's equation from which the oscillation condition is derived, causing a random shift of the oscillation frequency, which can be seen as a perturbation of the oscillating signal. As said, this mechanism affects the spectrum of the oscillator closest to the oscillation frequency. For noise components at larger frequency offset from the (noiseless) oscillation frequency, the noise can be seen as an input signal to a nonlinear circuit under a large periodic excitation, in this case the self-oscillation; it is therefore frequency converted between all harmonics of the oscillating signal, including the DC component. This mechanism is the same as that taking place in mixers; however, it is particularly

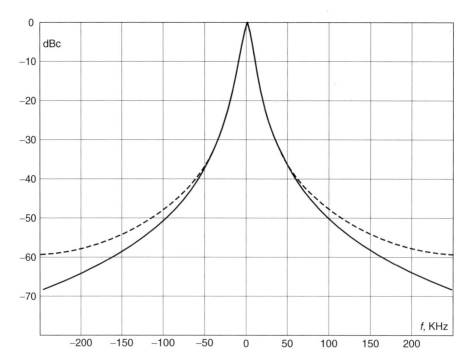

Figure 5.30 Widening of the spectral line at the oscillation frequency caused by noise

disturbing because of the strong $\dfrac{1}{f}$ noise component in active devices being up-converted from very low frequency to near the oscillation frequency. Both mechanisms can be implemented in a nonlinear simulator and are accurately predicted, provided noise sources are accurately modelled.

Frequency conversion from baseband to near the carrier is essentially a second-order nonlinear effect (see Section 1.3.1). Therefore, its magnitude depends very much on the importance of second-order nonlinearities. A qualitative measure of second-order nonlinearities is the amplitude of the rectification of the oscillating signal, generating an increase of the DC bias currents and voltages. Therefore, a low-noise oscillator should exhibit a fairly constant DC bias current with and without oscillation, to within a few percent of the static unperturbed current. Generation of current and voltage harmonics has been treated in detail in Chapter 4, and the reader can refer to that approach for a correct procedure while designing an oscillator as in Section 5.3. It can be noted that, while third-order (or odd-order in general) nonlinearities are required for the saturation of the signal in an amplifier, and therefore for the achievement of the steady-state amplitude of the oscillation, second-order nonlinearities can be avoided by careful design without affecting the proper performances of the oscillator, especially when emphasis is put on noise instead of efficiency.

Another parameter related to the noise performances of the oscillator is the oscillation frequency shift with bias voltages. A first qualitative indicator is the frequency

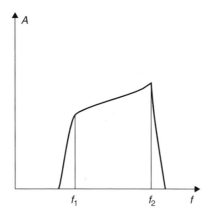

Figure 5.31 Oscillation frequency shift from operating bias to suppression of oscillation

shift from normal operating bias and the reduced bias causing oscillation extinction [28] (Figure 5.31).

The larger the shift, the noisier the oscillator. This approach, however, can be used for accurate, quantitative evaluation of the phase noise in an oscillator. The pushing factor is defined as the shift in oscillation frequency caused by a change in bias voltage:

$$K = \frac{\Delta f_{\text{osc}}}{\Delta V_{\text{bias}}} \tag{5.123}$$

It can be computed from a nonlinear model with a DC analysis, or directly measured by changing the bias voltage and measuring the oscillation frequency or measured by insertion of white noise at low frequency from the bias network of the transistor [29]. Once the pushing factor is known, the single-sideband noise is computed as

$$L(f) = 20 \log \left(\frac{K \cdot \Delta V_{\text{noise}}}{\sqrt{2} \cdot f} \right) \tag{5.124}$$

where f is the offset frequency from the oscillation frequency and ΔV_{noise} is the noise voltage present at the control node.

The assumption underlying the approach is that the low-frequency noise experiences the same up-conversion mechanisms as DC voltage and current; therefore, if the spectral characteristics of the low-frequency noise are known, its up-conversion as a sideband of the spectral line of the oscillating signal can also be computed. In other words, no low-frequency dispersion is assumed to take place in the active device. This is a limiting assumption, but it is experimentally verified to an acceptable degree of accuracy, at least for some types of active devices. An example of typical phase noise for a dielectric resonator oscillator (DRO) oscillator is shown in Figure 5.32.

Another important point is the correct identification of the control node causing the shift in oscillation frequency. For FET devices, this is usually the gate node; however,

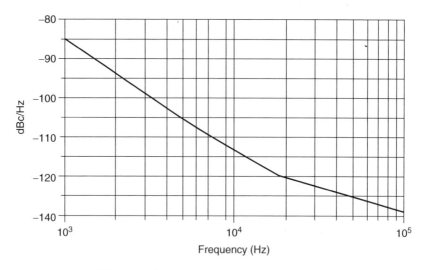

Figure 5.32   Typical phase noise for a DRO

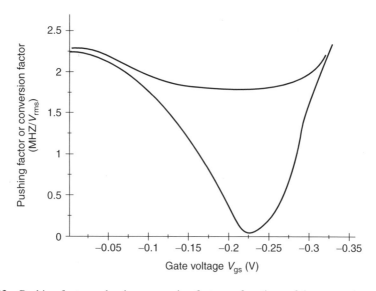

Figure 5.33   Pushing factor and noise conversion factor as functions of the gate voltage in an FET

experimental evidence suggests that control is not limited to it. It can be seen that for some values of the gate bias voltage the pushing factor is zero; however, the phase noise at the same gate bias is reduced but does not vanish (Figure 5.33) [29].

As stated above, the spectral characteristics of the low-frequency noise must be known in order to predict its up-conversion by the pushing factor. However, low-frequency noise is dependent on bias current and voltage, but these are modulated by the large oscillating signal. Therefore, the spectrum of the low-frequency noise changes

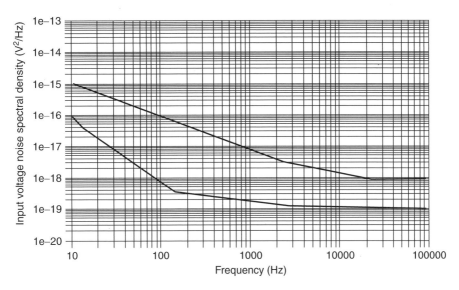

Figure 5.34 Equivalent input low-frequency noise for a transistor under oscillation (1) and at rest (2)

under the effect of the oscillation [30]. As an example, the measured noise of a BJT under oscillation and at the operating point in stable conditions is shown in Figure 5.34 [30].

Let us now describe some arrangements for phase noise analysis within nonlinear analysis methods. The conversion noise is analysed in the same way as in mixers, and the description of the analysis algorithm is not repeated here. The modulation noise is treated in more detail in the following.

The experimental behaviour of low-frequency noise and the single-sideband noise are sketched in Figure 5.35 for a typical case, where the abscissa is the actual frequency for low-frequency noise and the offset frequency from the oscillation frequency for the phase noise. Three regions are clearly visible for phase noise: closest to the carrier the noise has a 30 dB/decade slope; at larger frequency offset, a region with 20 dB/decade slope is present that becomes a flat noise floor for large offset frequencies.

The 30 dB/decade region is due to modulation of the oscillating signal by the $\frac{1}{f}$ noise, which is predominant at very low frequency, and therefore at very small offset frequencies from the carrier. 10 dB/decade are contributed by the $\frac{1}{f}$ dependency of the baseband noise; 20 dB/decade are contributed by the modulation mechanism. In fact, the noise power modulates the frequency of the oscillating signal; therefore, phase is modulated with a $\frac{1}{\omega}$ law, which becomes $\frac{1}{\omega^2}$ for noise power. After the knee voltage, the white noise has no frequency dependence, and only the 20 dB/decade of the modulation mechanism remains. When the conversion noise is predominant, noise modulates the phase directly and no additional contribution from the conversion mechanism to frequency dependence is introduced; therefore the noise spectrum is flat.

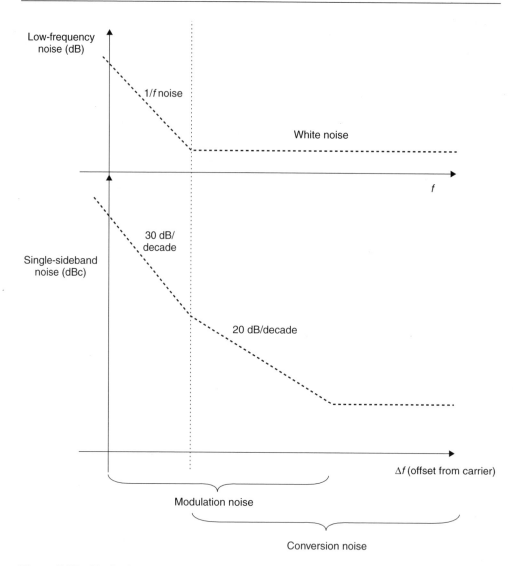

Figure 5.35 Single-sideband noise spectrum of a typical microwave oscillator as a function of the offset frequency

Harmonic balance is currently the most common analysis method for nonlinear circuits at microwave frequencies; modern algorithms can handle noise in oscillators quite generally. Conversion noise is usually modelled by means of the conversion matrix, very much as in mixers; the method will be described in Chapter 7. Conversion noise is present only in oscillators, and its algorithm within a harmonic balance environment is briefly described hereafter.

First, the noiseless oscillator is analysed, and the unperturbed solution is found. The values of the phasors of the electrical unknown quantities are found at fundamental frequency and all the harmonics, as described above. Also, frequency is found as an additional unknown of the problem (eq. (5.114)). The solving system is

$$\vec{I}_{\text{NL}}(\vec{V}) + \ddot{\vec{Y}} \cdot \vec{V} + \vec{I}_{\text{o}} = \vec{0} \tag{5.125}$$

where, in fact, the vector of unknown voltages is modified as in eq. (5.66), and the other vectors accordingly. The vector of the excitation currents now includes only the Norton equivalents of DC bias voltages and/or currents. Once the noiseless solution $\vec{V}_0$ is found, the noise is added as additional Norton equivalent excitation currents at the ports connecting the linear and nonlinear subcircuits (Figure 5.36).

The eq. (5.125) is perturbed around the noiseless solution:

$$\vec{I}_{\text{NL}}(\vec{V}_0) + \left. \frac{\partial \vec{I}_{\text{NL}}(\vec{V})}{\partial \vec{V}} \right|_{\vec{V}=\vec{V}_0} \cdot \delta\vec{V} + \ddot{\vec{Y}} \cdot (\vec{V}_0 + \delta\vec{V}) + \vec{I}_{\text{o}} + \vec{I}_{\text{noise}}$$

$$= \left. \frac{\partial \vec{I}_{\text{NL}}(\vec{V})}{\partial \vec{V}} \right|_{\vec{V}=\vec{V}_0} \cdot \delta\vec{V} + \ddot{\vec{Y}} \cdot \delta\vec{V} + \vec{I}_{\text{noise}} = \vec{0} \tag{5.126}$$

This is a non-autonomous problem that is solved with standard iterative methods, so that the perturbation phasors and the frequency shift $\delta\vec{V}$ are found. The noise sources are in fact modelled as modulated sinusoids at carrier harmonics, with random pseudo-sinusoidal phase and amplitude modulation laws of frequency ω. This results in

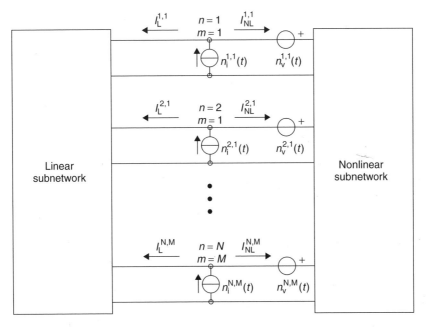

Figure 5.36 Norton equivalent of the noise currents at the ports connecting linear and nonlinear subcircuits

frequency fluctuations with a mean-square value proportional to the available power of the noise sources. The associated mean-square phase fluctuations are proportional to the available noise power divided by ω^2. Some formulations of the harmonic balance problem require a special treatment of the derivatives in eq. (5.126) because their value is close to zero, and the solution method requires their inversion (see Chapter 1) [31], while others are immune from this problem [32].

Other approaches allow the noise performance evaluation, as for instance by means of direct time-domain numerical integration [33] or Volterra series expansion [34]; however, so far the harmonic balance approach has proved to be quite successful. Lately, an envelope analysis harmonic balance approach has been proposed, with promising results: the noise is straightforwardly introduced as an (random) envelope-modulating signal (see Chapter 1) [35]. Also, a general analytical formulation has been proposed that includes both modulation and conversion mechanisms as particular cases [36]; however, its perspectives for implementation do not look very promising because of numerical ill-conditioning of some formulae. A thorough comparison of the different approaches can be found in [37].

5.7 BIBLIOGRAPHY

[1] K. Kurokawa, 'Some basic characteristics of broadband negative resistance oscillator circuits', *Bell Syst. Tech. J.*, **48**, 1937–1955, 1969.

[2] G.R. Basawapatna, R.B. Stancliff, 'A unified approach to the design of wide-band microwave solid-state oscillators', *IEEE Trans. Microwave Theory Tech.*, **MTT-27**(5), 379–385, 1979.

[3] G.D. Vendelin, *Design of Amplifiers and Oscillators by the S-parameter Method*, Wiley, New York (NY), 1982.

[4] R.D. Martinez, R.C. Compton, 'A general approach for the S-parameter design of oscillators with 1 and 2-port active devices', *IEEE Trans. Microwave Theory Tech.*, **40**(3), 569–574, 1992.

[5] M.Q. Lee, S.-J. Yi, S. Nam, Y. Kwon, K.-W. Yeom, 'High-efficiency harmonic loaded oscillator with low bias using a nonlinear design approach', *IEEE Trans. Microwave Theory Tech.*, **47**(9), 1670–1679, 1999.

[6] R.W. Jackson, 'Criteria for the onset of oscillation in microwave circuits', *IEEE Trans. MTT*, **40**(3), 566–569, 1992.

[7] K.M. Johnson, 'Large signal GaAs MESFET oscillator design', *IEEE Trans. Microwave Theory Tech.*, **27**(3), 217–227, 1979.

[8] D.J. Esdale, M.J. Howes, 'A reflection coefficient approach to the design of one-port negative impedance oscillators', *IEEE Trans. Microwave Theory Tech.*, **29**(8), 770–776, 1981.

[9] K. Kurokawa, 'Noise in synchronised oscillators', *IEEE Trans. Microwave Theory Tech.*, **MTT-16**(4), 234–240, 1968.

[10] R.J. Gilmore, F.J. Rosenbaum, 'An analytic approach to optimum oscillator design using S-parameters', *IEEE Trans. Microwave Theory Tech.*, **31**(8), 633–639, 1983.

[11] C. Rauscher, 'Large-signal technique for designing single-frequency and voltage-controlled GaAs FET oscillators', *IEEE Trans. Microwave Theory Tech.*, **MTT-29**(4), 293–304, 1984.

[12] K.L. Kotzebue, 'A technique for the design of microwave transistor oscillators', *IEEE Trans. Microwave Theory Tech.*, **MTT-32**, 719–721, 1984.

[13] T.J. Brazil, J.C. Scanlon, 'A nonlinear design and optimisation procedure for GaAs MESFET oscillators', *IEEE MTT-S Int. Symp. Dig.*, 1987, pp. 907–910.

[14] Y. Xuan, C.M. Snowden, 'A generalised approach to the design of microwave oscillators', *IEEE Trans. Microwave Theory Tech.*, **35**(12), 1340–1347, 1987.

[15] Y. Xuan, C.M. Snowden, 'Optimal computer-aided design of monolothic microwave integrated oscillators', *IEEE Trans. Microwave Theory Tech.*, **37**(9), 1481–1484, 1989.

[16] Y. Mitsui, M. Nakatani, S. Mitsui, 'Design of GaAs MESFET oscillator using large-signal S-parameters', *IEEE Trans. Microwave Theory Tech.*, **25**(12), 981–984, 1977.

[17] E.W. Bryerton, W.A. Shiroma, Z.B. Popovic, 'A 5-GHz high-efficiency class-E oscillator', *IEEE Microwave Guided Wave Lett.*, **6**(12), 441–443, 1996.

[18] M. Prigent, M. Camiade, G. Pataut, D. Reffet, J.M. Nebus, J. Obregon, 'High efficiency free running class-F oscillator', *IEEE MTT-S Int. Symp. Dig.*, 1995, pp. 1317–1320.

[19] V. Rizzoli, A. Costanzo, A. Neri, 'Harmonic-balance analysis of microwave oscillators with automatic suppression of degenerate solution', *Electron. Lett.*, **28**(3), 256–257, 1992.

[20] C.R. Chang, M.B. Steer, S. Martin, E. Reese, 'Computer-aided analysis of free-running oscillators', *IEEE Trans. Microwave Theory Tech.*, **39**(10), 1735–1745, 1991.

[21] B.D. Bates, P.J. Khan, 'Stability of multifrequency negative-resistance oscillators', *IEEE Trans. Microwave Theory Tech.*, **MTT-32**, 1310–1318, 1984.

[22] V. Rizzoli, A. Costanzo, C. Cecchetti, 'Numerical optimisation of microwave oscillators and VCOs', *IEEE MTT-S Int. Microwave Symp. Dig.*, 1993, pp. 629–632.

[23] V. Rizzoli, A. Neri, 'Harmonic-balance analysis of multitone autonomous nonlinear microwave circuits', *IEEE MTT-S Int. Microwave Symp. Dig.*, 1991, pp. 107–110.

[24] K.K.M. Cheng, J.K.A. Everard, 'A new and efficient approach to the analysis and design of GaAs MESFET microwave oscillators', *IEEE MTT-S Int. Microwave Symp. Dig.*, 1990, pp. 1283–1286.

[25] L.O. Chua, Y.S. Tang, 'Nonlinear oscillation via volterra series', *IEEE Trans. Circuits Syst.*, **CAS-29**(3), 150–167, 1982.

[26] L.O. Chua, C.Y. Ng, 'Frequency-domain analysis of nonlinear systems: formulation of transfer functions', *Electron. Circuits Syst.*, **3**(6), 257–269, 1979.

[27] Y. Hu, J.J. Obregon, J.-C. Mollier, 'Nonlinear analysis of microwave FET oscillators using volterra series', *IEEE Trans. Microwave Theory Tech.*, **MTT-37**(11), 1689–1693, 1989.

[28] R.G. Rogers, 'Theory and design of low phase noise microwave oscillators', *Proc. 42nd Annual Frequency Control Symposium*, 1988, pp. 301–303.

[29] J. Verdier, O. Llopis, R. Plana, J. Graffeuil, 'Analysis of noise up-conversion in microwave field-effect transistor oscillators', *Trans. Microwave Theory Tech.*, **44**(8), 1478–1483, 1996.

[30] M. Regis, O. Llopis, J. Graffeuil, 'Nonlinear modelling and design of bipolar transistors ultra-low phase-noise dielectric-resonator oscillators', *IEEE Trans. Microwave Theory Tech.*, **46**(10), 1589–1593, 1998.

[31] W. Anzill, P. Russer, 'A general method to simulate noise in oscillators based on frequency domain techniques', *IEEE Trans. Microwave Theory Tech.*, **41**(12), 2256–2263, 1993.

[32] V. Rizzoli, F. Mastri, D. Masotti, 'General noise analysis of nonlinear microwave circuits by the piecewise harmonic-balance technique', *IEEE Trans. Microwave Theory Tech.*, **42**(5), 807–819, 1994.

[33] G.R. Olbrich, T. Felgentreff, W. Anzill, G. Hersina, P. Russer, 'Calculation of HEMT oscillator phase noise using large signal analysis in time domain', *IEEE MTT-S Symp. Dig.*, 1994, pp. 965–968.

[34] C.-L. Chen, X.-N. Hong, B.X. Gao, 'A new and efficient approach to the accurate simulation of phase noise in microwave MESFET oscillators', *Proc. IEEE MTT-S IMOC '95 Conf.*, 1995, pp. 230–234.

[35] E. Ngoya, J. Rousset, D. Argollo, 'Rigorous RF and microwave oscillator phase noise calculation by envelope transient technique', *IEEE MTT-S Symp. Dig.*, 2000, pp. 91–94.

[36] E. Mehrshahi, F. Farzaneh, 'An analytical approach in calculation of noise spectrum in micro-wave oscillators based on harmonic balance', *IEEE Trans. Microwave Theory Tech.*, **48**(5), 822–831, 2000.

[37] A. Suarez, S. Sancho, S. VerHoeye, J. Portilla, 'Analytical comparison between time- and frequency-domain techniques for phase-noise analysis', *IEEE Trans. Microwave Theory Tech.*, **50**(10), 2353–2361, 2002.

6

Frequency Multipliers and Dividers

6.1 INTRODUCTION

This introduction describes the main topics in the design of frequency multipliers, with special emphasis on active ones.

The generation of high-frequency sinusoidal signals is a key functionality, often required in microwave and millimetre-wave systems such as, for instance, transmitters/receivers. The performances of microwave solid-state oscillators, however, degrade with increasing frequency. In order to overcome this problem, high-performance oscillators operating at a lower frequency are often employed followed by frequency multipliers with good conversion efficiency and output power. Solid-state frequency multipliers can be realised both in passive and active configurations, that is, employing passive devices with reactive or resistive nonlinearities (diodes under reverse or forward bias respectively) or active devices (MESFET, HEMT, HBT) biased in a strongly nonlinear operating region. Active multipliers offer the advantage over passive ones of exhibiting conversion gain rather than losses, eliminating or reducing the need for a high-frequency amplifier after the multiplier; their bandwidth can also be made to be fairly wide. The availability of active devices exhibiting a conversion gain well into the millimeter-wave region with non-negligible bandwidth is actually pushing towards the active solution. Moreover, this choice has the obvious advantage of allowing functional integration in a single technology if a monolithic implementation is attempted.

The intrinsic nonlinear nature of frequency conversion requires the use of nonlinear design methodologies, both on the side of accurate and efficient nonlinear models and algorithms, and on the side of clear optimum design conditions and procedures. This is especially true for monolithic solutions, with the aim of reducing unnecessary design time and increasing the possibility of first-time success of an optimised design.

To this moment however, while the availability and accuracy of general nonlinear models and analysis algorithms ensures the prediction of the performances of the active

Nonlinear Microwave Circuit Design   F. Giannini and G. Leuzzi
© 2004 John Wiley & Sons, Ltd   ISBN: 0-470-84701-8

multiplier with sufficient dependability for practical applications, not everything is clear from the methodological point of view. In recent years, simplified approaches have shed some light on the basic mechanisms of harmonic generation; however, more complex considerations are to be made when full-nonlinear effects are taken into account.

It is to be pointed out that the main design goals for a frequency multiplier are on the one hand high conversion gain from the input signal at fundamental (oscillator) frequency to the converted (multiplied) frequency, and on the other hand a low DC power consumption. The relative importance of these two quantities obviously depends on the particular application. Bandwidth is also important in the case when a variable-frequency signal (and therefore a voltage-controlled fundamental oscillator) is required by the system. Another issue is reliability, which could be impaired by the excitation of potentially dangerous nonlinearities of the actual device. Stability must also be checked, as in any high-gain microwave circuit, with the additional complication of strongly non-linear operations.

In the following, a short description of passive multipliers is first presented. This issue has been investigated in detail in the past, and many theoretical and experimental results are available. Then, active multipliers are presented, which are rapidly gaining popularity in applications but which still lack a comprehensive treatment and a generally agreed design methodology.

6.2 PASSIVE MULTIPLIERS

Passive multipliers are extensively used for their simplicity. The basic principles and some examples are shown.

Passive multipliers are popular for the simplicity of their structure, for the reliability of the nonlinear frequency-multiplying element, usually a diode, and for the very high maximum frequency of operations. Quite naturally, the frequency multiplication cannot yield any conversion gain, but only losses; this is partly compensated by the low or zero DC power consumption. The cascading of an amplification stage can balance the power budget, but requires two circuits for the complete treatment of the signal. Both the circuits are reasonably well established now, and a reliable design can be performed; however, unnecessary complication of the circuitry results, compared to an active implementation. Only the passive multiplier circuit is described in the following, as the amplifier is a standard linear, quasi-linear or high-efficiency amplifier at the multiplied frequency.

Passive multipliers can be classified as resistive or capacitive (or reactive, in general) types. In the first case, the frequency-multiplying mechanism is the strong nonlinearity of the conduction current in the diode. In the second case, the frequency-multiplying mechanism is the nonlinear nature of the reactance of the diode, typically the junction capacitance. In this latter case, the depletion capacitance in reverse bias is used as non-linear reactance in order to avoid the conduction current present when the diffusion capacitance is not negligible. However, especially at high frequency, both mechanisms are found to contribute to frequency multiplication [1]. A great variety of diode structures have been developed, especially for very high frequencies, that can reach the THz range [2–4]; many structures have a back-to-back arrangement and a symmetric C-V characteristic, that allow zero-bias operations and efficient frequency tripling [5].

Resistive multipliers [6, 7], in principle, have infinite bandwidth, given the non-frequency-dependent nature of resistive nonlinearities. However, the associated junction reactance and the reactive parasitic elements of the diode imply a frequency-dependent behaviour of the element. Moreover, matching networks will further limit the bandwidth. Nonetheless, a significant bandwidth can be achieved in practice [8]. On the other hand, power dissipation is always present in the nonlinear element, which imposes a lower limit to frequency conversion losses. It can be demonstrated [6, 7, 9, 10] that for the conversion gain the Manley–Rowe relation holds:

$$G_C \leq \frac{1}{n^2}$$

where n is the frequency multiplication factor. Therefore, resistive multipliers are not suitable when conversion losses are a critical issue, especially for triplers or quadruplers. Unfortunately, losses are important especially at high frequency, where frequency multipliers are most useful.

Resistive multipliers are efficiently used within balanced configurations. Singly balanced or doubly balanced arrangements have intrinsic fundamental-frequency and odd-harmonic frequency rejection, and, therefore, reduced need for filtering networks. Therefore, bandwidths in excess of one octave can be achieved [8], though conversion loss is quite high (in the order of 10 dB or more).

Reactive multipliers [9–14], in principle, can have zero conversion losses. This requires proper reactive loading at all frequencies other that the input (fundamental) frequency and the output (multiplied) frequency; an obvious consequence is that the multiplier has a narrowband, given the strong frequency dependence of reactive impedances. Moreover, when operating at high frequencies, it is difficult to exactly control the values of reactances because of the high influence of parasitics. It is therefore difficult to exactly tune a reactive multiplier.

As mentioned above, the general structure of a frequency multiplier requires large-signal matched terminations at fundamental (input) frequency and multiplied (output) frequency so that most input power reaches the nonlinear device and most output power is transferred to the load. All other frequencies are usually referred to as idler frequencies; their loads must be optimised for optimum multiplication and transfer of output power to the load. In principle, reactive loads at idler frequencies are desirable because they do not dissipate active power. Their actual value must be determined under large-signal conditions, and in general requires detailed optimisation. However, short-circuits usually yield stability of operation, and efficient multiplication, because of the high current circulation in the diode also at idler frequency.

In practice, only the first two or three harmonics will be controlled, resulting in a conversion loss usually higher than the theoretical minimum. Normal conversion loss values for a reactive (narrowband) doubler are around 6 to 9 dB, and above 10 dB for a tripler [10].

6.3 ACTIVE MULTIPLIERS

The general requirements for an active multiplier are described, together with some general design guidelines.

6.3.1 Introduction

Frequency multiplication is performed by generation of upper harmonics of a fundamental-frequency input signal; this is obtained by excitation of strong nonlinearities within the semiconductor device. In the case of active devices, these are usually the nonlinearities in the output current characteristics [15–21]. In particular, current clipping caused by pinching off the channel is most used because of the high distortion introduced, the low or zero-bias current dissipation and high reliability of operation; this is roughly equivalent to Class-B or Class-C operations in an amplifier and similarly yields a conversion gain in the order of 6 to 9 dB less than the fundamental-frequency gain, at least in principle [16]. Other strong nonlinearities are less practical: for instance, current clipping by gate-channel junction forward conduction requires Class-A biasing and therefore high DC power dissipation, and causes possible gate junction damages; however, gain is higher than that near pinch-off. Another very strong nonlinearity, that is gate-drain or channel breakdown, is never used because of reliability problems, in addition to the necessity for high bias voltage. The effect of nonlinearities other than the transconductance modulation has been investigated [16, 20, 22, 23, 24], leading to the conclusion that a possible alternative is the modulation of the output conductance (see also Chapter 7). Other non-linearities, for example, those due to nonlinear gate-source and gate-drain capacitances, are less effective for harmonic generation, giving a minor contribution to frequency multiplication. However, the presence of reactive feedback elements within the active device can not only enhance the harmonic generation but also give rise to instability problems and must be taken into account for comprehensive and reliable multiplier design.

Terminations presented to the active device belong to two basic types: at the signal input and output ports, that is, gate/base port at fundamental frequency and drain/collector port at the desired harmonic frequency; and the other ports, usually limited to the second- or third-harmonic frequency, while ports at higher frequencies are shorted or opened. For the first type, conjugate match allows maximum power transfer, and, therefore, most efficient conversion; since the device operates in nonlinear regime, large-signal impedances (or equivalent) must be used. For the second type, often called idler ports, reactive terminations are a natural choice for minimising power dissipation. Terminations at harmonic frequencies at the input port are sometimes neglected (i.e. shorted), assuming that the contribution of nonlinearities in the input mesh is minor; in the following, it is shown that this is not usually true. Terminations at the output port other than at multiplied frequency, on the other hand, play an important role: in principle, they must be as close as possible to a short circuit for two basic reasons. First, it is important that the most effective nonlinearity, that is the transconductance, be fully exploited and, therefore, that the load line be as vertical as possible in the plane of the output characteristics. Second, for high output power the output voltage swing must cover the full range between the knee or saturation voltage and the breakdown voltage, that is, the active region. If other frequency components are present, the multiplied frequency component of the output voltage must be reduced consequently in order that the total voltage does not exceed the limits of the active region. These simple considerations are essentially confirmed also by a full-nonlinear analysis.

In the case in which the nonlinearity responsible for the frequency multiplication is the modulated output conductance, the fundamental-frequency termination must be close to an open circuit. This is easily understood by remarking that the output conductance near the knee voltage is modulated by the output voltage (see Chapter 7), and therefore a large fundamental-frequency voltage swing must be ensured. Similar considerations as above apply for the other terminations.

The analysis and design of active multipliers is presented in the following, first by simplified piecewise-linear approaches that allow a first insight into the harmonic generation principle and give indications on efficient operating regions. By these means, a first qualitative design can be performed: regions of optimum design are found and approximate values for the terminating networks are derived. Then, full-nonlinear analysis is described for inclusion of detailed nonlinear effects that mandate some care in the nonlinear optimisation of the actual circuit prior to fabrication.

6.3.2 Piecewise-linear Analysis

In this paragraph, a piecewise-linear model and analysis is used to derive the main features and behaviour of an active frequency multiplier. General guidelines are available that help the designer in the first phase of the circuit design.

Simplified nonlinear methods are useful especially at an early design stage when a general understanding of the basic principles and mechanisms must be gained in order to assess the basic structure of the circuit and the expected performances. They are also a useful means to evaluate the performances of the active device and are of help in its selection. Such an approach makes use of a simplified device model and is based on reasonable assumptions on the frequency multiplication mechanisms. In general, its results qualitatively agree with a full-nonlinear approach, thus providing a valuable tool for a quick preliminary analysis and a reasonable starting point for full-nonlinear analysis.

A simplified model of the active device (in this case a field-effect transistor) is shown in Figure 6.1, where the only nonlinearity is the output current source. In general, many elements contribute to the generation of upper harmonics, as for example, gate-source and gate-drain capacitances. However, the main harmonic-generating effect is usually provided by output current nonlinearities [16, 18, 19, 21]. For an accurate analysis, a detailed description of the output current dependence on the controlling voltages should be used; however, the main effects are retained, while at the same time allowing a simple analytical treatment when the current is described by a simple piecewise-linear model. An example is shown in Figure 6.2, where the constant transconductance in the region between pinch-off and forward gate-junction conduction is also evidenced.

This model allows a very easy design if the assumption is made that the load line does not reach the ohmic and breakdown regions, that is, if the operating voltage is

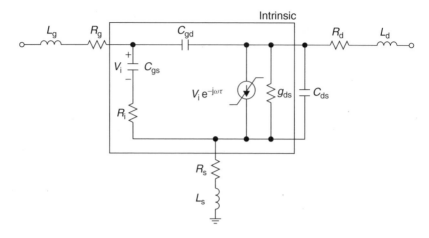

Figure 6.1 A simplified model for a field-effect transistor

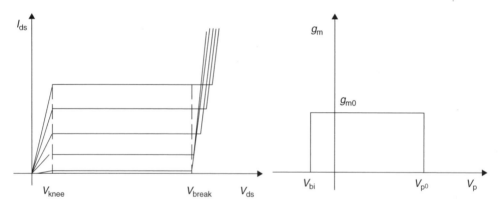

Figure 6.2 A piecewise-linear constant-transconductance model for the output characteristics

always higher than the knee voltage and smaller than the breakdown voltage. This can be ensured by proper drain voltage biasing and by suitable choice of the output load. Then, the output current depends on the gate voltage only. Therefore, the current waveform is known when the gate bias voltage and signal amplitude are known; it has the shape of a truncated sinusoid, as shown in Figure 6.3.

The harmonic content of the current waveform of Figure 6.3 is easily and analytically computed, allowing the direct determination of the optimum gate bias and gate signal amplitude for maximum amplitude of the desired harmonic. Then, the optimum output bias and load is consequently determined. As said above, all output harmonics must be shorted except the desired output harmonic; this assumption allows the direct determination of the optimum output load: it is the resistance that maximises the voltage swing within the hard nonlinear limits imposed by breakdown (upper limit) and knee voltage (lower limit).

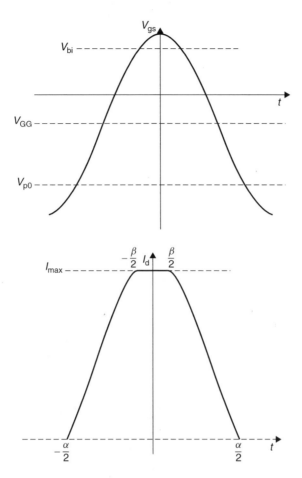

Figure 6.3 Input voltage and output current waveform for a simplified piecewise-linear model

For the sake of clarity, we will first describe multiplier operations when the output current is clipped only by the pinch-off of the channel (Class C–A); then, the case of a symmetric clipping by pinch-off and forward gate-junction conduction is described for frequency tripling (Class-A); and finally, the general case of asymmetric upper and lower clipping will be described. For all the cases, analytic formulae are given.

In the former case, and under the mentioned hypotheses, the current has the explicit expression

$$I_d(t) = \frac{I_{max}}{1 - \cos\left(\frac{\alpha}{2}\right)} \cdot \left[\cos(\omega t) - \cos\left(\frac{\alpha}{2}\right)\right] \quad \text{if} \quad |\omega t| \le \frac{\alpha}{2}$$

$$I_d(t) = 0 \qquad\qquad\qquad\qquad\qquad \text{otherwise} \qquad\qquad (6.1)$$

where the maximum amplitude is normalised to the maximum channel current. The waveform and the conduction angle α are shown in Figure 6.4.

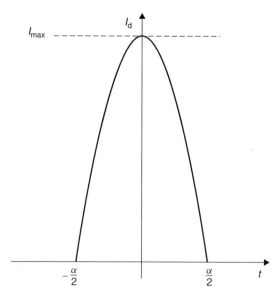

Figure 6.4 Output current waveform in the case of clipping due to pinch-off only

The amplitude of the DC component and of the first three current harmonics of the output current are explicitly given by the following formulae [25]:

$$I_{d,0} = \frac{I_{max}}{2\pi} \cdot \frac{2 \sin\left(\frac{\alpha}{2}\right) - \alpha \cdot \cos\left(\frac{\alpha}{2}\right)}{1 - \cos\left(\frac{\alpha}{2}\right)}$$

$$I_{d,1} = \frac{I_{max}}{2\pi} \cdot \frac{\alpha - \sin(\alpha)}{1 - \cos\left(\frac{\alpha}{2}\right)}$$

$$I_{d,2} = \frac{I_{max}}{6\pi} \cdot \frac{3 \sin\left(\frac{\alpha}{2}\right) - \sin\left(\frac{3\alpha}{2}\right)}{1 - \cos\left(\frac{\alpha}{2}\right)}$$ (6.2)

$$I_{d,3} = \frac{I_{max}}{6\pi} \cdot \frac{\sin(\alpha) \cdot [1 - \cos(\alpha)]}{1 - \cos\left(\frac{\alpha}{2}\right)}$$

or, approximately, for $\alpha \le \pi$ [19]

$$I_{d,0} \cong I_{max} \cdot \frac{2\alpha}{\pi^2}$$

$$I_{d,n} \cong I_{max} \cdot \frac{4\alpha}{\pi^2} \cdot \left| \frac{\cos(n\alpha)}{1 - \left(\frac{2n\alpha}{\pi}\right)^2} \right| \qquad n \ge 1$$ (6.3)

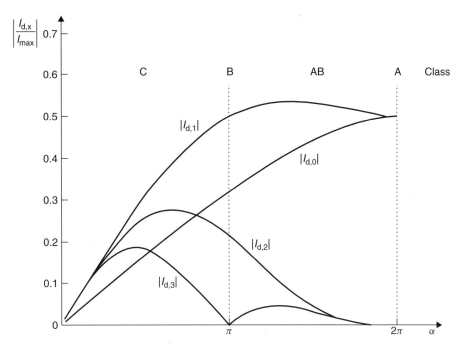

Figure 6.5 Normalised amplitudes of the output current harmonics vs current conduction angle α

The normalised amplitudes of the DC and first three current harmonics as a function of the conduction angle α are also shown in Figure 6.5.

Clearly, for the best conversion gain, a Class-C bias must be selected. For instance, for optimum doubler operations, an operating angle of approximately 126° must be selected that yields a second-harmonic current amplitude about one-fourth (0.27) of the maximum output current. For a frequency tripler, an operating angle of approximately 75° must be selected for a third-harmonic current amplitude equal to about one-sixth (0.185) of the maximum output current.

For maximum output power, it is desirable that the peak output current value be equal to the maximum channel current I_{max}; in this case, the conduction angle is related to the gate bias voltage V_{GG} and to the sinusoidal input signal amplitude $\hat{V}_{\text{gs}}$ by (Figure 6.6)

$$V_{\text{GG}} = \frac{V_{\text{bi}} \cdot \cos\left(\frac{\alpha}{2}\right) + V_{\text{po}}}{1 + \cos\left(\frac{\alpha}{2}\right)} \qquad \hat{V}_{\text{gs}} = \frac{V_{\text{bi}} - V_{\text{po}}}{1 + \cos\left(\frac{\alpha}{2}\right)} \tag{6.4}$$

The optimum output load at third harmonic must be such that the output voltage swing $V_{\text{ds},n}$ be as large as possible but not so large as to cause gate-drain breakdown or to drive the operating point into the ohmic region:

$$R_{\text{L},n} = \frac{V_{\text{ds},n}}{I_{\text{d},n}} = \frac{V_{\text{breakdown}} - V_{\text{knee}}}{2 \cdot I_{\text{d},n}} \tag{6.5}$$

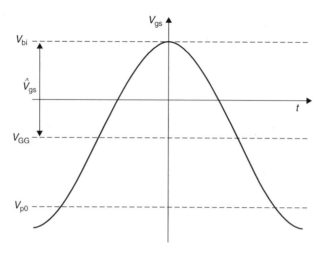

Figure 6.6   The input voltage waveform for maximum output current

This is true provided all other output voltage harmonics are shorted; otherwise, the total output voltage swing would be greater, and the load should be reduced, with consequent reduction of the output power. In Figure 6.7, the load curves for a doubler (a) and a tripler (b) with the optimum circulation angle α are shown; the corresponding waveforms are shown in Figure 6.8.

This approach describes in a simple and straightforward way the behaviour of a Class-C-like frequency multiplier and gives significant indications on optimum bias, operating power and loading. As already mentioned, Class-C biasing avoids reliability concerns caused by gate-junction forward conduction and leads to very low DC power consumption. However, gain is not high because the transistor is off for a large part of the period; moreover, output voltage and power are limited by gate-drain breakdown.

Similar results are obtained if a slightly more realistic behaviour of the transconductance is assumed. For instance, a quartic function can be assumed instead of a linear dependence for the output current above pinch-off, in order to account for a non-stepwise transconductance:

$$I_{\mathrm{d}}(t) = I_{\max} \cdot \left[1 - \left(\frac{2\,\omega t}{\alpha} \right)^2 \right]^2 \quad \text{if} \quad |\omega t| < \frac{\alpha}{2}$$

$$I_{\mathrm{d}}(t) = 0 \qquad\qquad\qquad\qquad \text{otherwise}$$

(6.6)

The relative waveform is shown in Figure 6.9; the amplitudes of the DC component and of the first three harmonics is given by [25]

$$I_{\mathrm{d},0} = I_{\max} \cdot \frac{4\alpha}{15\pi}$$

$$I_{\mathrm{d},1} = \frac{16 I_{\max}}{\pi \cdot \left(\dfrac{\alpha}{2} \right)^4} \cdot \left[3 - \left(\frac{\alpha}{2} \right)^2 \cdot \sin\left(\frac{\alpha}{2} \right) - \frac{3\alpha}{2} \cdot \cos\left(\frac{\alpha}{2} \right) \right]$$

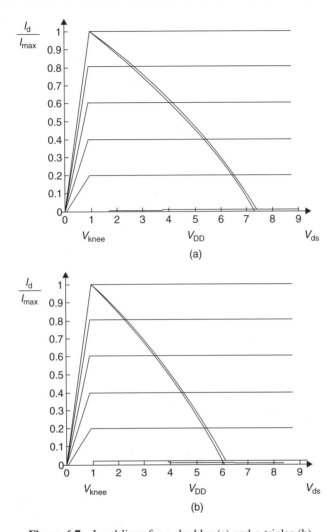

Figure 6.7 Load lines for a doubler (a) and a tripler (b)

$$I_{d,2} = I_{max} \cdot \frac{(48 - 16\alpha^2) \cdot \sin(\alpha) - 48\alpha \cdot \cos(\alpha)}{2\pi\alpha^4}$$

$$I_{d,3} = \frac{16 I_{max}}{\pi \cdot \left(\frac{3\alpha}{2}\right)^4} \cdot \left\{ \left[1 - \frac{1}{3}\left(\frac{3\alpha}{2}\right)^2 \right] \cdot \sin\left(\frac{3\alpha}{2}\right) - \frac{3\alpha}{2} \cdot \cos\left(\frac{3\alpha}{2}\right) \right\} \quad (6.7)$$

and its dependence on the circulation angle is shown in Figure 6.10. The behaviour is qualitatively very similar, indicating a weak dependence on the details of the waveform.

The described approach has been described for a very simplified model; however, the qualitative results are valid for more realistic devices as well. As an example, let us consider the Angelov model for an HEMT (see Chapter 3); the output characteristics

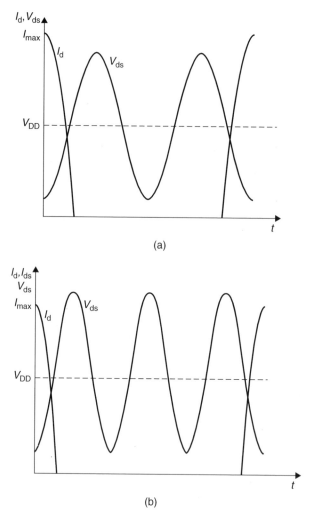

Figure 6.8 Output current and voltage waveforms for a doubler (a) and a tripler (b)

are shown in Figure 6.11 (a), and the drain current and the transconductance versus gate voltage for $V_{ds} = 2$ V are also plotted in the same figure (b), with the bell-shaped transconductance clearly depicted. For the analysis, an arbitrary but reasonable value of pinch-off voltage $V_{po} = -1$ V has been defined, and the maximum current is arbitrarily selected as the current corresponding to maximum transconductance I_{pk} (see Figure 6.11) for a constant $V_{ds} = 2$ V. The DC and the first three relative harmonic amplitudes are plotted in Figure 6.12: a clear agreement with the idealised cases of piecewise-linear characteristics is obtained.

The case of frequency triplers based on symmetric clipping in Class-A operations is now described [26, 27]; the input voltage and output current waveforms are shown in Figure 6.13.

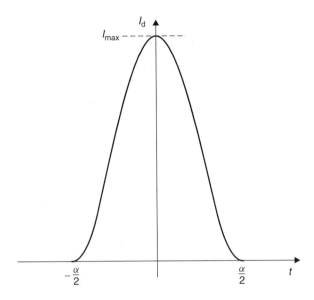

Figure 6.9 Output current waveform in the case of a quartic function

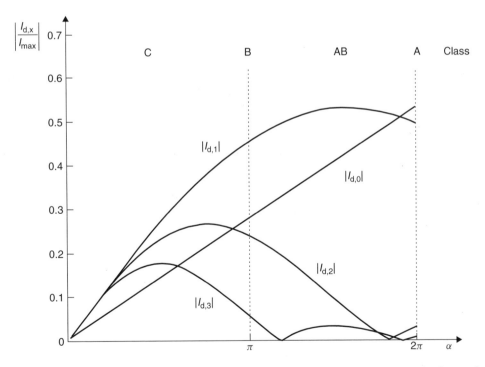

Figure 6.10 Normalised amplitudes of the output current harmonics vs current conduction angle α for a quartic function

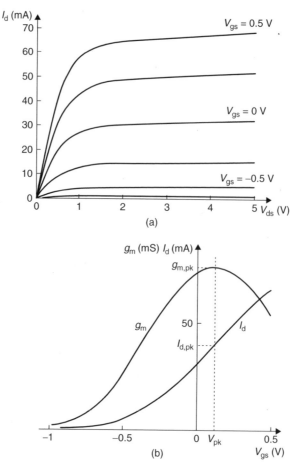

Figure 6.11 Output characteristics of an HEMT, as modelled by means of the Angelov model (a) and current and transconductance vs gate voltage for $V_{ds} = 2$ V (b)

In this case, the output current approaches a square wave for high amplitudes of the input signal because of the symmetric clipping by channel pinch-off (at zero current) and by forward conduction of the gate junction (at maximum current). It is well known that a square wave has only odd-harmonic content, whose amplitude ideally is

$$I_{d,n} = \frac{2}{n\pi}(-1)^n \cdot I_{max} \qquad n \text{ odd}$$
$$I_{d,n} = 0 \qquad\qquad\qquad n \text{ even}$$

(6.8)

In this case, the optimum load for maximum output power is

$$R_{L,3} = \frac{V_{ds,3}}{I_{d,3}} = \frac{3\pi \cdot (V_{breakdown} - V_{knee})}{2I_{max}}$$

(6.9)

For a tripler, the normalised amplitude of the third-harmonic current can have a maximum normalised amplitude of approximately 2.1. All harmonics at outputs other than

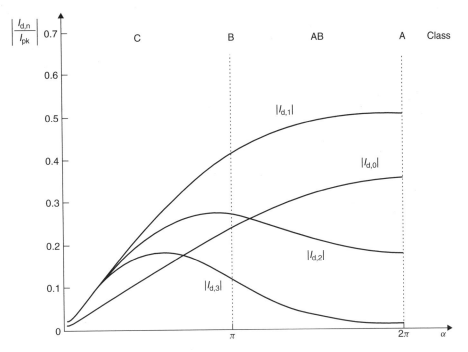

Figure 6.12 Normalised amplitudes of the output current harmonics vs current conduction angle for the HEMT in Figure 6.11

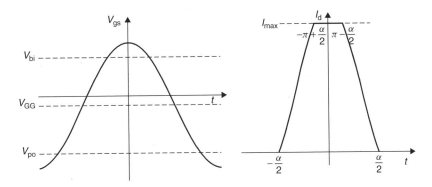

Figure 6.13 Input voltage and output current waveform for symmetric clipping

the third must be shorted, as in the previous case; however, even-order current harmonics are low (ideally zero) because of the symmetry of the clipping, and the relevant loads do not affect the behaviour of the circuit very much. In practice, the current is not an ideal square wave, and the relative amplitudes of the current harmonic can be computed (Figure 6.14).

For full overdrive, the maximum theoretical DC-to-third harmonic conversion efficiency is approximately 2.2. All harmonics at outputs other than the third must be shorted, as in the previous case; however, even-order current harmonics are low because of the

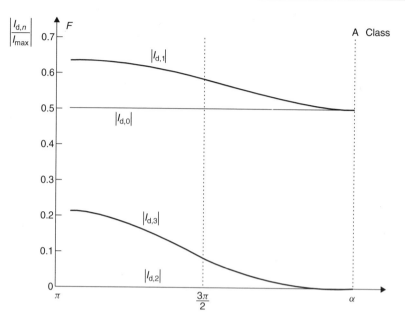

Figure 6.14 Normalised amplitudes of the output current harmonics vs current conduction angle for Class-A symmetric clipping

symmetry of the clipping, and the relevant loads do not affect the behaviour of the circuit very much. The harmonics of the input signal must not necessarily be short-circuited, as long as the input is significantly overdriven.

A relatively larger current and output power is obtained in this case with respect to the Class-C case at the expense of a more critical reliability and a higher DC power consumption. However, devices with lower breakdown can be used, and higher conversion gain is usually available owing to Class-A biasing.

The general case of non-symmetric clipping is now described by the use of a similar approach [28]. In this case, input voltage and output current waveforms are as in Figure 6.3. The coefficients of the harmonic currents are

$$
\begin{aligned}
I_{d,1} = -\frac{2}{\pi} \cdot \frac{\cos\left(\frac{\alpha}{2}\right)}{\cos\left(\frac{\beta}{2}\right) - \cos\left(\frac{\alpha}{2}\right)} \left[\sin\left(\frac{\alpha}{2}\right) - \sin\left(\frac{\beta}{2}\right) \right] \\
+ \frac{1 - \dfrac{\alpha + \beta}{\pi}}{\cos\left(\frac{\beta}{2}\right) - \cos\left(\frac{\alpha}{2}\right)} - \frac{1}{2\pi} \cdot \frac{\sin\left(\frac{\alpha}{2}\right) + \sin\left(\frac{\beta}{2}\right)}{\cos\left(\frac{\beta}{2}\right) - \cos\left(\frac{\alpha}{2}\right)} \\
+ \frac{2\sin\left(\frac{\beta}{2}\right)}{\pi} \quad n = 1
\end{aligned}
\tag{6.10}
$$

$$I_{d,n} = -\frac{2\cos\left(\dfrac{\alpha}{2}\right)}{n\pi} \cdot \frac{\sin\left(\dfrac{n\alpha}{2}\right) - \sin\left(\dfrac{n\beta}{2}\right)}{\cos\left(\dfrac{\beta}{2}\right) - \cos\left(\dfrac{\alpha}{2}\right)}$$

$$+ \frac{1}{(n-1)\pi} \cdot \frac{\sin\left[\dfrac{(n-1)\alpha}{2}\right] - \sin\left[\dfrac{(n-1)\beta}{2}\right]}{\cos\left(\dfrac{\beta}{2}\right) - \cos\left(\dfrac{\alpha}{2}\right)}$$

$$+ \frac{1}{(n+1)\pi} \cdot \frac{\sin\left[\dfrac{(n+1)\alpha}{2}\right] - \sin\left[\dfrac{(n+1)\beta}{2}\right]}{\cos\left(\dfrac{\beta}{2}\right) - \cos\left(\dfrac{\alpha}{2}\right)}$$

$$+ \frac{2}{n\pi}\sin\left(\dfrac{n\beta}{2}\right) \qquad n > 1 \tag{6.11}$$

By varying the gate bias voltage and the amplitude of the gate input voltage, all possible combinations of upper and lower current clipping are obtained. The results are conveniently presented as contour plots of constant second- or third-harmonic current amplitude, as shown in Figure 6.15. The plot is symmetric with respect to the bisector $\alpha = \beta$; the points on this line represent square waves with duty cycle from 0 ($\alpha = \beta = 0$) to 1 ($\alpha = \beta = 2\pi$). The point midway on this line for $\alpha = \beta = \pi$ is a symmetric

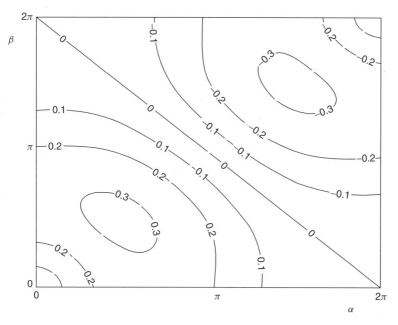

Figure 6.15 Contour plots of the second-harmonic current amplitude as a function of current circulation angles

square wave (duty cycle equal to 1) that has no second-harmonic component. The plot is antisymmetric with respect to the line $\alpha + \beta = 2\pi$, which represents symmetric clipping from a Class-A bias point, with no second-harmonic component. It is easily seen that the plot in Figure 6.5 is obtained for $\beta = 0$. An interesting result is that a region with a second-harmonic component higher than 0.3 is found for asymmetric clipping around a square waveform with duty cycle equal to 1/4 (or symmetrically, equal to 3/4); for this value, the second-harmonic component is equal to $\dfrac{1}{\pi} = 0.318$.

So far, only the nonlinear element in the equivalent circuit in Figure 6.1 has been considered, that is, the piecewise-linear controlled current source. The input signal has been applied directly to the control electrode (gate, in our case), and its bias level and amplitude have been optimised in order to provide optimum current waveforms, with the highest content of the desired harmonic. Terminations and output power have been considered at the terminals of the controlled current source as well. The rest of the equivalent circuit, that is the linear elements, have been neglected so far. However, the other elements of the equivalent circuit of the active device play a strong role in the determination of the optimum loads and performances, both at idler and at input/output ports; they can also determine situations where instabilities occur. This is easily seen even by a simplified analysis including the complete equivalent circuit of the active device as shown in Figure 6.1, with a piecewise-linear output current source.

The general structure of a frequency multiplier is shown in Figure 6.16, where the input and output matching networks can provide independent terminations at each harmonic frequency, at least in principle. Practical filtering structures can in fact provide a wide range of impedances with a relatively simple physical layout [20].

The active device is modelled by means of the simplified equivalent circuit shown in Figure 6.1. The signal reaching the internal gate port is easily computed by linear analysis; the harmonic content of the drain current is computed by means of the explicit expressions listed above. While not strictly necessary, it is also assumed that the output current is driven only by the fundamental frequency of the input gate voltage, for simplification of the analysis; this implies that the effect of higher harmonics at the terminal controlling the current source is neglected.

The frequency doubler is biased at the pinch-off gate voltage for Class-AB or Class-C operations. The input port at fundamental frequency is conjugately matched for

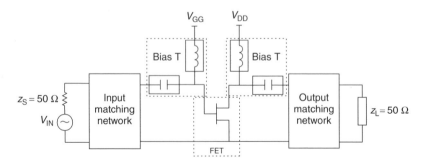

Figure 6.16 The general structure of a frequency multiplier

maximum power transfer into the device and so is the output port at second-harmonic frequency. The terminations at the output port at fundamental frequency and at the input port at second-harmonic frequency (idler ports) are now investigated; given the assumption above, their effects can be studied independent of each other. We assume reactive terminations at both idler ports for minimum dissipated power into the terminations and for input–output isolation. Also, reactive terminations reflect the signal at idler ports back into the active device; this contribution can generate destructive or constructive interference [29, 30], and can be exploited for increasing the conversion gain.

First, the phase of the reactive termination at the output port at fundamental frequency is swept; the conversion gain is plotted in Figure 6.17.

In particular, a region where the doubler becomes unstable is apparent [17, 31–33]. This range of values corresponds to the unstable region of the Smith Chart of the fundamental-frequency output termination for a potentially unstable transistor ($k < 1$); for its evaluation, the linear equivalent circuit must be computed with an averaged transconductance, as from the expressions above [17, 20, 32, 33]. This behaviour is caused by the internal feedback provided by the gate-source capacitance and the source resistance and inductance, which induces a positive feedback for a range of values of the terminations as in any linear circuit. This mechanism has nothing to do with the harmonic-generating waveform shaping since it is due to fundamental-harmonic linear behaviour of the active device. If the active device is unconditionally stable, the unstable area is not present; however, this is hardly likely to happen, since the active device must have enough linear gain at the second-harmonic frequency for providing conversion gain, and therefore it is usually potentially unstable at fundamental frequency.

Clearly, any conversion gain can be designed by choosing the output load sufficiently close to the unstable region. However, as in any linear amplifier, this is by no means a wise

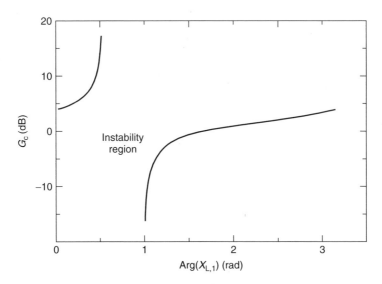

Figure 6.17 Conversion gain as a function of the phase of the reactive termination at output port at fundamental frequency using the simplified model of the active device

choice for practical circuit design. A suitable stability margin must therefore be selected in order to ensure stability and also for bandwidth considerations [17, 20, 32, 33].

The phase of the reactive termination at the input port at second-harmonic frequency is now swept. The conversion gain is shown in Figure 6.18. A less dramatic variation of the conversion gain is apparent [17, 30, 32].

In fact, the weaker dependence of the conversion gain on the phase of the termination, and the absence of unstable regions, depends on the low gain of the active device at second-harmonic frequency. This in turn is due not only to the higher value of the frequency with respect to the fundamental frequency but also to the nonlinear conditions of operation, where the transconductance is modulated by the fundamental-frequency large signal, reducing the incremental gain at second-harmonic frequency. Nonetheless, the possibility of an instability also at the higher frequency cannot be ruled out [33].

The effect of the termination at the input port at second-harmonic frequency can also be seen as providing a constructive or destructive interference [29, 30]; when the conversion gain is maximum, the conductive part of the equivalent output admittance at second-harmonic frequency (signal output port) is minimum, ensuring maximum available output power [32–34].

The curve in Figure 6.18 is fairly independent of the termination at fundamental frequency: when the value of the latter is varied, the curve is shifted upwards or downwards, corresponding to a higher or lower effective transconductance and conversion efficiency [17, 33].

The qualitative results obtained with the simplified model not only yield a comprehension of the basic mechanisms in a frequency multiplier but also allow a preliminary design to be performed. An accurate design, however, requires a full-nonlinear simulation, as presented in the next paragraph.

6.3.3 Full-nonlinear Analysis

Full-nonlinear analysis, based on the harmonic balance algorithm and on accurate models, is presented in this paragraph, leading to a comprehensive methodology for multiplier design.

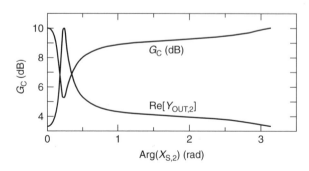

Figure 6.18 Conversion gain as a function of the phase of the reactive termination at input port at second-harmonic frequency using the simplified model of the active device

Frequency multipliers, and especially doublers, have been extensively studied by means of full-nonlinear analysis algorithms and models [15, 17, 20, 21, 30, 32–39], and also by means of advanced measurement set-up [40] and extensive fabrication of test circuits [30]. The results are in general agreement with what has been described in the preceding paragraph, with a better quantitative agreement to measurements. Moreover, other nonlinearities can be studied, in addition to the pure transconductance, provided that a complete and accurate nonlinear model is used, as described in Section 6.3.1 above.

The general structure of a single-device active frequency doubler can be represented as in Figure 6.16. As said before, the active device is biased in a strongly nonlinear region in order to have an output signal very rich in harmonics, including the desired multiple of the input (fundamental) frequency. The internal drain–source terminals are also shown in the figure, where the current waveform behaves approximately as described in the previous paragraph. We can assume that a pinch-off bias (Class-AB or Class-C) is used.

The active device is modelled by means of a fully nonlinear equivalent circuit model, as shown in Figure 6.19. The matching networks terminated by the external 50 − Ω loads are represented by the frequency-dependent terminations $Z_s(n\omega)$ and $Z_L(n\omega)$: the determination of their optimum values is the main goal of the design procedure. We assume that they are reactive at idler ports, and complex loads at signal input and output ports. The input and output matching networks act as impedance transformers at the desired frequencies or frequency bands.

The analysis is limited to fundamental and second-harmonic frequencies, since they determine the basic properties and performances of the doubler. It may also happen that

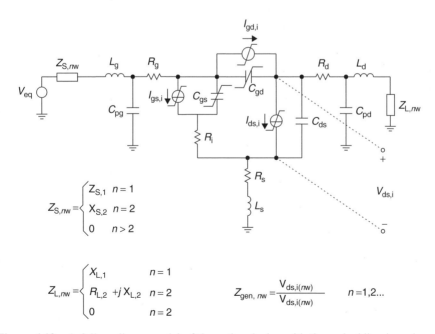

Figure 6.19 A full-nonlinear model of the active device with the embedding impedances

the second harmonic is so high in frequency that the third one is out of the technological control of the designer. Higher frequencies are therefore considered to be short-circuited in this analysis, both at the input and at the output of the multiplying device.

Given the nonlinear nature of the circuit, all results depend on the power level of the signal; however, it turns out that the results do not change much, at least qualitatively, for different power levels within a reasonable range.

In order to better understand the internal mechanisms of the frequency doubler, we consider the nonlinear active device as a frequency-converting network, as shown in Figure 6.20. The electrical ports at the left-hand side of the active device correspond to the gate physical port at fundamental and second-harmonic frequency respectively, and, similarly, the right-hand-side electrical ports correspond to the drain physical port at the same frequencies. A wave impinging the network at any port will cause a wave to come out of all ports; therefore, in general, all four loads ($\Gamma_{S,1}$, $\Gamma_{S,2}$, $\Gamma_{L,1}$ and $\Gamma_{L,2}$) affect the frequency conversion and also the stability of the frequency-converting circuit. In the following, the waves and the behaviour of the four-port frequency-converting network are defined as in Section 3.4.3, as the nonlinear scattering functions.

First, the conjugate matching of the signal input and output ports, that is, the gate port at fundamental frequency and the drain port at second-harmonic frequency, are verified to be an optimum condition by nonlinear analysis [41], or large-signal vector measurement (see Chapter 2). As an example, the conversion gain for a frequency doubler is shown in Figure 6.21 as a function of the output termination (amplitude and phase), and in Figure 6.22 as a function of the input termination (amplitude and phase). The maxima of the surfaces, indicating maximum output power, correspond to the condition of conjugate match at the two ports. In these analyses, when the load at a port is changed, the matching at the other port is readjusted, in order to have consistent results. While not being a theoretical proof, these results seem to confirm what the intuition suggests by extrapolation of linear concepts.

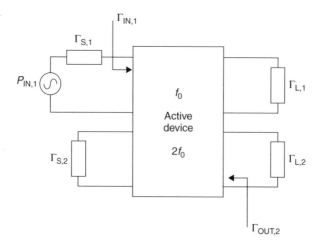

Figure 6.20   The active device represented as a frequency-converting network

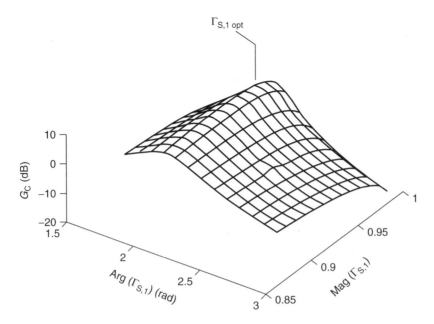

Figure 6.21 Conversion gain as a function of the termination at the gate port at fundamental frequency (input port)

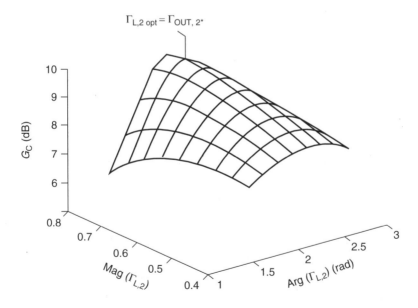

Figure 6.22 Conversion gain as a function of the termination at the drain port at second-harmonic frequency (output port)

The terminations at the idler ports are now investigated by keeping one of them fixed at a typical value and scanning the other one within the whole Smith Chart; for completeness, complex values of the terminations are also considered. The input termination at fundamental frequency (signal input port) and the output termination at second-harmonic frequency (signal output port) are readjusted for conjugate matching at each step. The plot of the conversion gain versus the termination at input (gate) port at second harmonic has a maximum near the edge of the Smith Chart (Figure 6.23).

This result confirms that a reactive load yields the best conversion performance and gives a quantitative estimation of the optimum value. From the point of view of the frequency-converting network in Figure 6.20, the maximum conversion gain is obtained when the output admittance of the network at the signal output port has its minimum value (Figure 6.24), as already mentioned above.

In this case, no instability problems are detected; this is due to the limited gain of the transistor at second-harmonic frequency, as mentioned above.

As an illustration of the effect of the second-harmonic input termination, the voltage and current waveforms at the intrinsic gate terminal and at the intrinsic drain terminal of the active device (see Figure 6.19) are shown in Figure 6.25 ((a) and (b), respectively) for a high-gain value of the second-harmonic input termination, and in Figure 6.26 for a low-gain value of the termination [42].

In the case of high-gain termination, there is a substantial second-harmonic content in the gate voltage and current, which induces a second-harmonic content in the drain current and voltage via linear amplification. In the case of low-gain termination, the second harmonic is shorted at the gate, and therefore there is no second-harmonic linear transfer to the drain via the transconductance.

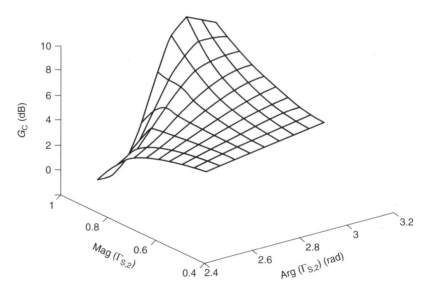

Figure 6.23 Conversion gain as a function of the termination at the gate port at second-harmonic frequency (idler port at input)

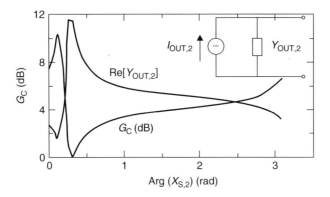

Figure 6.24 Conversion gain and real part of the output admittance at the signal output port as a function of the phase of the reactive termination at the gate port at second-harmonic frequency (idler port at input)

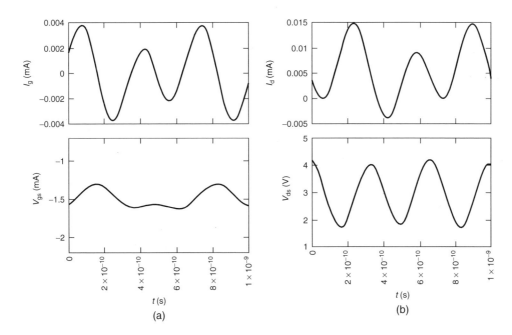

Figure 6.25 Voltage and current waveforms at the intrinsic gate terminal (a) and at the intrinsic drain terminal (b) for a value of the second-harmonic input termination corresponding to a high conversion gain

The fundamental-frequency output termination $\Gamma_{L,1}$ (idler port at output) is now scanned within the whole Smith Chart for completeness; at each step, the terminations at the signal input and output ports are readjusted for large-signal conjugate matching, and also the idler termination at input port is set at each step to the optimum value as described above. A region for the termination is found where the conversion gain increases, until the harmonic balance analysis does not converge any more (Figure 6.27).

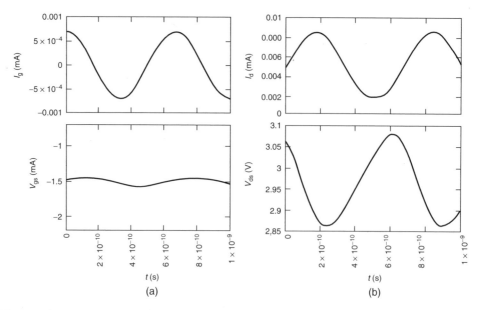

Figure 6.26   Voltage and current waveforms at the intrinsic gate terminal (a) and at the intrinsic drain terminal (b) for a value of the second-harmonic input termination corresponding to a low conversion gain

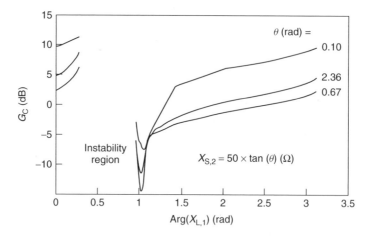

Figure 6.27   Conversion gain as a function of the phase of the reactive termination at the drain port at fundamental frequency (idler port at output) for three values of the phase of the reactive load at the gate port at second-harmonic frequency (idler port at input)

As an illustration of the effect of the fundamental-frequency output termination, the voltage and current waveforms at the intrinsic gate terminal and at the intrinsic drain terminal of the active device (see Figure 6.19) are shown in Figure 6.28 ((a) and (b), respectively) for a high-gain value of the fundamental-frequency output termination (at the left of the instability region in Figure 6.27), and in Figure 6.29 for a low-gain value of the termination (at the right of the instability region in Figure 6.27) [42]. In the case of high-gain termination, the drain current waveform is clipped by the pinch-off, generating a strong second-harmonic component; the fundamental-frequency output voltage is almost shorted, and the second-harmonic output voltage component is quite large. The amplitude of the input gate voltage must be large, in order to drive the almost-shorted output with a substantial current. In the case of low-gain termination, the fundamental-frequency output termination is close to an open circuit, causing an early saturation of the output voltage. Therefore, the gate voltage swing is very small, and so is the output current swing. No clipping and very little second-harmonic content is present in the drain current waveform, and consequently also in the drain voltage. In both cases, terminations have been selected in order to have no second-harmonic content in the gate signal, for the sake of illustration of the harmonic generation in the output by means of the current-clipping mechanism.

It can be seen from Figure 6.27 that the conversion gain first decreases when approaching the instability region from the 'right', and then rises again toward instability [17, 32]; a decrease of the conversion gain is also seen from the simplified model (Section 6.3.2, Figure 6.17), where no successive increase is present when approaching the instability region. A possible explanation is deduced from the waveforms in Figures 6.28 and 6.29. When approaching the instability region from the 'left', that is from the short circuit, the circuit tends to oscillate at second harmonic; when coming from the 'right', that is from the open circuit, the oscillation starts as a fundamental-frequency oscillation, with very low second-harmonic content, that generates a second-harmonic signal when the voltage amplitude becomes large. The simplified model, with the piecewise-linear transconductance model of the controlled current source, cannot

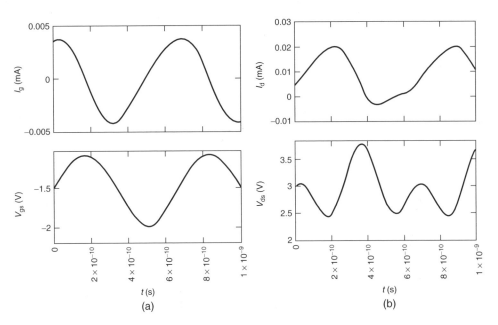

Figure 6.28 Voltage and current waveforms at the intrinsic gate terminal (a) and at the intrinsic drain terminal (b) for a high conversion gain

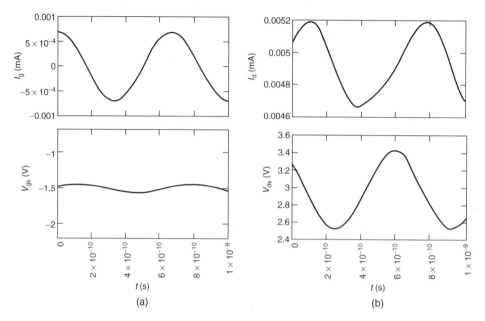

Figure 6.29 Voltage and current waveforms at the intrinsic gate terminal (a) and at the intrinsic drain terminal (b) for a low conversion gain

account for distortion because of a large voltage amplitude, and, therefore, the rise of the second-harmonic component does not show up in the analysis.

From what has been said, it is clear that stability may be a concern in the design of a frequency doubler. As a first approach, small-signal stability circles at either frequencies are a useful indication of potentially dangerous terminations [20, 33]. However, a more convenient formal approach can be done by means of the nonlinear scattering functions (see Chapter 3) [43], which relate small incremental perturbations of the large-signal waves by means of a linear matrix, in a similar way as the scattering parameters relate small perturbations of a large DC bias signal (Chapter 2). The need for the use of incremental parameters from a large-signal regime is based on the fact that a device under compression does not have the same gain properties as in small-signal conditions (Figure 6.30(a)), and that frequency conversion within the device creates paths for the signal that could induce a positive feedback across different frequencies (Figure 6.30(b)). It can be seen that critical situations can take place even when the linear parameters indicate a stable design [44].

So far, only frequency doublers have been considered with full-nonlinear analysis. For higher-order frequency multipliers (triplers, etc.), similar considerations can be made, leading to a similar design approach [26, 27, 45–47].

6.3.4 Other Circuit Considerations

As seen above, the harmonic-generating element in an active doubler is mainly the controlled current source; this is a resistive nonlinearity that has an infinite bandwidth, in

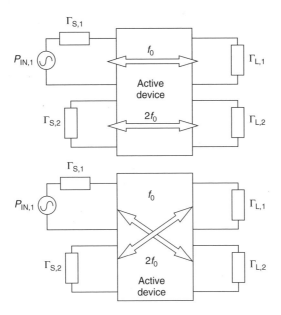

Figure 6.30   Signal paths within the circuit including a frequency-converting active device

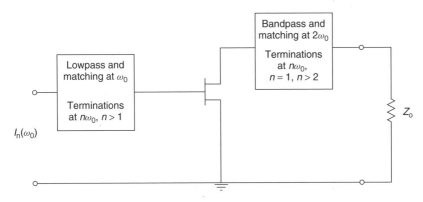

Figure 6.31   General structure of an active single-ended frequency multiplier

principle. The actual bandwidth that can be obtained is limited by the reactances of the transistor and by the limited capability of external networks to synthesise the required loads for good frequency doubling. When properly designed, rather wide bandwidths can be obtained. The general structure of a frequency multiplier is shown in Figure 6.31.

The input network provides large-signal matching at fundamental frequency and terminates the harmonics with reactive terminations that ensure optimum conversion gain and isolation from the output. Similarly, the output network ensures large-signal matching at second-harmonic frequency and terminates the fundamental frequency and the higher-order harmonics with reactive terminations, in order to reject the fundamental and higher-order harmonics, and ensure isolation from the input. Bandwidths in the order of 40% can be obtained at the expense of conversion gain [20, 48, 49].

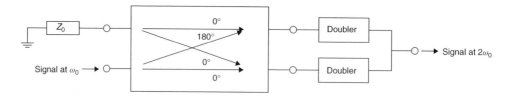

Figure 6.32 A balanced frequency doubler

A frequency doubler can take advantage from a balanced configuration [31, 50–58]. Two identical single-ended doublers are driven out of phase by a 180° coupler, and their outputs are combined in-phase, for example, by a simple T-junction (Figure 6.32).

The fundamental-frequency signal and all the odd-order harmonics are 180° out-of-phase at the output, and therefore cancel; the second-harmonic signal and all even-order harmonics are in-phase at the output, and combine. Such an arrangement, therefore, ensures intrinsic isolation between input and output without the need for filters. Conversion gain is the same as for the single-ended doubler, and the output power is 3 dB higher, provided that a correspondingly higher input power is supplied; no matching improvement is obtained.

6.4 FREQUENCY DIVIDERS – THE REGENERATIVE (PASSIVE) APPROACH

In this paragraph, the operating principle of regenerative frequency dividers are described, together with a stability analysis.

Frequency dividers can be classified into two main types: regenerative dividers, where the power is converted from the fundamental-frequency input signal to the fractional-frequency signal by a passive nonlinear device, and oscillating dividers, where an oscillator at the fractional frequency is phase locked by the input signal at fundamental frequency, corresponding to a harmonic frequency of the oscillator. The latter type is treated in Chapter 8 together with other injection-locked circuits, while the former type is described hereafter.

The general structure of a regenerative frequency divider is shown in Figure 6.33 in which a frequency divider-by-two is shown [59–61]. The input pumping signal is fed to a nonlinear device, usually a reverse-biased diode, where frequency conversion takes place. An input filter prevents the frequency-converted signal to bounce back towards the signal source, while an output filter prevents the input signal to reach the load. The filters also provide matching in order to allow maximum power transfer from input to output.

The diode can be analysed by means of the conversion matrix, as described in Chapter 8; however, a reduced formulation will be used here for the case of a frequency halver [61] for better clarity. The circuit can be seen as two linear subnetworks connected by a frequency-converting nonlinear element. At fundamental and subharmonic (fractional) frequencies, the circuit is as in Figure 6.34.

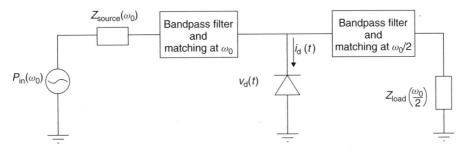

Figure 6.33   The structure of a regenerative frequency divider-by-two

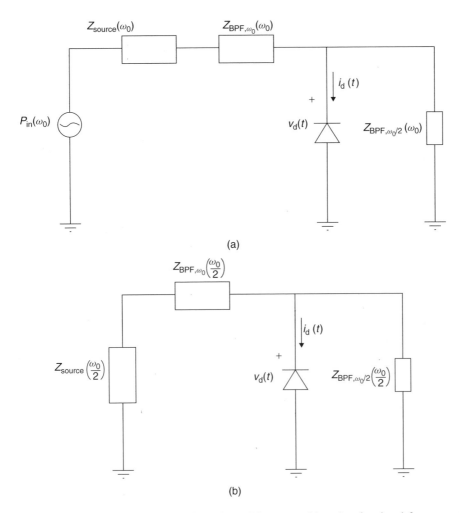

Figure 6.34   The frequency divider at fundamental frequency (a) and at fractional frequency (b)

The pumping signal provides a large sinusoidal voltage at ω_0 in the form

$$v_{d,LO}(t) = V_{LO} \cdot \sin(\omega_0 t) \tag{6.12}$$

for which we assume zero phase and zero DC bias. The capacitance of the diode can be expanded in Fourier series, assuming a simple expression for the junction capacitance:

$$C(t) = \frac{C_{j0}}{\sqrt{1 - \dfrac{V_{LO} \cdot \sin(\omega_0 t)}{v_{bi}}}}$$

$$\cong C_{j0}\left(1 + \frac{V_{LO}}{2v_{bi}} \cdot \sin(\omega_0 t) + \frac{3}{16}\left(\frac{V_{LO}}{v_{bi}}\right)^2 \cdot (1 - \cos(2\,\omega_0 t)) + \ldots\right) \tag{6.13}$$

If a small signal at fractional frequency $\dfrac{\omega_0}{2}$ is present in the circuit,

$$v_{ss}(t) = v_{ss} \cdot \sin\left(\frac{\omega_0}{2}t\right) \tag{6.14}$$

the small-signal current in the diode is

$$i_d(t) = \frac{d\left(C(t) \cdot v_{ss} \cdot \sin\left(\dfrac{\omega_0}{2}t\right)\right)}{dt}$$

$$\cong \frac{d\left(\begin{array}{c} C_{j0}\cdot\left(1 + \dfrac{3}{16}\left(\dfrac{V_{LO}}{v_{bi}}\right)^2\right) \cdot v_{ss}\cdot\sin\left(\dfrac{\omega_0}{2}t\right) \\[2mm] + C_{j0} \cdot \dfrac{V_{LO}}{2v_{bi}} \cdot v_{ss} \cdot \sin(\omega_0 t) \cdot \sin\left(\dfrac{\omega_0}{2}t\right) + \ldots \end{array}\right)}{dt} \tag{6.15}$$

Equation (6.15) gives rise to a conversion-matrix-like expression. The component at fractional frequency of the small-signal current is

$$i_d(t) \cong \frac{\omega_0}{2} \cdot C_{j0} \cdot \left(1 + \frac{3}{16}\left(\frac{V_{LO}}{v_{bi}}\right)^2\right) \cdot v_{ss} \cdot \cos\left(\frac{\omega_0}{2}t\right) - \frac{\omega_0}{2} \cdot C_{j0} \cdot \frac{V_{LO}}{4v_{bi}} \cdot v_{ss} \cdot \sin\left(\frac{\omega_0}{2}t\right) \tag{6.16}$$

The first term is capacitive:

$$C_d = C_{j0} \cdot \left(1 + \frac{3}{16}\left(\frac{V_{LO}}{v_{bi}}\right)^2\right) \tag{6.17}$$

while the second is real and out-of-phase with respect to the voltage; therefore, it is a negative conductance:

$$G_d \cong -\frac{\omega_0}{2}C_{j0} \cdot \frac{V_{LO}}{4v_{bi}} \tag{6.18}$$

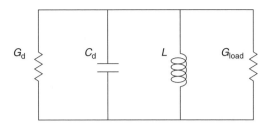

Figure 6.35 The small-signal equivalent circuit of the frequency divider at fractional frequency

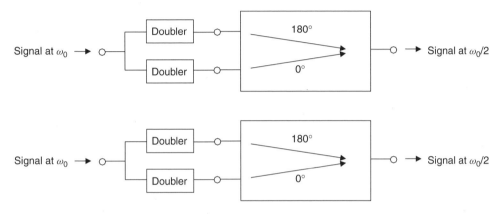

Figure 6.36 Balanced frequency divider-by-two

The equivalent circuit at fractional frequency $\dfrac{\omega_0}{2}$ corresponding to the circuit in Figure 6.34(b) is therefore as in Figure 6.35.

The circuit must resonate at $\dfrac{\omega_0}{2}$, that is, the inductance must resonate the diode capacitance. Moreover, in order for the subharmonic signal to be self-sustained in the circuit, the load admittance must dissipate less power than the equivalent diode negative resistance generates, converting it from the pump signal. As seen in Chapter 5, it must be

$$G_{\mathrm{d}} < -G_{\mathrm{load}} \quad \text{or} \quad |G_{\mathrm{d}}| > G_{\mathrm{load}} \tag{6.19}$$

If this is true, the subharmonic signal grows until the negative conductance starts decreasing for the effect of higher-order terms in eq. (6.14), and an equilibrium is reached.

The frequency divider-by-two can be arranged in a balanced configuration using a balun [62] (Figure 6.36) for intrinsic isolation between input and output. The filters as in Figure 6.33 can now be omitted, or at least greatly simplified, and the circuit can therefore have a much larger bandwidth.

6.5 BIBLIOGRAPHY

[1] K. Benson, M.A. Frerking, 'Theoretical efficiency for triplers using nonideal varistor diodes at submillimetre wavelengths', *IEEE Trans. Microwave Theory Tech.*, **MTT-33**(12), 1367–1374, 1985.

[2] J. Grajal de la Fuente, V. Krozer, F. Maldonado, 'Modelling and design aspects of millimetre-wave Schottky varactor frequency multipliers', *IEEE Microwave Guided Lett.*, **8**(11), 387–389, 1998.

[3] A. Rydberg, H. Grönqvist, E. Kollberg, 'Millimetre and submillimetre-wave multipliers using quantum barrier varactor (QBV) diodes', *IEEE Electron. Device Lett.*, **11**(9), 373–375, 1990.

[4] D. Choudhury, M.A. Frerking, P.D. Batelaan, 'A 200 GHz tripler using a single barrier varactor', *IEEE Trans. Microwave Theory Tech.*, **MTT-41**(4), 595–599, 1993.

[5] K. Krishnamurti, R.G. Harrison, 'Analysis of symmetric-varactor frequency triplers', *IEEE MTT-S Int. Symp. Dig.*, 1993, pp. 649–652.

[6] C.H. Page, 'Frequency conversion with positive nonlinear resistors', *J. Res. Natl. Bureau Stnd.*, **56**, 173–179, 1956.

[7] C.H. Page, 'Harmonic generation with ideal rectifiers', *Proc. IRE*, **46**, 1738–1740, 1958.

[8] S.A. Maas, Y. Ryu, 'A broadband, planar, monolithic resistive frequency doubler', *IEEE MTT-S Int. Symp. Dig.*, 1994, pp. 175–178.

[9] P. Penfield, P.P. rafuse, *Varactor Applications*, MIT Press, Cambridge (MA), 1962.

[10] M. Faber, *Microwave and Millimetre-wave Diode Frequency Multipliers*, Artech House, Norwood (MA), 1995.

[11] C.C.H. Tang, 'An exact analysis of varactor frequency multipliers', *IEEE Trans. Microwave Theory Tech.*, **MTT-14**, 210–212, 1966.

[12] M. Frerking, J. East, 'Novel semiconductor varactors', *Proc. IEEE*, 1992.

[13] T.C. Leonard, 'Prediction of power and efficiency of frequency doublers exhibiting a general nonlinearity', *Proc. IEEE*, **51**, 1135–1139, 1963.

[14] J.O. Scanlan, 'Analysis of varactor harmonic generators', in *Advances in Microwaves*, Vol. 2, L. Young (Ed.), Academic Press, New York (NY), 1967, pp. 87–105.

[15] M.S. Gupta, R.W. Laton, T.T. Lee, 'Performance and design of microwave FET harmonic generators', *IEEE Trans. Microwave Theory Tech.*, **MTT-29**(3), 261–263, 1981.

[16] A. Gopinath, J.B. Rankin, 'Single-gate MESFET frequency doublers', *IEEE Trans. Microwave Theory Tech.*, **MTT-30**(6), 869–875, 1982.

[17] C. Rauscher, 'High-frequency doubler operation of GaAs field-effect transistors', *IEEE Trans. Microwave Theory Tech.*, **MTT-31**(6), 462–473, 1983.

[18] R. Gilmore, 'Concept in the design of frequency multipliers', *Microwave J.*, **30**(3), 129–139, 1987.

[19] S.A. Maas, *Nonlinear Microwave Circuits*, Artech House, Norwood (MA), 1988.

[20] M. Borg, G.R. Branner, 'Novel MIC bipolar frequency doublers having high gain, wide bandwidth and good spectral performance', *IEEE Trans. Microwave Theory Tech.*, **MTT-39**(12), 1936–1946, 1991.

[21] D.G. Thomas, G.R. Branner, 'Optimisation of active microwave frequency multiplier performance utilising harmonic terminating impedances', *IEEE Trans. Microwave Theory Tech.*, **MTT-44**(12), 2617–2624, 1996.

[22] J. Golio, *Microwave MESFETs and HEMTs*, Artech House, Boston (MA), 1991.

[23] E. Camargo, R. Soares, R.A. Perichon, M. Goloubkoff, 'Sources of nonlinearity in GaAs MESFET frequency multipliers', *IEEE MTT-S Int. Symp. Dig.*, 1983, pp. 343–345.

[24] E. Camargo, F. Correra, 'A high-gain GaAs MESFET frequency quadrupler', *IEEE MTT-S Int. Symp. Dig.*, 1987, pp. 177–180.

[25] P. Colantonio, *Metodologie di progetto per amplificatori di potenza a microonde*, Doctoral Thesis, University of Roma Tor Vergata, Roma (Italy), 1999.

[26] H. Fudem, E.C. Niehenke, 'Novel millimetre-wave active MMIC tripler', *IEEE MTT-S Int. Symp. Dig.*, 1998, pp. 387–390.

[27] G. Dow, L. Rosenheck, 'A new approach for mm-wave generation', *Microwave J.*, **26**, 147–162, 1983.

[28] E. O'Ciardha, S.U. Lidholm, B. Lyons, 'Generic-device frequency-multiplier analysis -a unified approach', *IEEE Trans. Microwave Theory Tech.*, **48**(7), 1134–1141, 2000.

[29] Y. Iyama, A. Iida, T. Takagi, S. Urasaki, 'Second-harmonic reflector-type high gain FET frequency doubler operating in K-band', *IEEE MTT-S Int. Symp. Dig.*, 1989, pp. 1291–1294.

[30] D.G. Thomas, G.R. Branner, 'Single-ended HEMT multiplier design using reflector networks', *IEEE Trans. Microwave Theory Tech.*, **MTT-49**(5), 990–993, 2001.

[31] T. Hirota, H. Ogawa, 'Uniplanar monolithic frequency doublers', *IEEE Trans. Microwave Theory Tech.*, **MTT-37**(8), 1249–1254, 1989.

[32] P. Colantonio, F. Giannini, G. Leuzzi, E. Limiti, 'Non linear design of active frequency doublers', *Int. J. RF Microwave Comput.-Aided Eng.*, **9**(2), 117–128, 1999.

[33] I. Schmale, G. Kompa, 'A stability-ensuring procedure for designing high conversion-gain frequency doublers', *IEEE MTT-S Int. Symp. Dig.*, 1998, pp. 873–876.

[34] J. Soares Augusto, M. Joao Rosario, J. Caldinhas Vaz, J. Costa Freire, 'Optimal design of MESFET frequency multipliers', *Proc. 23rd European Microwave Conf.*, Madrid (Spain), Sept. 1993, pp. 402–405.

[35] S. El-Rabaie, J.A.C. Stewart, V.F. Fusco, J.J. McKeown, 'A novel approach for the large-signal analysis and optimisation of microwave frequency doublers', *IEEE MTT-S Int. Symp. Dig.*, 1988, pp. 1119–1122.

[36] C. Guo, E. Ngoya, R. Quere, M. Camiade, J. Obregon, 'Optimal CAD of MESFETs frequency multipliers with and without feedback', *1988 IEEE MTT-S Symp. Dig.*, Baltimore (MD), June 1988, pp. 1115–1118.

[37] C. Fager, L. Landén, H. Zirath, 'High output power, broadband 28–56 GHz MMIC frequency doubler', *IEEE MTT-S Int. Symp. Dig.*, 2000, pp. 1589–1591.

[38] R. Gilmore, 'Design of a novel frequency doubler using a harmonic balance algorithm', *IEEE MTT-S Int. Symp. Dig.*, 1986, pp. 585–588.

[39] R. Gilmore, 'Octave-bandwidth microwave FET doubler', *Electron. Lett.*, **21**(12), 532–533, 1985.

[40] R. Larose, F.M. Ghannouchi, R.G. Bosisio, 'Multi-harmonic load-pull: a method for designing MESFET frequency multipliers', *IEEE Military Comm. Conf.*, 1990, pp. 455–469.

[41] P. Colantonio, F. Giannini, G. Leuzzi, E. Limiti, 'On the optimum design of microwave frequency doublers', *IEEE MTT-S Int. Symp. Dig.*, 1995, pp. 1423–1426.

[42] M. Tosti, *Progettazione ottima di moltiplicatori di frequenza a microonde*, *Laurea* Dissertation, Univ. Roma Tor Vergata, Roma (Italy), 1999.

[43] J. Verspecht, P. Van Esch, 'Accurately characterizing of hard nonlinear behavior of microwave components by the nonlinear network measurement system: introducing the nonlinear scattering functions', *Proc. INNMC '98*, Duisburg (Germany), Oct. 1998, pp. 17–26.

[44] G. Leuzzi, F. Di Paolo, J. Verspecht, D. Schreurs, P. Colantonio, F. Giannini, E. Limiti, 'Applications of the non-linear scattering functions for the non-linear CAD of microwave circuits', *Proc. IEEE ARFTG Symp.*, 2002.

[45] J.P. Mima, G.R. Branner, 'Microwave frequency tripling utilising active devices', *IEEE MTT-S Int. Symp. Dig.*, 1999, pp. 1048–1051.

[46] G. Zhao, 'The effects of biasing and harmonic loading on MESFET tripler performance', *Microwave Opt. Tech. Lett.*, **9**(4), 189–194, 1995.

[47] G. Zhang, R.D. Pollard, C.M. Snowden, 'A novel technique for HEMT tripler design', *IEEE MTT-S Int. Symp. Dig.*, 1996, pp. 663–666.

[48] J.H. Pan, 'Wideband MESFET microwave frequency multiplier', *IEEE MTT-S Int. Symp. Dig.*, 1978, pp. 306–308.

[49] A.M. Pavio, S.D. Bingham, R.H. Halladay, C.A. Sapashe, 'A distributed broadband monolithic frequency multiplier', *IEEE MTT-S Int. Symp. Dig.*, 1988, pp. 503–504.

[50] I. Angelov, H. Zirath, N. Rorsman, H. Grönqvist, 'A balanced millimetre-wave doubler based on pseudomorphic HEMTs', *IEEE Int. Symp. Dig.*, 1992, pp. 353–356.

[51] R. Bitzer, 'Planar broadband MIC balanced frequency doublers', *IEEE MTT-S Int. Symp. Dig.*, 1991, pp. 273–276.

[52] J. Fikart, Y. Xuan, 'A new circuit structure for microwave frequency doublers', *IEEE Microwave Millimetre-wave Circuit Symp. Dig.*, 1993, pp. 145–148.

[53] W.O. Keese, G.R. Branner, 'In-depth modelling, analysis and design of balanced active microwave frequency doublers', *IEEE MTT-S Int. Symp. Dig.*, 1993, pp. 562–565.

[54] M. Abdo-Tuko, R. bertenburg, 'A balanced Ka-band GaAs FET MMIC frequency doubler', *IEEE Trans. Microwave Guided Wave Lett.*, **4**, 217–219, 1994.

[55] M. Cohn, R.G. Freitag, H.G. Henry, J.E. Degenford, D.A. Blackwell, 'A 94 GHz MMIC tripler using anti-parallel diode arrays for idler separation', *IEEE MTT-S Int. Symp. Dig.*, 1994, pp. 736–739.

[56] O. von Stein, J. Sherman, 'Odd-order MESFET multipliers with broadband, efficient, low spurious response', *IEEE MTT-S Int. Symp. Dig.*, 1996, pp. 667–670.

[57] R. Stancliff, 'Balanced dual-gate GaAs FET frequency doublers', *IEEE MTT-S Int. Symp. Dig.*, 1981, pp. 143–145.

[58] T. Hiraoka, T. Tokumitsu, M. Akaike, 'A miniaturised broad-band MMIC frequency doubler', *IEEE Trans. Microwave Theory Tech.*, **MTT-38**(12), 1932–1937, 1990.

[59] F. Sterzer, 'Microwave parametric subharmonic oscillations for digital computing', *Proc. IRE*, **47**, 1317–1324, 1959.

[60] L.C. Upadhyayula, S.Y. Narayan, 'Microwave frequency dividers', *RCA Rev.*, **34**, 595–607, 1973.

[61] G.R. Sloan, 'The modelling, analysis and design of filter-based parametric frequency dividers', *IEEE Trans. Microwave Theory Tech.*, **MTT-41**(2), 224–228, 1993.

[62] R.G. Harrison, 'A broad-band frequency divider using microwave varactors', *IEEE Trans. Microwave Theory Tech.*, **MTT-25**, 1055–1059, 1977.

7

Mixers

7.1 INTRODUCTION

In this introduction, the basic principles of mixing circuits are introduced.

Mixers are based on an intrinsically nonlinear operation, that is, multiplication of a reference signal from the local oscillator by the input signal, with consequent amplitude multiplication and frequency shifting. However, if the reference signal from the local oscillator is constant in both amplitude and frequency, and the input signal is small enough not to generate higher-order products other than multiplication, the result is a linear frequency shifting of the input signal. The multiplication can be seen in different ways: for instance, introducing a switch in series to the input signal we get (Figures 7.1 and 7.2):

$$s(t) = \frac{1}{2} + \sum_n S_n \cdot \sin(\omega_{LO} \cdot t) \quad v_{in}(t) = V_{in} \cdot \sin(\omega_{in} t) \tag{7.1}$$

$$v_{out}(t) = v_{in}(t) \cdot s(t) = \frac{v_{in}(t)}{2}$$

$$+ V_{in} \sum_n \frac{S_n}{2} (\cos((\omega_{LO} - \omega_{in}) \cdot t) - \cos((\omega_{LO} + \omega_{in}) \cdot t)) \tag{7.2}$$

The spectrum of the output voltage v_{out} is as in Figure 7.3 (for comparison, see Figure 1.42 in Section 1.4).

The wanted frequency component is extracted by means of a filter. However, the switch is usually realised by means of a nonlinear device, for example a diode, commanded by a large series voltage source at the local oscillator frequency, as in Figure 7.4.

This implies that the spectrum of the output voltage is rather as in Figure 7.5, in which large components appear at the local oscillator frequency and at all its harmonics (see Figure 1.44 in Section 1.4, and derivation therein).

It is much more difficult in this case to suppress the large, unwanted frequency components by means of a filter. This is the reason why special arrangements are so

Nonlinear Microwave Circuit Design F. Giannini and G. Leuzzi
© 2004 John Wiley & Sons, Ltd ISBN: 0-470-84701-8

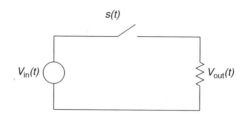

Figure 7.1   The mixer as a switch

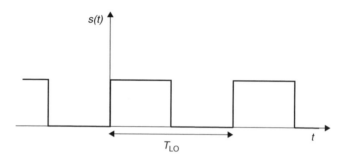

Figure 7.2   The switching function

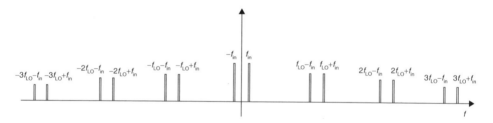

Figure 7.3   The spectrum of the output voltage of an ideal mixer

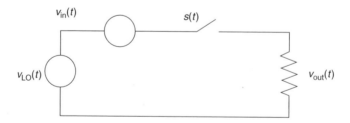

Figure 7.4   A more realistic arrangement for a mixer

popular, where some components are suppressed (or rather, attenuated) exploiting the symmetry properties of balanced mixer. These will be considered in some detail in the following (Section 7.2.2).

As shown in Section 1.4, the input signal is simply multiplied by a switch function $s(t)$ only if the switch function is not affected by the input signal itself. This is no more

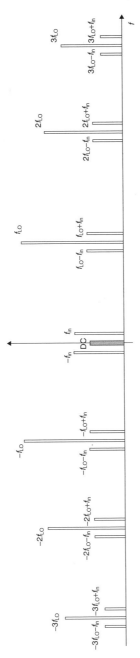

Figure 7.5 The spectrum of the output voltage of a more realistic mixer

true when the input signal becomes large, and distortion and intermodulation arise; this is also the factor determining the upper limit of the dynamic range of the mixer. We will see this case in more detail in the following (Section 7.4). At low levels, noise determines the lower limit of the dynamic range. For a correct evaluation of the noise level in a mixer, the nonlinear behaviour must be taken into account: this will be done in more detail in Section 7.5.

7.2 MIXER CONFIGURATIONS

In this paragraph, the main types of mixers that differ for the type of mixing nonlinearity and for the symmetry of the configuration are described.

7.2.1 Passive and Active Mixers

Mixers have traditionally relied on diodes as the nonlinear mixing element. In this case, the typical configuration is shown in Figure 7.6.

The input signal is the RF, while the output signal is the IF in the case of a downconverter; *vice versa* in the case of an upconverter. The input network provides the optimum terminations to the LO and IN signals and filters the OUT signal generated by the nonlinearity in the diode, in order to ensure minimum conversion losses and maximum isolation between the input and output ports. It must also provide isolation between the LO and the IN ports in order to avoid interference. More dangerously, the large LO signal could saturate the output of the IN amplifier stage, when present. Similarly, the output network provides optimum loading for the OUT signal and stops the IN and LO signals. The practical design and realisation of the filtering structures can be problematic, especially when the frequency of an unwanted large signal (typically the LO fundamental or low-harmonic frequency) lies very close to the input or output frequency that requires a good match. As we will see in the following, a balanced structure can suppress, or rather attenuate, an unwanted spectral line, easing the design of the filtering and matching networks.

In the case of the diode, the main nonlinearity is the I/V exponential characteristic, which presents a differential resistance ranging from nearly open circuit when

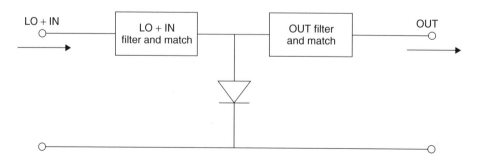

Figure 7.6   The general structure of a diode mixer

reverse biased to a very low value when forward biased. The junction capacitance has a much smaller variation range and its contribution to mixing is much less important; it can be considered constant, and neglected for approximate analysis. A large LO signal drives the diode into forward and reverse bias for the largest part of the signal period, making the diode work very much as the ideal switch in Figure 7.4. A small forward bias current, bringing the diode at the edge of forward conduction, allows the LO signal to effectively switch it between almost short circuit (forward conduction) and almost open circuit (reverse bias) even for low amplitudes of the LO signal itself, thus enhancing the mixer performances; however, the need for a path for the bias current may complicate the layout and degrade the performances.

Active mixers make use of three-terminal devices such as MESFETs, HEMTs, HBTs or BJTs as nonlinear mixing elements, providing also some gain or at least reduced losses. Different nonlinearities are exploited depending on which terminal the large LO signal is fed to; however, the predominant nonlinear element is always the drain or collector current source, while capacitances provide a minor contribution. The output I/V characteristics of an FET are shown in Figure 7.7 in which the load curves corresponding to different modes of operation are indicated. The parameters modulated by the LO signal are the transconductance and the output conductance, that is, the derivatives of the I/V curves with respect to gate and drain voltage respectively:

$$g_m = \left.\frac{\partial I_d}{\partial V_{gs}}\right|_{V_{ds}=\text{const.}} \qquad g_d = \left.\frac{\partial I_d}{\partial V_{ds}}\right|_{V_{gs}=\text{const.}} \tag{7.3}$$

The load line 1 in Figure 7.7 corresponds to a gate mixer, where the main nonlinearity is the transconductance, modulated by the LO signal applied to the gate, with the drain voltage fairly constant. The input signal is applied to the gate as well, while the output signal is taken at the drain port, as shown in Figure 7.8.

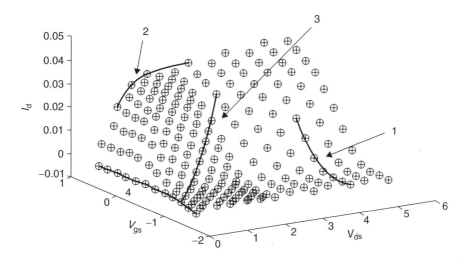

Figure 7.7 Load lines on the output I/V characteristics of an FET corresponding to different operation modes

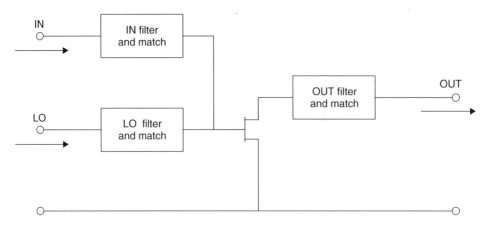

Figure 7.8   The general structure of a gate mixer

The LO signal modulates the transconductance, and therefore the gain of the common-source amplifier for the IN signal, from zero below pinch off to the maximum value along the load line. The behaviour is very much like that of a switch with gain. In Figure 7.7, the load line has a constant V_{ds} voltage path, implying a short-circuit drain termination at the LO fundamental frequency and harmonics; this is discussed in some detail below, together with the terminations at the IN and OUT frequencies.

This configuration does not provide any intrinsic isolation between LO and IN signals and has a very bad isolation between LO and OUT ports since the already large LO signal is further amplified by the FET into the OUT port. The IN signal is also amplified by the FET, but its amplitude is relatively smaller and is more easily filtered out at the OUT port. The LO and IN ports are isolated from the OUT signal because of the low reverse gain of the FET. This configuration is likely to provide a conversion gain if properly terminated; however, it is also prone to instability if the gain is exceedingly large.

The load line 2 in Figure 7.7 corresponds to a drain mixer, where the main nonlinearities are the transconductance and the output conductance, modulated by an LO signal applied to the drain, with the gate voltage fairly constant. The input signal is applied to the gate, while the output signal is taken at the drain port, as shown in Figure 7.9.

The LO signal modulates the transconductance and the output conductance of the FET, and therefore the gain of the common-source amplifier for the IN signal, while switching between the saturated and ohmic regions of the characteristics. The behaviour is again like that of a switch with gain. In Figure 7.7, the load line has a constant V_{gs} voltage path, implying a short-circuit gate termination at the LO fundamental frequency and harmonics.

This configuration does not provide any intrinsic isolation between LO and OUT signals and has a bad isolation between IN and both OUT and LO ports since the IN signal is amplified by the FET. The IN port is isolated from the LO and OUT signals because of the low reverse gain of the FET. It is likely to provide a conversion gain if properly terminated; however, it is also prone to instability if the gain is large.

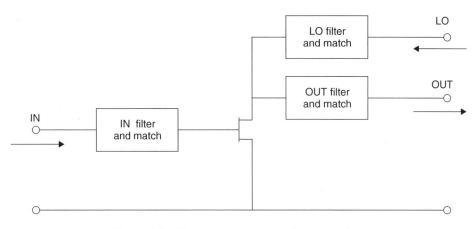

Figure 7.9   The general structure of a drain mixer

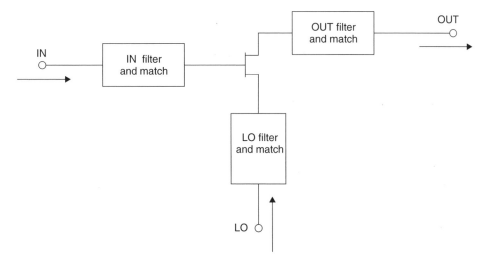

Figure 7.10   The general structure of a source mixer

The load line 3 in Figure 7.7 corresponds to a source mixer, where the main nonlinearities are the transconductance and the output conductance, modulated by an LO signal applied to the source, with the gate and drain voltages fairly constant. The input signal is applied to the gate, while the output signal is taken at the drain port, as shown in Figure 7.10.

The LO signal modulates the transconductance and the output conductance of the FET and therefore the gain of the amplifier for the IN signal. The behaviour is again like that of a switch with gain. In Figure 7.7, the load line has a constant V_{gd} voltage path, implying short-circuit gate and drain termination at the LO fundamental frequency and harmonics.

This configuration does not provide any intrinsic isolation between LO and OUT signals and has a bad isolation between IN and both OUT and LO ports. The IN port is

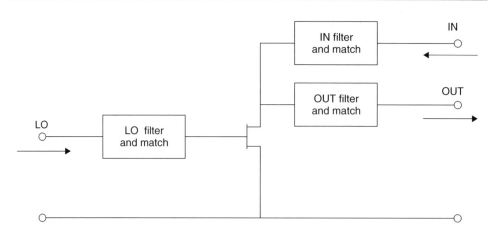

Figure 7.11 The general structure of a resistive (channel) mixer

isolated from both the LO and the OUT signal because of the low reverse gain of the FET. It is likely to provide a conversion gain if properly terminated.

The load line 4 in Figure 7.7 corresponds to what could be called a channel mixer since the main nonlinearity is the channel conductance, modulated by an LO signal applied to the gate, with zero-drain bias. It is known as resistive mixer because the FET has no drain bias (cold FET), and therefore has no gain. The input signal is applied to the drain, while the output signal is taken at the drain or source port, as shown in Figure 7.11.

The LO signal modulates the channel (output) conductance of the FET, making the FET behave as a time-variant resistance when seen from the drain port. In Figure 7.7, the load line has a constant V_{ds} voltage path, implying short-circuit drain termination at the LO fundamental frequency and harmonics.

This configuration provides a moderate isolation between LO and both IN and OUT signals: on the one hand, the FET does not have any gain, but on the other hand, the gate-channel capacitance is high at zero-drain voltage, providing non-negligible coupling. No intrinsic isolation is provided between IN and OUT ports. No gain is provided because of the cold FET; however, very linear conversion is ensured by the superior linearity of the output conductance in the ohmic region compared to the linearity of transconductance and output conductance in the regions of operations described above. Therefore, this configuration is especially valuable for low-intermodulation applications.

7.2.2 Symmetry

As already mentioned above, symmetric or antisymmetric pairing of identical basic mixers provides an effective means to suppress or, more realistically, attenuate some unwanted frequency components in the spectra of the input and output signals. The suppression is especially needed for the large local oscillator signal, which could saturate or seriously reduce the performances of an amplifier stage, but it is important for components with smaller amplitude also. Intermodulation within external systems of these unwanted components with the wanted signals can produce spurious signals interfering with the normal behaviour of the systems themselves. Filters alone could not provide the necessary attenuation because of fabrication tolerances or limited quality factors, because of narrow transition bands between the passband and the suppressed band or because of the unpractically large size of the required filtering network.

Several different arrangements are available to the designer; the basic ones are described in the following in a qualitative way [1]. The basic principle requires that two identical nonlinear elements are each fed with the superposition of the same LO and IN signals, but with different phases; the output signals are then summed up in the load. Each nonlinearity generates spectral lines as in Figure 7.5, some of which are in-phase and therefore are summed up in the load, some others are out-of-phase and therefore cancel in the load; the phase of each line depends on the order of the line itself. In order to generate identical signals with different phases, couplers are used. The most common ones are the hybrid coupler providing (ideally) identical amplitude and 90° phase difference between the output ports when the signal is fed by either of the input ports, and the delta/sigma coupler providing (ideally) identical amplitude and phase at the two output ports when the signal is fed from the sigma port, and identical amplitude and 180° phase difference between the two output ports when the signal is fed from the delta port (see Figure 7.12).

Let us illustrate the point by means of a simplified representation, preserving the symmetry properties of nonlinearities and couplers and neglecting the amplitudes of the spectral lines. Let us consider only the resistive part of the response of the nonlinear mixing device (a diode, in this example) and expand the current in power series of the input voltage (Section 1.3.1). The amplitudes of the coefficients of the power series are arbitrarily set to 1, and only their sign is retained, in order to keep track of the phase of

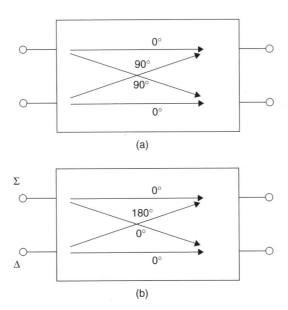

(a)

(b)

Figure 7.12   Schematic representation of the hybrid coupler (a) and of the delta/sigma coupler (b)

each term; this will be done for all amplitudes in the following. For the two diodes in Figure 7.13 (a) and (b), the currents are therefore expressed as in eqs. (7.4a) and (7.4b) respectively:

$$i = v + v^2 + v^3 + \cdots \tag{7.4a}$$

$$i = -v + v^2 - v^3 + \cdots \tag{7.4b}$$

Let us now illustrate an arrangement with a pair of diodes in anti-parallel configuration at the output ports of a delta/sigma coupler as in Figure 7.14, with their currents entering the output node.

The voltages at diodes (a) and (b) are

$$v_a = v_{LO} + v_{IN} \tag{7.5a}$$

$$v_b = -v_{LO} + v_{IN} \tag{7.5b}$$

The corresponding currents are

$$i_a = (v_{LO} + v_{IN}) + (v_{LO} + v_{IN})^2 + (v_{LO} + v_{IN})^3 + \cdots \tag{7.6a}$$

$$i_b = -(-v_{LO} + v_{IN}) + (-v_{LO} + v_{IN})^2 - (-v_{LO} + v_{IN})^3 + \cdots \tag{7.6b}$$

Expanding the binomials and subtracting the two currents we get the output current:

$$i_a = (v_{LO} + v_{IN}) + (v_{LO}^2 + v_{LO}v_{IN} + v_{IN}^2)$$
$$+ (v_{LO}^3 + v_{LO}^2 v_{IN} + v_{LO}v_{IN}^2 + v_{IN}^3) + \cdots \tag{7.7a}$$

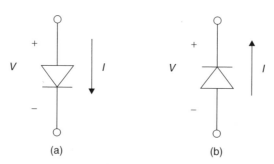

(a)   (b)

Figure 7.13   Voltage and current in the diodes

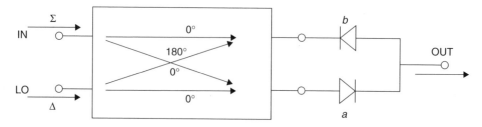

Figure 7.14   A singly balanced mixer with LO rejection at the output

$$i_b = (v_{LO} - v_{IN}) + (v_{LO}^2 - v_{LO}v_{IN} + v_{IN}^2)$$

$$+ (v_{LO}^3 - v_{LO}^2 v_{IN} + v_{LO}v_{IN}^2 - v_{IN}^3) + \cdots \tag{7.7b}$$

$$i_{OUT} = i_a - i_b = v_{IN} + v_{LO}v_{IN} + v_{LO}^2 v_{IN} + v_{IN}^3 + \cdots \tag{7.8}$$

Some terms are out-of-phase and cancel; some others are in-phase and combine in the output load. Remembering the considerations in Section 1.3.1, we see that the first and fourth terms are components at the input frequency and at its third-harmonic frequency; the latter can be neglected, given the small amplitude of the input signal. The second term is the mixed signal (see Introduction above), and provides the two sidebands of the local oscillator signal. The third term is the mixing of the input signal with the second harmonic of the local oscillator and with its rectified DC term. The former product can be used for subharmonic mixing, in the case that a local oscillator at a sufficiently high frequency be not available; it must otherwise be rejected by the output filter. The second product is an additional term at the input frequency. Then, there are higher-order terms that can be neglected to a first approximation. The local oscillator with its harmonics is cancelled by the symmetry of the configuration; the other unwanted terms can be rejected by filtering, with much greater ease than in a single-diode mixer. The situation is shown in Figure 7.15 for an upconverting mixer, where the combined and cancelled terms are shown as solid and dotted bars respectively.

The singly balanced mixer in Figure 7.14 therefore has intrinsic isolation between the local oscillator port and the output port; it also has an isolation between input port and local oscillator port. No isolation is provided between input and output ports.

The cancellation of the LO oscillator at the output has an intuitive explanation. Referring to Figure 7.16 and recalling the symmetry of the arrangement, it is apparent that the LO current closes its path without entering the output branch (a) during the first

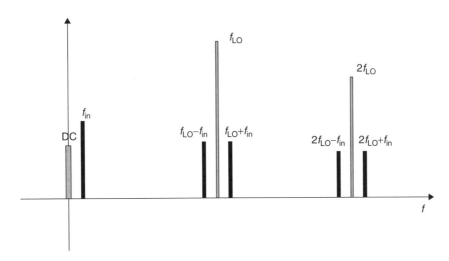

Figure 7.15 Combined (solid) and cancelled spectral lines (dotted) for the singly balanced mixer in Figure 7.14

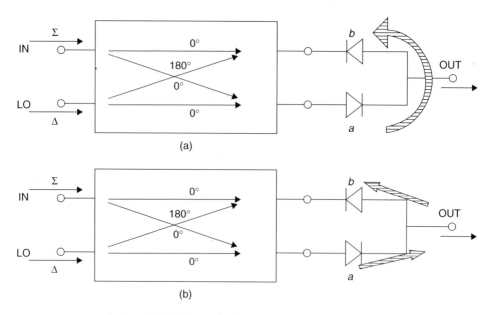

(a)

(b)

Figure 7.16   The paths for LO (a) and in currents (b)

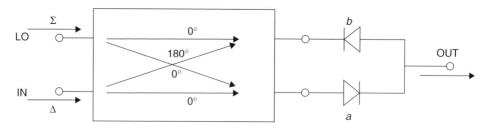

Figure 7.17   A singly balanced mixer with IN rejection at the output

half-period; it is blocked by the diodes during the second half-period. The input current on the other hand enters the output branch in order to close the path, through the upper arm of the coupler during the first half-period, and through the lower arm of the coupler during the second half-period.

Let us now interchange the input and local oscillator ports, as in Figure 7.17. It is easy to see that the output current is

$$i_{OUT} = i_a - i_b = v_{LO} + v_{LO}v_{IN} + v_{LO}v_{IN}^2 + v_{LO}^3 + \cdots \qquad (7.9)$$

The first and fourth terms are components at the local oscillator frequency and at its third-harmonic frequency. The second term is the mixed signal (see Introduction above) and provides the two sidebands of the local oscillator signal. The third term is the mixing of the second harmonic of the input signal and of its rectified DC term with the local oscillator. The former product can be neglected given the small amplitude of the input

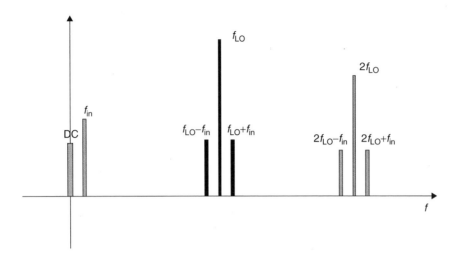

Figure 7.18 Combined (solid) and cancelled spectral lines (dotted) for the singly balanced mixer in Figure 7.17

signal; the latter product is an additional term at the frequency of the local oscillator. Then, there are higher-order terms that can be neglected to a first approximation. The input signal is cancelled by the symmetry of the configuration, while the local oscillator is still present. The situation is shown in Figure 7.18, in which the combined and cancelled terms are shown as solid and dotted bars respectively.

The singly balanced mixer in Figure 7.16 therefore has intrinsic isolation between the input port and the output port; it also has isolation between input port and local oscillator port. No isolation is provided between local oscillator and output ports.

By similar derivation, it can be seen that a hybrid coupler with anti-parallel single-diode mixers provides isolation between input and output ports only if the single-diode mixers are well matched; interchanging the input and local oscillator ports has no effect, given the symmetry of the coupler; and the output spectrum is as shown in Figure 7.19.

A peculiar and useful characteristic of the singly balanced mixers described above is the rejection of the AM noise from the local oscillator. This is easily seen by letting $v_{IN} = 0$ and replacing $v_{LO} \rightarrow v_{LO} + v_{noise}$ in eq. (7.5). It is easily seen that the noise is rejected at the output.

A subharmonically pumped mixer is a circuit that exploits the second harmonic of the local oscillator for mixing with the input signal. A simple balanced configuration that does not require a coupler is shown in Figure 7.20.

By carrying out the derivation as above, with the local oscillator signal and input signal fed in-phase to the two anti-parallel diodes, the output spectrum is as shown in Figure 7.21.

The peculiar features of this arrangement are the very simple circuit scheme without couplers; the low conversion losses (in case of diodes) due to the suppression of

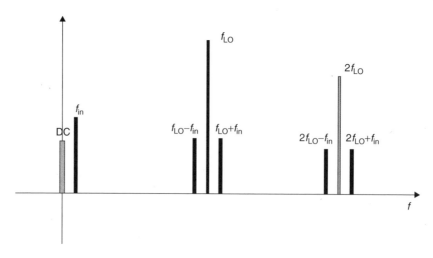

Figure 7.19 Combined (solid) and cancelled spectral lines (dotted) for a singly balanced mixer with a hybrid coupler

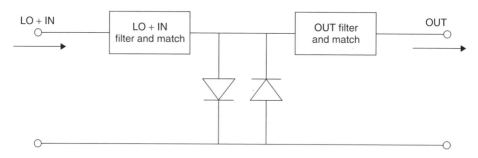

Figure 7.20 A subharmonically pumped mixer with an anti-parallel pair of single-diode mixers

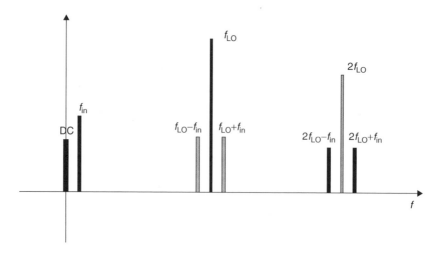

Figure 7.21 Combined (solid) and cancelled spectral lines (dotted) for the subharmonically pumped mixer in Figure 7.20

fundamental-frequency mixing products; and the large separation between input, local oscillator and output frequencies, which eases the suppression of unwanted components by filtering.

Other balanced schemes, in particular doubly balanced mixers, can be analysed by similar means and are widely treated in the literature (e.g. [1]).

7.3 MIXER DESIGN

In this paragraph, the optimum loading conditions for the local oscillator and for the small signals at the converted frequencies are described on the basis of large-signal considerations as in Section 7.2.1 and on the small-signal linear time-variant frequency-converting representation (the conversion matrix).

A mixer fulfils the conditions described in Section 1.4: a large signal (the local oscillator) pumps the nonlinear elements into large-signal regime, generating a number of harmonics and modulating its differential admittance. When a small input signal is superimposed, a whole spectrum of converted signals is generated as sidebands of the harmonics of the local oscillator. If the input signal is small, the conversion is linear, and the mixer can be seen as a linear frequency-converting n-port network, where the number of ports equals the number of physical ports (e.g. for a single-diode mixer, only one port) times the number of non-negligible harmonics of the local oscillator. The spectrum of the small signal is shown in Figure 7.22 and the small-signal equivalent network of the mixer is represented in Figure 7.23, corresponding to the admittance representation in eq. (1.152) in Chapter 1, repeated here for convenience, in a general form [2–7]:

$$\vec{I} + \vec{Y} \cdot \vec{V} = 0 \tag{7.10}$$

where

$$\vec{I} = \begin{bmatrix} I_N \\ \cdot \\ I_0 \\ \cdot \\ I_{-N} \end{bmatrix} \quad \vec{V} = \begin{bmatrix} V_N \\ \cdot \\ V_0 \\ \cdot \\ V_{-N} \end{bmatrix} \tag{7.11}$$

and the subscript indicates the frequency as shown in Figure 7.22.

For mixer design, the nonlinear pumping must first be applied, and the behaviour of the nonlinear element determined by means of a large-signal analysis or measurement; a few possible arrangements have been described above in Section 7.2.1. Once this is done, the conversion matrix is computed and the optimum values for the embedding admittances at the port frequencies must be determined. In fact, two ports are the actual input and output ports of the mixer: in the case of an upconverting mixer as in Figure 7.22, the input frequency is $f_{in} = f_0$ and the output frequency is $f_{out} = -f_{-1}$. We remember that in Chapter 1, the use of negative frequencies for the lower sidebands has been introduced for the sake of simplification in the notation. It must be remembered also that the phasors with negative index are the conjugate of the ordinary phasors since they correspond to negative frequencies; the relevant impedances must be conjugated accordingly.

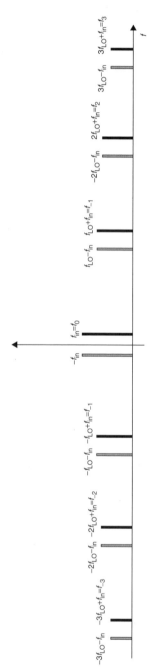

Figure 7.22   The spectrum of the frequency-converted signals in a mixers

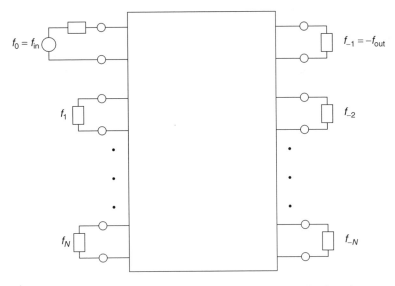

Figure 7.23   The small-signal linear equivalent network of a mixer

In principle, the embedding admittance at all converted frequencies contribute to the conversion gain of the mixer, as in any linear n-port network; in practice, only a few ports other than the input and output ones can provide an improvement to the performances of the circuit, among which the image frequency $f_{image} = f_1$ [1] is especially important. Therefore, as a first approximation, all ports other than input and output can be terminated with matched loads, or short circuits, without unacceptably degrading the conversion performances of the mixer. Therefore, we can reduce the n-port network to a two-port network by means of standard reduction techniques [1, 8] and treat the latter as a standard two-port network. We can therefore define a stability factor, stability circles and a maximum available gain or maximum stable gain [7–11]. Diode mixers invariably show unconditional stability, while active mixers can be unstable.

Let us come back to the preliminary large-signal analysis or tuning when the mixer is driven only by the local oscillator. First of all, it is worth remarking that the embedding admittances at the converted frequencies can be unknown during the large-signal analysis or tuning of the mixer under local oscillator pumping since only the harmonic frequencies of the local oscillator affect the large-signal steady state of the mixer. The embedding admittances are designed during the successive step, the proper termination of the linear network. At least in principle a linear parameter such as, for example, the maximum available gain of the linear reduced two-port frequency-converting network can be used as a quality factor during large-signal tuning. In practice, more empirical considerations are preferred. The load line as shown in Figure 7.7 must modulate the nonlinear parameters for the selected configuration in such a way as to ensure not only maximum conversion but also stability. Therefore, a nearly switching behaviour of the FET must be ensured, with the load line extending across regions with highly different behaviour. From a numerical point of view, the fundamental-frequency component of the modulated parameter (e.g. transconductance) must be maximised for maximum conversion gain. On the other hand, stability must be ensured: in particular, case 1 for a FET (gate mixing) seems to be the most problematic configuration. Short circuits at the drain termination at LO fundamental frequency and harmonics seems to be the best compromise in terms of conversion gain and stability [1, 9, 12, 13]; it also improves the isolation between the local oscillator and the output. As a general rule, the LO port must be conjugately matched for best use of the LO power.

As mentioned, the large-signal analysis is properly performed by means of non-linear analysis algorithms (see Chapter 1). In the case that a specific mixer configuration and a specific device have been selected, it is possible to perform approximated analysis that yields quite accurate results with significantly reduced numerical effort, or even with analytical expressions. For instance, explicit formulae have been derived for HEMT gate mixers, assuming that only the gate–source capacitance, the transconductance and the output conductance are nonlinear [14]. Similarly, analytic formulae have been derived for an MESFET drain mixer under the assumption that only transconductance and output conductance behave nonlinearly [15]. Resistive 'channel' mixers also have been analysed by assuming a simple nonlinear circuit for an FET, where the only nonlinearity is the channel resistance [16].

Once the large-signal analysis is performed, the small-signal equivalent circuit is available for conversion optimisation. The typical strategy for the spurious terms consists of shorting the unwanted terminations, for example, output load at input signal frequency and *vice versa*. This approach tends to improve isolation between unwanted ports, to improve stability and to minimise noise generation. The input and output ports at the corresponding frequencies are conjugately matched for the input and output match and for maximum power transfer from the input to the output and maximum conversion gain. This can be practically achieved by making use of simple expressions derived from the equivalent circuit of the active device [1].

7.4 NONLINEAR ANALYSIS

In this paragraph, the nonlinear techniques for the analysis of mixers are described; in particular, an extension of the Volterra series is described for the prediction of intermodulation distortion.

In principle, mixers can be analysed by any nonlinear method that can manage two tones as input signals, one of which is very large (the local oscillator) and the other is very small (the input signal). Therefore, time-domain direct integration or harmonic/spectral balance methods are suitable algorithms, since they can handle very strong nonlinearities and two-tone analysis. Volterra series, as has been described in Chapter 1, is not suitable, because it is limited to mild nonlinear problems. However, there are numerical problems with the above-mentioned methods when they are applied to the analysis of a mixer. In practice, the numerical noise generated by truncations and approximations in the nonlinear analysis of a large signal (the LO) is comparable to the small input and output signals. A better approach consists of separating the two analyses: first, the large local oscillator signal is analysed by means of any of the methods seen, for example, in Chapter 1, for a single-tone input. Then, the small input signal is added as a small perturbation.

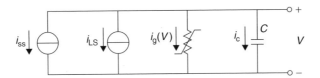

Figure 7.24   The example circuit for the calculation of the conversion matrix

This approach has been described in detail in Chapter 1 and is known as the conversion matrix approach, or sometimes it is referred to as large-signal/small-signal analysis. Within this approach, the input signal causes a perturbation of the large-signal trajectory determined by the large signal (the local oscillator). The perturbation is assumed to be so small that its effect can be linearised; for instance, in our example circuit (Figure 7.24) the current in the nonlinear conductance can be expressed as (eq. (1.133) and eq. (1.134), combined here for convenience)

$$i_g(t) = i_{g,LS}(t) + i_{g,ss}(t) \cong i_{g,LS}(t) + \left. \frac{di_g(v)}{dv} \right|_{v=v_{LS}(t)} \cdot v_{ss}(t) + \cdots$$

$$= i_{g,LS}(t) + g_{ss}(t) \cdot v_{ss}(t) + \cdots \tag{7.12}$$

By this approach, the frequency-conversion properties of the circuit are easily predicted; they are generated by the periodic time-dependence of the linearised conductance. The resulting spectrum is depicted in Figure 1.44 in Chapter 1. The circuit behaves as a linear time-dependent circuit, in which the small-signal incremental parameters (as e.g. the nonlinear conductance in Figure 7.24) are dependent on a time-variant bias voltage. However, eq. (7.12) holds only for small amplitudes of the input signal, where the circuit is linear with respect to the small input signal; therefore, it cannot predict phenomena such as distortion in the frequency conversion when the input signal has a moderately large amplitude or intermodulation between two moderately large input signals. These phenomena are due to the fact that the higher-order terms of the Taylor series in (7.12) cannot be neglected any more when the input signal is not so small. These phenomena are important because they limit the dynamic range of a mixer as a linear frequency-converting circuit, as much as they limit the dynamic range of a linear amplifier biased at a static operating point. In the case of a two-tone input signal, the spectrum of the output signal becomes as shown in Figure 7.25, in which only the portion of the spectrum around the fundamental component of the local oscillator is shown.

For the prediction of these nonlinear phenomena, the series expansion as in (7.12) is no more sufficient, and additional terms must be included [17–23]. The equation becomes

$$i_g(t) = i_{g,LS}(t) + i_{g,ss}(t) \cong i_{g,LS}(t) + \left. \frac{di_g(v)}{dv} \right|_{v=v_{LS}(t)} \cdot v_{ss}(t)$$

$$+ \frac{1}{2} \left. \frac{d^2 i_g(v)}{dv^2} \right|_{v=v_{LS}(t)} \cdot v_{ss}^2(t) + \frac{1}{6} \left. \frac{d^3 i_g(v)}{dv^3} \right|_{v=v_{LS}(t)} \cdot v_{ss}^3(t) + \cdots$$

$$= i_{g,LS}(t) + g_1(t) \cdot v_{ss}(t) + g_2(t) \cdot v_{ss}^2(t) + g_3(t) \cdot v_{ss}^3(t) + \cdots \tag{7.13}$$

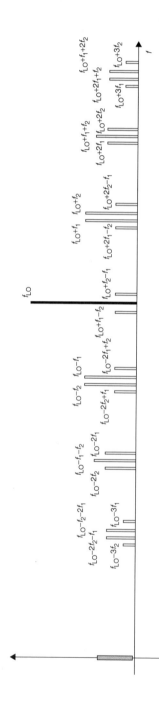

Figure 7.25 The spectrum of the output signal in the case of a moderately large two-tone input signal

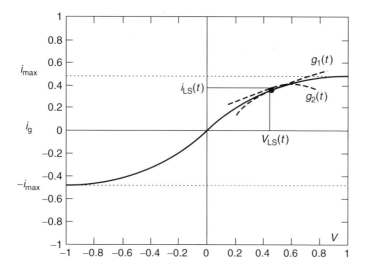

Figure 7.26 Graphical illustration of the first two coefficients in eq. (7.13)

Equation (7.13) is equivalent to eq. (1.44); however, in this case, the coefficients of the Taylor series expansion are dependent on time since the dynamic operating point varies with the instantaneous value of the local oscillator waveform. The first and second derivatives of the nonlinear current with respect to the voltage are shown in Figure 7.26 at a generic operating point, for the sake of illustration.

In the case of our example circuit, they have the following expressions:

$$g_1(v) = \frac{di_g(v)}{dv} = g \cdot \left(1 - tgh^2\left(\frac{g \cdot v}{i_{max}}\right)\right)$$

$$g_2(v) = \frac{d^2 i_g(v)}{dv^2} = \frac{2g^2}{i_{max}} \cdot tgh\left(\frac{g \cdot v}{i_{max}}\right) \cdot \left(tgh^2\left(\frac{g \cdot v}{i_{max}}\right) - 1\right) \qquad (7.14)$$

When driven by a large-signal periodic waveform, the conductances are expressed as

$$g_1(t) = g_1(v_{LS}(t)) = \sum_{m=-\infty}^{\infty} G_{1,m} \cdot e^{jm\omega_{LS}t} \qquad g_2(t) = g_2(v_{LS}(t)) = \sum_{m=-\infty}^{\infty} G_{2,m} \cdot e^{jm\omega_{LS}t}$$

$$(7.15)$$

and give rise to frequency conversion. Therefore, the output signal can still be expressed as a Volterra series, but now the nuclei are time-dependent; frequency conversion takes place for the higher-order components of the spectrum also.

Kirchhoff's equation for the small perturbation signal reads as (see eq. (1.138))

$$i_{ss}(t) + C \cdot \frac{dv_{ss}(t)}{dt} + g_1(t) \cdot v_{ss}(t) + g_2(t) \cdot v_{ss}^2(t) + \cdots = 0 \qquad (7.16)$$

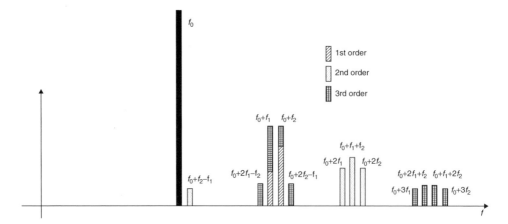

Figure 7.27   Contribution of the different orders to the spectrum of the output signal

The small-signal voltage can therefore be expanded in Volterra series as (see eq. (1.53))

$$v_{ss}(t) = v_{ss,1}(t) + v_{ss,2}(t) + \cdots \tag{7.17}$$

The contributions of the orders of the series are shown in Figure 7.27 for the upper sideband of the fundamental-frequency component of the local oscillator.

By replacing eq. (7.17) in (7.16) in orders and keeping only the first-order terms, we get

$$i_{ss}(t) + C \cdot \frac{dv_{ss,1}(t)}{dt} + g_1(t) \cdot v_{ss,1}(t) = 0 \tag{7.18}$$

where

$$i_{ss}(t) = I^{(1)} \cdot \sin(\omega_1 t) + I^{(2)} \cdot \sin(\omega_2 t) \tag{7.19}$$

is the two-tone excitation. The first-order nucleus is simply the conversion matrix; the first-order voltages are expressed as

$$v_{ss,1}^{(1)}(t) = \sum_{m=-\infty}^{\infty} V_{ss,1,m}^{(1)} \, e^{j(m\omega_{LS}+\omega_1)t}$$

$$v_{ss,1}^{(2)}(t) = \sum_{m=-\infty}^{\infty} V_{ss,1,m}^{(2)} \, e^{j(m\omega_{LS}+\omega_2)t} \tag{7.20}$$

where the superscript refers to the frequency of the input tone. The first-order Kirchhoff's equation can be written in the frequency domain as

$$\vec{I}_{ss,1} + \vec{C} \cdot \vec{V}_{ss,1} + \vec{G}_1 \cdot \vec{V}_{ss,1} = 0 \quad \vec{I}_{ss,2} + \vec{C} \cdot \vec{V}_{ss,2} + \vec{G}_1 \cdot \vec{V}_{ss,2} = 0 \tag{7.21}$$

where the vectors of voltage and current phasors and the conversion matrices are defined as in Section 1.4. The equation splits into two independent equations because of the linearity of the first order, by definition.

Once the first-order Kirchhoff's equation has been solved and the first-order terms have been found, the second-order Kirchhoff's equation can be written as

$$C \cdot \frac{dv_{ss,2}(t)}{dt} + g_1(t) \cdot v_{ss,2}(t) + g_2(t) \cdot v_{ss,1}^2(t) = 0 \tag{7.22}$$

The spectrum of the second-order terms (see Figure 7.26) is deduced from the known term:

$$v_{ss,1}^2(t) = (v_{ss,1}^{(1)}(t) + v_{ss,1}^{(2)}(t))^2 \tag{7.23}$$

where the first-order terms are expressed as in eq. (7.20). The second-order Kirchhoff's equation can be written in the frequency domain as

$$\vec{C} \cdot \vec{V}_{ss,2} + \vec{G}_1 \cdot \vec{V}_{ss,2} + \vec{G}_2 \cdot \vec{V}_{ss,1}^2 = 0 \tag{7.24}$$

where the vectors of voltage phasors and the second-order conversion matrix are defined as for the first-order terms. The second-order voltage phasors are found from the linear system of equations (7.24). Recursively, all higher-order terms are found, similar to the standard Volterra series analysis, and the weak nonlinear behaviour of the mixer with respect to the two-tone input signal is found. Therefore, the compression, distortion and intermodulation effects of the mixer are found, and the dynamic range is determined.

7.5 NOISE

In this paragraph, the noise behaviour of mixers is described with reference to its nonlinear origin and characteristics.

As a first approach, noise can be considered as a small signal, whose spectral statistical properties are known. The behaviour of a mixer with respect to external noise is then, to a first approximation, the same as with respect to any small deterministic signal present at its ports. We remark that a mixer can be seen from the point of view of an external small signal as a linear frequency-converting n-port network, where each port corresponds to a physical port at a sideband frequency (Figure 7.23). Therefore, the noise present at the output port due to external sources is the superposition of the input noise at all ports, each multiplied by the relevant conversion loss (Figure 7.28). If the external noise sources are uncorrelated, as for instance in the case that they are due to a lossy external linear circuit, the converted contributions are uncorrelated and add in power [24].

However, to this noise of external origin, all circuits add noise generated by the internal elements of the circuits themselves; this added noise is the origin of the noise figure being greater than 1. If the frequency components of the noise of internal origin at the various mixer ports are uncorrelated, then the noise is treated as the uncorrelated external noise. However, this assumption is true only for some of the internal noise sources.

Noise in semiconductor devices has different physical origins that account for the different behaviours that are encountered; for instance, shot noise and thermal noise are

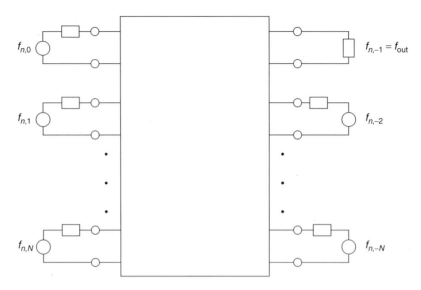

Figure 7.28 Frequency conversion of the external noise by the mixer

essentially white, while $\dfrac{1}{f}$ noise has the dependence on frequency, as its name implies. $\dfrac{1}{f}$ noise is rarely of importance in mixers, unless at very low frequency of operation. Thermal noise is independent of the bias state of the device and behaves very much as in a linear resistive device. Shot noise, on the other hand, depends on the current level within the device; its mean-square current can be written as

$$\overline{i^2}_{shot} = 2qI \cdot B \tag{7.25}$$

where q is the electron charge, I is the current through the device and B is the bandwidth. If the device is biased at a static operating point, the spectrum of the shot noise is white, and its components at different frequencies are uncorrelated.

However, if the active device is driven by a large-signal periodic current, as in a mixer pumped by the local oscillator signal, then the shot noise components generated at different frequencies are correlated since the noise-generating phenomenon is modulated at the local oscillator frequency. The correlation coefficients between the components at two different ports m and n are the harmonics of the local oscillator current [5, 6, 25–30]:

$$C_{m,n} = 2qI_{m-n} \cdot B \tag{7.26}$$

The relation (7.26) is intuitively justified as being equivalent to a conversion-matrix term that connects two frequency components at different ports.

Once generated internally with a degree of correlation due to current modulation, the noise signal is frequency-converted as much as the external noise. However, when superimposing at the output port the converted contributions from the various ports, their original correlation must be accounted for, and they do not simply add in power.

In case other noise-generating mechanisms are present in the device, they can simply be superimposed, provided the generating mechanisms are uncorrelated; otherwise, they must be included in the formulation [31].

This fundamentally nonlinear internal noise correlation is such that the noise properties of a mixer cannot be represented by a simple resistive model; in particular, noise level is increased with respect to a statically biased semiconductor device. Given the formulation above, prediction of correct noise levels are easily obtained once the large-signal local oscillator is known, and the conversion matrix is computed. Obviously, the physical mechanisms at the origin of the noise, which determine the correlation properties of the internal noise at the mixer ports, must be known. This is by no means a trivial problem; however, many noise sources are known well enough for effective noise prediction within nonlinear analysis methods such as, for example, harmonic balance.

7.6 BIBLIOGRAPHY

[1] S.A. Maas, *Microwave Mixers*, Artech House, Norwood (MA), 1993.

[2] H.C. Torry, C.H. Whitmer, *Crystal Rectifiers*, McGraw-Hill, New York (NY), 1948.

[3] C.F. Edwards, 'Frequency conversion by means of a nonlinear admittance', *Bell Syst. Tech. J.*, **35**, 1403–1416, 1956.

[4] S. Egami, 'Nonlinear-linear analysis and computer-aided design of resistive mixers', *IEEE Trans. Microwave Theory Tech.*, **MTT-22**, 270–275, 1974.

[5] D.N. Held, A.R. Kerr, 'Conversion loss and noise of microwave and millimetre-wave mixers: part 1 - theory', *IEEE Trans. Microwave Theory Tech.*, **MTT-26**, 49–55, 1978.

[6] A.R. Kerr, 'Noise and loss in balanced and subharmonically pumped mixers: part 1 - theory', *IEEE Trans. Microwave Theory Tech.*, **MTT-27**, 938–943, 1979.

[7] S.A. Maas, 'Theory and analysis of GaAs mixers', *IEEE Trans. Microwave Theory Tech.*, **MTT-32**, 1402–1406, 1984.

[8] B. Schüppert, 'A fast and reliable method for computer analysis of microwave mixers', *IEEE Trans. Microwave Theory Tech.*, **34**(19), 110–119, 1986.

[9] M.J. Rosario, J. Costa Freire, 'Design techniques for MESFET mixers for maximum conversion gain', *IEEE Trans. Microwave Theory Tech.*, **38**(12), 1972–1979, 1990.

[10] Y.H. Liew, J. Joe, 'RF and IF ports matching circuit synthesis for a simultaneous conjugate-matched mixer using quasi-linear analysis', *IEEE Trans. Microwave Theory Tech.*, **50**(9), 2056–2062, 2002.

[11] F. Giannini, G. Leuzzi, E. Limiti, F. Morgia, 'Optimum nonlinear design of active microwave mixers', *Proc. GAAS 2000*, Paris (France), Oct. 2000.

[12] P. Harrop, T.A.C.M. Clasem, 'Modelling of a FET mixer', *Electron. Lett.*, **14**(12), 369–370, 1978.

[13] C. Camacho-Peñalosa, C.S. Aitchison, 'Analysis and design of MESFET gate mixers', *IEEE Trans. Microwave Theory Tech.*, **MTT-35**, 643–652, 1987.

[14] Y.K. Kwoon, D. Pavlidis, 'Phasor diagram analysis of millimetre-wave HEMT mixers', *IEEE Trans. Microwave Theory Tech.*, **MTT-43**(9), 2165–2167, 1995.

[15] J.C.H. Cayrou, M. Gayral, J. Graffeuil, J.F. Sautereau, 'Simple expression for conversion gain of MESFET drain mixers', *Electron. Lett.*, **29**(17), 1514–1516, 1993.

[16] S. Peng, 'A simplified method to predict the conversion loss of FET resistive mixers', *IEEE MTT-S Symp. Dig.*, 1997, pp. 857–860.

[17] L.M. Orloff, 'Intermodulation analysis of crystal mixers', *Proc. IEEE*, **52**, 173–179, 1964.

[18] J.G. Gandiner, A.M. Yousif, 'Distortion performance of single-balanced diode modulators', *Proc. Inst. Electron. Eng.*, **117**(8), 1609–1614, 1970.

[19] R.B. Swerdlow, 'Analysis of intermodulation noise in frequency converters by volterra series', *IEEE Trans. Microwave Theory Tech.*, **MTT-26**(4), 305–313, 1978.

[20] S.A. Maas, 'Two-tone intermodulation in diode mixers', *IEEE Trans. Microwave Theory Tech.*, **MTT-35**(3), 307–314, 1987.

[21] S. Peng, P. McCleer, G.I. Haddad, 'Nonlinear models for intermodulation analysis of FET mixers', *IEEE Trans. Microwave Theory Tech.*, **MTT-43**, 1037–1045, 1995.

[22] J.A. Garcia, M.L. De la Fuente, J.C. Pedro, N.B. Carvalho, Y. Newport, A. Mediavilla, A. Tazon, 'Time-varying volterra-series analysis of spectral regrowth and noise power ratio in FET mixers', *IEEE Trans. Microwave Theory Tech.*, **49**(3), 545–549, 2001.

[23] J. Kim, Y. Kwon, 'Intermodulation analysis of dual-gate FET mixers', *IEEE Trans. Microwave Theory Tech.*, **MTT-50**(6), 1544–1555, 2002.

[24] V. Rizzoli, F. Mastri, C. Cecchetti, 'Computer-aided noise analysis of MESFET and HEMT mixers', *IEEE Trans. Microwave Theory Tech.*, **MTT-37**(9), 1401–1410, 1989.

[25] C. Dragone, 'Analysis of thermal shot noise in pumped resistive diodes', *Bell Syst. Tech. J.*, **47**, 1883–1902, 1968.

[26] A.A.M. Saleh, *Theory of Resistive Mixers*, M.I.T. Press, Cambridge (MA), 1971.

[27] V. Rizzoli, A. Neri, 'State of the art and present trends in nonlinear microwave CAD techniques', *IEEE Trans. Microwave Theory Tech.*, **36**, 343–365, 1988.

[28] V. Rizzoli, C. Cecchetti, A. Lipparini, F. Mastri, 'General-purpose harmonic-balance analysis of nonlinear microwave circuits under multitone excitation', *IEEE Trans. Microwave Theory Tech.*, **MTT-36**(12), 1650–1660, 1988.

[29] V. Rizzoli, 'General noise analysis of nonlinear microwave circuits by the piecewise harmonic balance technique' *IEEE Trans. Microwave Theory Tech.*, **MTT-42**(5), 807–819, 1994.

[30] S. Heinen, J. Kunish, I. Wolff, 'A unified framework for computer-aided noise analysis of linear and nonlinear microwave circuits', *IEEE Trans. Microwave Theory Tech.*, **MTT-39**(12), 2170–2175, 1991.

[31] V. Rizzoli, F. Mastri, D. Masotti, 'Advanced piecewise-harmonic-balance noise analysis of nonlinear microwave circuits with application to Schottky-barrier diodes', *IEEE MTT-S Int. Symp. Dig.*, 1992, pp. 247–250.

8

Stability and Injection-locked Circuits

8.1 INTRODUCTION

In this introduction, the main circuits in which an externally forced signal and an autonomous oscillatory signal coexist are described.

So far, circuits in which either an externally forced signal or an autonomous, internally generated oscillatory signal is present have been described. The former group includes power amplifiers, frequency doublers and mixers; for these circuits, the frequency of operation is determined by the external excitation, and the primary job of the designer is to ensure good match and high gain for the incident signal. The latter group includes oscillators, and the primary job of the designer is to ensure the existence of the signal and the stability of its frequency and amplitude. In this chapter, circuits in which an externally forced signal coexists with an internally generated, oscillatory signal, whose existence and properties must be ensured by proper design, are described. The circuits described in this chapter are injection-locked oscillators, non-regenerative frequency dividers and multipliers and self-oscillating mixers. A typical feature of these circuits is that the oscillation takes place in a circuit in nonlinear regime, requiring an extension of the oscillation and stability criteria described for basically linear circuits in the Chapter 5. This is done preliminarily in this chapter, with a short description of the onset of instabilities in nonlinear circuits. The relevant stability criterion can be used for determining whether a nonlinear steady state is stable or if it is an unstable solution of the Kirchhoff's equations, only numerically existent but never reached by the circuit.

8.2 LOCAL STABILITY OF NONLINEAR CIRCUITS IN LARGE-SIGNAL REGIME

In this paragraph, the stability criterion for nonlinear circuits in large-signal regime is introduced, together with the main numerical techniques for its evaluation.

Nonlinear Microwave Circuit Design   F. Giannini and G. Leuzzi
© 2004 John Wiley & Sons, Ltd   ISBN: 0-470-84701-8

In Chapter 5, the criterion for stability or oscillation has been given for linear circuits. In fact, linear electronic circuit always include nonlinear elements, nonlinearly biased by a large DC source in a suitable operating region and linearised around the static bias point. A basic assumption for linear dynamic operations is that the alternate signals superimposed to the static bias be small enough for the dynamic operating point to keep within a region around the static bias point where linearisation holds. If this is the case, the nonlinear element (the active device) can be replaced by a linearised equivalent, for example, a small-signal equivalent circuit. Since the bias point is static, or time-invariant, the linearised representation is time-invariant as well and is studied by means of the usual techniques for linear circuits. In the case of oscillators, the stability properties of a small alternate signal perturbing the steady state are studied in order to suppress or allow the onset of oscillations.

Nonlinear circuits or elements operating under large-signal regime can be considered as driven by time-variant large sources that include both the static bias sources and the large alternate signal source such that the operating point is dynamically driven through regions wherein the behaviour of the active element cannot be linearised. In other words, the alternate signal superimposed to the static bias point is so large that the dynamic load curve of the active device extends to regions where the device behaves nonlinearly. Therefore, the alternate signal cannot be seen as a small, linear perturbation of a static bias point; rather, the sum of the bias source and alternate signal must be considered simultaneously as a single large source nonlinearly driving the active device. This is the steady-state regime whose stability we want to investigate by studying the properties of a small perturbation to this steady state. Linearisation around the dynamic operating point is still possible but yields a time-variant linearised representation; therefore, small perturbations of the nonlinear regime is expected to have a different behaviour from perturbations in a linear(ised) time-invariant circuit.

The case of a small alternate signal superimposed to a large steady-state periodic signal has already been described in Section 1.4 and applied in the Chapter 7, where the small signal includes the input signal and all the frequency-converted signals. In the case of a mixer, the small-signal circuit is a linear(ised) incremental forced circuit, as said above. In the case of nonlinear stability analysis, the small signal is a perturbation of the steady state, whose stability is thus investigated; the small signal therefore has no external excitation, and the linearised incremental network is in fact an autonomous circuit at the sideband frequencies, whose self-oscillations must be investigated. The conversion-matrix formalism is still used, but the small-signal time-variant problem is now autonomous.

Linear periodic time-variant systems have been studied by Floquet two centuries ago by means of an analytical approach [1, 2]. We will now see how its results are equivalent to those of a conversion-matrix approach. Let us study a linear second-order differential equation with periodic coefficients in the form

$$x''(t) + a(t) \cdot x'(t) + b(t) \cdot x(t) = 0 \tag{8.1}$$

where $a(t)$ and $b(t)$ are periodic:

$$a(t + T) = a(t) \quad b(t + T) = b(t) \tag{8.2}$$

Floquet states that the solution has the form

$$x(t) = X_1 \cdot p_1(t) \cdot e^{\mu_1 \cdot t} + X_2 \cdot p_2(t) \cdot e^{\mu_2 \cdot t} \tag{8.3}$$

where

$$p_1(t + T) = p_1(t) \quad p_2(t + T) = p_2(t) \tag{8.4}$$

are periodic functions with the same periodicity of the coefficients, and where the Floquet coefficients μ_1 and μ_2 are

$$\mu_1 = v_1 + j\frac{2\pi \cdot k}{T} \quad k_1 = \pm 1,\ \pm 2, \ldots \quad \mu_2 = v_2 + j\frac{2\pi \cdot k}{T} \quad k_2 = \pm 1, \pm 2, \ldots \tag{8.5}$$

where v_1 and v_2 are complex numbers. The multiple values of the Floquet coefficients correspond in fact to the same behaviour of the solution. The Floquet coefficients as well as the coefficients X_1 and X_2 are found from the boundary conditions for the specific problem.

Let us apply this result to a series resonant RLC circuit in which the resistance and the capacitance are nonlinear, driven by a periodic local oscillator signal with frequency $\omega_0 = \dfrac{2\pi}{T}$. Let also assume that the periodic steady state has been found by any nonlinear analysis method. Linearisation around the periodic steady state yields a linear time-variant circuit as in Figure 8.1 (see Section 1.4),

where $R_{\text{active}}(t)$ and $C(t)$ are known periodic functions:

$$R_{\text{active}}(t + T) = R_{\text{active}}(t) \quad C(t + T) = C(t) \tag{8.6}$$

because of the periodicity of the local oscillator. Kirchhoff's voltage law equation reads as

$$R(t) = R_{\text{passive}} + R_{\text{active}}(t) \quad i(t) = C(t) \cdot \frac{dv_\text{c}(t)}{dt}$$

$$L \cdot \frac{di(t)}{dt} + R(t) \cdot i(t) + v_\text{c}(t) = 0 \tag{8.7}$$

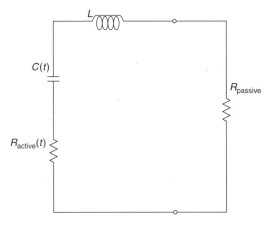

Figure 8.1   A time-variant linearised RLC series resonant circuit

$$\frac{d^2 v_c(t)}{dt^2} + \frac{R(t)}{L} \cdot \frac{dv_c(t)}{dt} + \frac{v_c(t)}{LC(t)} = 0 \qquad (8.8)$$

The general solution is

$$v_c(t) = V_1 \cdot v_{p1}(t) \cdot e^{\mu_1 t} + V_2 \cdot v_{p2}(t) \cdot e^{\mu_2 t} \qquad (8.9)$$

where

$$v_{p1}(t + T) = v_{p1}(t) \quad v_{p2}(t + T) = v_{p2}(t) \qquad (8.10)$$

while V_1 and V_2 are constants depending on the boundary conditions.

It is easily seen that eq. (8.8) reduces to the usual equation of an RLC circuit if the resistance and the capacitance are constant.

The periodic functions $v_{p1}(t)$ and $v_{p2}(t)$ can be expanded in Fourier series as

$$v_{p1}(t) = \sum_n V_{1,n} \cdot e^{jn\omega_0 t} \quad v_{p2}(t) = \sum_n V_{2,n} \cdot e^{jn\omega_0 t} \qquad (8.11)$$

and so also the linearised resistance and capacitance:

$$R(t) = R_{\text{passive}} + \sum_n R_{\text{active},n} \cdot e^{jn\omega_0 t} \quad C(t) = \sum_n C_n \cdot e^{jn\omega_0 t} \qquad (8.12)$$

or more conveniently,

$$\frac{R(t)}{L} = \sum_n A_n \cdot e^{jn\omega_0 t} \quad \frac{1}{L \cdot C(t)} = \sum_n B_n \cdot e^{jn\omega_0 t} \qquad (8.13)$$

Replacing the above expressions in Kirchhoff's equation and truncating the series expansions, we get

$$\frac{d^2 \left(\sum_n V_n \cdot e^{(\mu + jn\omega_0 t)} \right)}{dt^2} + \sum_m A_m \cdot e^{jm\omega_0 t} \cdot \frac{d \left(\sum_n V_n \cdot e^{(\mu + jn\omega_0 t)} \right)}{dt}$$

$$+ \sum_m B_m \cdot e^{jm\omega_0 t} \cdot \sum_n V_n \cdot e^{(\mu + jn\omega_0 t)} = 0 \qquad (8.14)$$

where A_m and B_m are known coefficients, and V_n and μ are unknown. This is in fact the conversion-matrix formalism (see Section 1.4). By balancing each harmonic term separately, we get a homogeneous system of equations in the coefficients V_n, whose determinant must be zero for a non-trivial solution. This last condition yields the value of the Floquet coefficients μ. In matrix form,

$$
\begin{bmatrix}
(\mu + jN\omega)^2 & \cdots & \cdots & \cdots & 0 \\
\cdots & \cdots & \cdots & \cdots & \cdots \\
\cdots & \cdots & \mu^2 & \cdots & \cdots \\
\cdots & \cdots & \cdots & \cdots & \cdots \\
0 & \cdots & \cdots & \cdots & (\mu - jN\omega)^2
\end{bmatrix}
\cdot
\begin{bmatrix}
V_N \\
\cdots \\
V_0 \\
\cdots \\
V_{-N}
\end{bmatrix}
$$

$$
+
\begin{bmatrix}
A_0 & \cdots & A_N & \cdots & A_{2N} \\
\cdots & \cdots & \cdots & \cdots & \cdots \\
A_{-N} & \cdots & A_0 & \cdots & A_N \\
\cdots & \cdots & \cdots & \cdots & \cdots \\
A_{-2N} & \cdots & A_{-N} & \cdots & A_0
\end{bmatrix}
\cdot
\begin{bmatrix}
(\mu + jN\omega) & \cdots & \cdots & \cdots & 0 \\
\cdots & \cdots & \cdots & \cdots & \cdots \\
\cdots & \cdots & \mu & \cdots & \cdots \\
\cdots & \cdots & \cdots & \cdots & \cdots \\
0 & \cdots & \cdots & \cdots & (\mu - jN\omega)
\end{bmatrix}
$$

$$
\cdot
\begin{bmatrix}
V_N \\
\cdots \\
V_0 \\
\cdots \\
V_{-N}
\end{bmatrix}
+
\begin{bmatrix}
B_0 & \cdots & B_N & \cdots & B_{2N} \\
\cdots & \cdots & \cdots & \cdots & \cdots \\
B_{-N} & \cdots & B_0 & \cdots & B_N \\
\cdots & \cdots & \cdots & \cdots & \cdots \\
B_{-2N} & \cdots & B_{-N} & \cdots & B_0
\end{bmatrix}
\cdot
\begin{bmatrix}
V_N \\
\cdots \\
V_0 \\
\cdots \\
V_{-N}
\end{bmatrix}
=
\begin{bmatrix}
0 \\
\cdots \\
0 \\
\cdots \\
0
\end{bmatrix}
\tag{8.15}
$$

In compact notation,

$$
\overset{\leftrightarrow}{D}(\mu) \cdot \vec{V} = 0 \tag{8.16}
$$

whence

$$
\det(\overset{\leftrightarrow}{D}(\mu)) = 0 \tag{8.17}
$$

As mentioned, the solution of eq. (8.17) yields the Floquet coefficients corresponding to the non-trivial solution. In general,

$$
\mu_i = \alpha_i + j\beta_i \tag{8.18}
$$

where i denotes the ith solution of eq. (8.17). If at least one of the Floquet coefficients has positive real part

$$
\alpha_i > 0 \tag{8.19}
$$

then the periodic steady state is unstable; otherwise, all solutions decay to zero for $t \to \infty$, and the periodic steady state is stable.

It is apparent that the Floquet coefficients have the role of the Laplace parameter $s = \alpha + j\omega$ and that the evolution of a small perturbation to a periodic regime can be studied in the Laplace domain by means of the conversion matrix. Stating it differently, when a small perturbation of the form

$$
v(t) = v_0 \cdot e^{st} \tag{8.20}
$$

is applied to the periodic large-signal steady state, the perturbation is frequency-converted to all the sidebands of the periodic large signal, generating components at the (complex) frequencies $jn\omega_0 \pm s$, with $n = 0, \ldots N$. Since no forcing term is present other than the small perturbation, its free evolution is ruled by the autonomous equation system

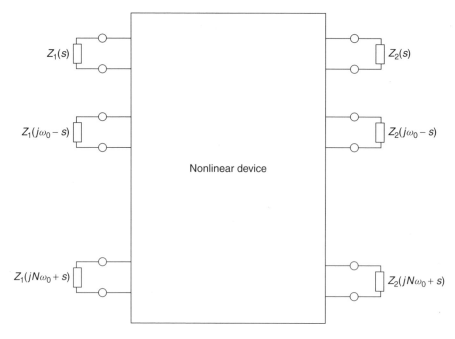

Figure 8.2   The multi-frequency network for the stability analysis of the periodic large-signal steady state of a nonlinear circuit including a two-port nonlinear active device

(8.15) or (8.16), which describes the behaviour of the time-variant, frequency-converting linear(ised) multi-port network as in Figure 8.2, where the two-port active device is the frequency-converting element.

If we rewrite system (8.15) or (8.16) as a function of Laplace parameter s, the network is stable or unstable depending on the sign of the real part of the solutions s_0 to the equation

$$\det(\overset{\leftrightarrow}{D}(s)) = 0 \tag{8.21}$$

As in a linear time-invariant oscillator, the stability criterion must be checked in order to find out whether the perturbation to the periodic large-signal steady state will stimulate the onset of oscillations or will decay exponentially to zero for $t \to \infty$. If the stability criterion for the network is satisfied, then the perturbation will decay to zero and the large-signal periodic steady state is stable; otherwise, a growing instability with complex frequency $s_0 = \alpha_0 + j\omega_0$ with $\alpha_0 > 0$ will establish itself in the circuit. The stability criterion is similar to the one described in Chapter 5 in linear conditions; only a multi-frequency network is involved, because of the time-dependence of the linear network.

The general stability criterion derived above, stating that a large-signal periodic regime is stable if the determinant of the conversion matrix has no zeroes with positive real part, can be verified by means of Nykvist's criterion [3]. The determinant (8.21) is computed for $s = j\omega$, and the function $f(\omega) = \det(\overset{\leftrightarrow}{D}(j\omega))$ as ω goes from $\omega = -\infty$ to $\omega = +\infty$ is plotted on the complex plane. The number of clockwise encirclements of the

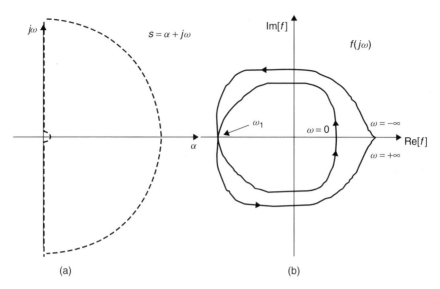

Figure 8.3   Contour in the plane of Laplace parameter (a) for the application of Nykvist's criterion (b)

origin of the complex plane gives the number of zeroes of the determinant with positive real part (Figure 8.3).

It is not actually necessary to compute the determinant in an infinite frequency range. First of all, it can be seen that for real signals,

$$f(\omega) = f^*(-\omega) \tag{8.22}$$

This condition ensures that the function $f(\omega)$ is symmetrical with respect to the real axis of the complex plane and that only positive frequencies must be computed. Moreover, given the periodicity of the large-signal steady state and assuming an infinite number of terms in the series expansion, the curve is periodic in the frequency ω and must be computed only from $\omega = 0$ to $\omega = \dfrac{\omega_0}{2}$ [3]. This can be intuitively explained by considering that an instability at a frequency in the range between two harmonics of the periodic large signal will be frequency-converted to all other sidebands, causing an encirclement of the origin in each frequency range between two neighbouring harmonics. In order to correctly compute the number of encirclements of the origin, it is necessary to multiply the determinant by the exponential term

$$e^{j\pi \frac{p_{max} \cdot \omega}{\omega_0}} \tag{8.23}$$

where p_{max} is the maximum order of derivatives with respect to time within the characteristic equations of the circuit elements (usually, $p_{max} = 1$). The exponential term has a zero for $s \to \infty$, removing the pole of the determinant [3].

Nykvist's plot also gives an estimation of the frequency of the unstable perturbation. At the frequency ω_1 where the determinant has zero imaginary part (see

Figure 8.3), the phase condition of Barkhausen's criterion is satisfied; its real part is negative, indicating the onset of an instability. Therefore, ω_1 is probably not far from the steady-state oscillation frequency of the perturbation (see Section 5.2). It must be remembered, however, that the nonlinear nature of the circuit does not allow any reliable prediction of the final state of the circuit when the perturbation grows up towards equilibrium.

Nykvist's criterion is very general and elegant and is suitable for implementation within general-purpose harmonic balance CAD programs. However the choice of the maximum order of the conversion matrix is somehow critical and can give rise to uncertainties in the stability determination. Moreover, it is an analysis-oriented method that does not provide any hint for stable or unstable circuit design. However, when used within a complete study of the parametric solutions of a nonlinear circuit, it turns out to be extremely powerful and versatile.

It must be remarked here that the stability criterion based on the conversion matrix is meaningful only when coupled to a harmonic or spectral balance algorithm. In the case that a time-domain algorithm is used, the solution is always stable, since the algorithm simulates the real-world behaviour of the circuit, even though with unavoidable approximations. This point is further elaborated below in Section 8.3.

Nykvist's criterion is equivalent to the criterion described in Section 5.2. In particular, it is equivalent to eq. (6.2), and following discussion, generalised to a time-variant linear network. A design-oriented approach is now described [4]. Let us assume that an instability must be designed at a given frequency ω_{osc}; the embedding impedances of the active device at the sideband frequencies $n\omega_0 \pm \omega_{\text{osc}}$ must be chosen in such a way that an oscillation or instability condition is fulfilled at any port of the linearised time-variant frequency-converting network shown in Figure 8.2. We remark that changing the embedding impedances at the sideband frequencies in principle does not affect the conversion matrix that is determined only by the large-signal periodic steady-state regime, with harmonic content at $\omega = n\omega_0$ only. For instance, if all but two embedding impedances are kept fixed, the conversion matrix can be reduced to a two-port (frequency-converting) network by standard conversion-matrix reduction methods [5]; let us select two ports at different frequencies for the design (Figure 8.4).

The resulting network, including a two-port active device, can now be designed as a linear oscillator at any of the two frequencies if the stability factor is less than 1. If this is the case, the input and output stability circles can be drawn and the input and output

Figure 8.4   The multi-frequency network in Figure 8.2 reduced to a frequency-converting two-port network

loads at the relevant frequencies can be chosen so to satisfy Barkhausen's criterion or the instability criterion as defined in Section 5.2. If the stability factor is greater than 1, then the large-signal state must be modified, for example, by increasing the amplitude of the large-signal source, or the bias point, or the loads at harmonic frequencies, in order to enhance the mixing properties of the active device in large-signal regime.

If the design is performed at two ports at different frequencies, the potential stability of the reduced network may be caused by the frequency-converting (out of diagonal) terms in the conversion matrix, which connect the input and the output of the network; if this is the case, the potential instability vanishes for decreasing amplitude of the large periodic signal at ω_0, since no frequency conversion takes place for a small amplitude of the local oscillator. The instability appears only for a sufficiently large amplitude of the periodic large signal.

In the opposite case that instability must be avoided, the loads of the two-port network must be chosen so that the circuit is stable for all perturbation frequencies $0 < \omega_{osc} < \infty$, very much as in linear amplifiers. This approach is quite general, but requires a modified formulation in the case that $\omega_{osc} = \dfrac{\omega_0}{2}$, because of the coincidence of upper and lower sidebands [6].

8.3 NONLINEAR ANALYSIS, STABILITY AND BIFURCATIONS

8.3.1 Stability and Bifurcations

In this paragraph, the parametric analysis of nonlinear circuits is introduced, which allows the detection of bifurcations and the determination of the stable and unstable regions of operations of the circuit.

In the previous paragraph, a criterion for the determination of the stability of a nonlinear regime has been described. The reason why this criterion has been introduced lies in the capability of the harmonic or spectral balance algorithm to yield a solution, even if the solution itself is not stable. A basic assumption of harmonic balance algorithms (see Section 1.3.2) is that the signal must be expanded in Fourier series, with one or more fundamental frequencies. If the actual solution includes any additional real or complex basic frequency, but this is not included in the expansion, the algorithm does not detect it. The solution, if any, is only mathematical, being physically unstable. A time-domain analysis on the other hand will not yield any unstable solution, always preferring the stable one. Therefore, harmonic balance analysis always requires a stability verification, typically of the Nykvist type. However, the possibility to find unstable solutions allows the designer to get a complete picture of the behaviour of the circuit.

A particularly illuminating approach requires the tracking of a solution as a function of a parameter of the circuit. In many cases, the value of a bias voltage or of an element of the circuit may determine the behaviour of the circuit, whether stable or not. By changing the value and checking the stability properties of the solution, the operating regions of the circuit are found. It is particularly important to detect the values of the parameter at which a stable solution becomes unstable, or *vice versa*. These particular values are called bifurcations because two solutions are found after the bifurcation, one of which is usually unstable.

As an example, let us consider again the shunt oscillator in Figure 5.6 in Chapter 5, which is repeated here in Figure 8.5. Let us consider the behaviour of the solution as a function of the parameter $G_{\text{tot}} = G_s + G_d$, that is, of the total conductance; we will first consider the linear solution.

Kirchhoff's equation for the circuit is

$$\frac{1}{L} \cdot \int v(t) \cdot t + G \cdot v(t) + C \cdot \frac{dv(t)}{dt} = 0 \tag{8.24}$$

The equation always has the trivial solution $v_{\text{DC}}(t) = 0$; it also has the non-trivial solution:

$$v_{\text{osc}}(t) = v_1 \cdot e^{(\alpha + j\omega) \cdot t} + v_2 \cdot e^{(\alpha - j\omega) \cdot t} \quad \alpha = -\frac{G_{\text{tot}}}{2C}$$

$$\omega = \sqrt{\left(\frac{G_{\text{tot}}}{2C}\right)^2 - \frac{1}{LC}} \quad \text{if} \quad G_{\text{tot}} < 2 \cdot \sqrt{\frac{C}{L}} \tag{8.25a}$$

$$v_{\text{osc}}(t) = v_1 \cdot e^{-\alpha_1 \cdot t} + v_2 \cdot e^{-\alpha_2 \cdot t} \alpha_1 = -\frac{G_{\text{tot}}}{2C} + \sqrt{\left(\frac{G_{\text{tot}}}{2C}\right)^2 - \frac{1}{LC}}$$

$$\alpha_2 = -\frac{G_{\text{tot}}}{2C} - \sqrt{\left(\frac{G_{\text{tot}}}{2C}\right)^2 - \frac{1}{LC}} \quad \text{if} \quad G_{\text{tot}} > 2 \cdot \sqrt{\frac{C}{L}} \tag{8.25b}$$

It easy to see that for $G_{\text{tot}} < 0$ an oscillatory solution growing in time is present; in this case, the DC solution is not stable, which is only mathematically possible. For positive values of the total conductance G_{tot}, a damped oscillatory solution or two exponentially decaying solutions are present in the circuit, indicating the stability of the DC solution. The values of the total conductance for which the nature of the solutions changes is the origin ($G_{\text{tot}} = 0$); an additional change takes place for $G_{\text{tot}} = 2 \cdot \sqrt{\frac{C}{L}}$, but no stable solution is involved in the change, which is not observable in steady-state conditions. The root locus can be plotted as in Figure 8.6.

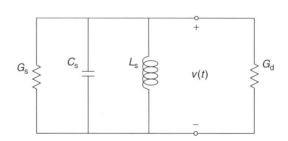

Figure 8.5   A parallel resonant circuit

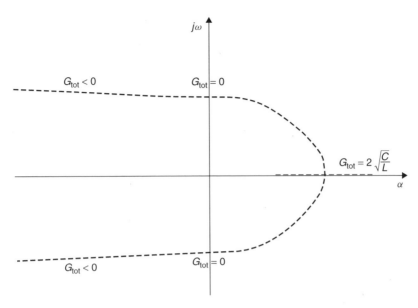

Figure 8.6 Root locus for the parallel resonant circuit in Figure 8.5

Correspondingly, the value of an electrical quantity in the circuit may be plotted as a function of the parameter; for instance, let us assume that the amplitude of the oscillation saturates to a finite value because of the nonlinearities of the circuit, so far neglected for stability analysis. Therefore, a plot of the amplitude of the oscillating voltage in the parallel resonant circuit after the oscillations have reached the equilibrium amplitude (see Section 5.3) looks as shown in Figure 8.7.

For $G_{tot} > 0$, eq. (8.24) has only one solution, which is the stable DC solution with $v_{osc}(t) = 0$. For $G_{tot} < 0$, eq. (8.24) has two solutions, of which the DC one with $v_{osc}(t) = 0$ is only mathematical because it is unstable, while the other with $v_{osc}(t) > 0$ is stable. The point $G_{tot} = 0$ is a bifurcation point. In this case, the generation of a new branch is caused by the sign change of the real part of the complex roots of the perturbation equation (8.24), or in other words by the onset of an autonomous oscillation; this is called a Hopf bifurcation. Other types of bifurcation are described below.

The circuit in Figure 8.5 becomes closer to real life if we consider it as the linearisation of a circuit including an active device, exhibiting a negative differential (small-signal) conductance in a given bias point range. By changing the bias, the total conductance may change sign and quench the oscillations. Therefore, the parameter that controls the bifurcation may be the bias voltage of the active device, for example, a tunnel diode.

Another example of control parameter is the gate bias voltage of an FET oscillator as described in Section 5.4, and shown in Figure 8.8.

The condition for the onset of the oscillation is

$$A(P_{in} = 0) \cdot \beta > 1 \tag{8.26}$$

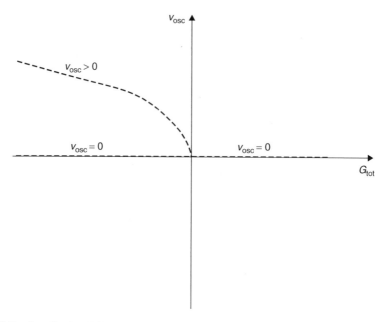

Figure 8.7 Amplitude of the oscillating voltage in the parallel resonant circuit in Figure 8.5

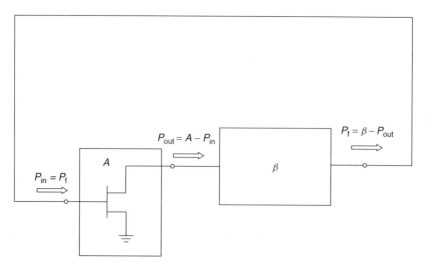

Figure 8.8 An oscillator based on an amplifier and a feedback network

If condition (8.26) is satisfied, the amplitude of the oscillation grows until the oscillation condition at equilibrium is fulfilled.

$$A(P_0) \cdot \beta = 1 \qquad (8.27)$$

The equilibrium is reached because the transistor has a gain compression for increasing operating power (see Section 5.3). The small-signal gain of the amplifier is usually controlled by the gate bias voltage, which therefore determines whether the condition for the onset of the oscillations is fulfilled or not, and the equilibrium operating power changes accordingly (Figure 8.9). The bifurcation diagram for the gate bias voltage is shown in Figure 8.10. The gate bias voltage has the same qualitative behaviour as shown in Figure 8.7.

Let us now treat the case in which the amplifier is biased near Class-B and the transistor has a transconductance increase for increasing gate bias voltage; the amplifier therefore exhibits a gain expansion at low input power, then a gain compression for higher input power when the limiting nonlinearities (forward gate junction conduction, breakdown, etc.) come into play (Figure 8.11).

This case lends itself to illustrating a different type of bifurcation; in the following, reference is made to Figures 8.11 and 8.12. Let us start the analysis with a gate bias

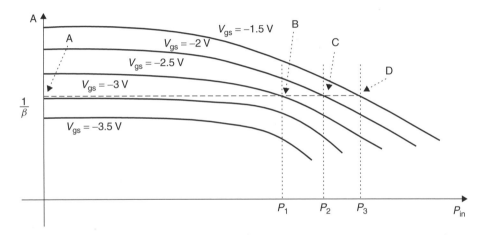

Figure 8.9 Power gain of an amplifier as a function of input power for different bias gate voltages, compared to the inverse of the attenuation of the feedback network

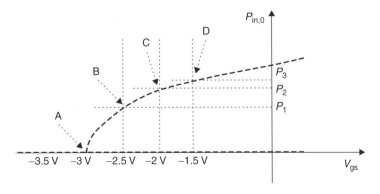

Figure 8.10 Bifurcation diagram for a feedback oscillator with the gate bias voltage as a control parameter; letters refer to points in Figure 8.9

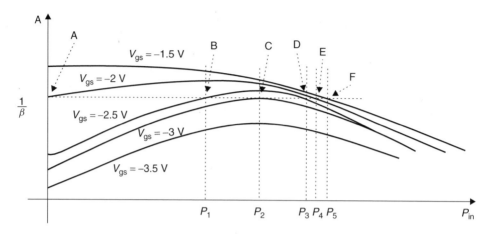

Figure 8.11 Power gain of an amplifier with gain expansion as a function of input power for different bias gate voltages, compared to the inverse of the attenuation of the feedback network

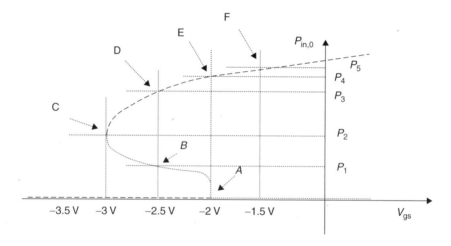

Figure 8.12 Bifurcation diagram for a feedback oscillator with the gate bias voltage as a control parameter when the amplifier has gain expansion; letters refer to points in Figure 8.11

voltage $V_{gs} = -3.5$ V; no oscillations are present in the circuit, which remains in the DC state. Then, let us increase the gate bias voltage; the small-signal condition for the onset of oscillation (8.26) is fulfilled for $V_{gs} = -2$ V, corresponding to point A, where also the equilibrium condition (8.27) is fulfilled. However, this equilibrium point is not stable, as described in Section 5.3. After a transient, the other equilibrium point E, which is reached is stable. If the gate bias voltage is further increased, the oscillation amplitude increases reaching point F and beyond.

Let us now come back towards decreasing gate bias voltages while the circuit is still oscillating. From point F, point E is first reached when $V_{gs} = -2$ V. Then, a further decrease of the gate bias voltage does not quench the oscillation, since a stable equilibrium exists, that is, point D for $V_{gs} = -2.5$ V. The circuit oscillates even though the small-signal start-up condition is not fulfilled, since the oscillation had started at higher gate bias voltages. Continuing along the path for decreasing leads the circuit to point C for $V_{gs} = -3$ V, where the oscillation is stopped. The path from point C to point A through point B is only mathematical, since point B and all other points along the path are unstable equilibrium points. The oscillator then exhibits hysteresis in its behaviour when the gate bias voltage is swept from pinch-off towards open channel and *vice versa*.

The bifurcation point C is called a turning-point or direct-type bifurcation; the bifurcation point A at which an unstable branch departs is a subcritical Hopf bifurcation, while the Hopf bifurcation described above (see point A in Figure 8.10) that gives rise to a stable oscillation is called supercritical. At the turning point C, coming from the point D and continuing towards point B, a real Laplace parameter, solution of the stability equation (8.21) becomes positive, causing the amplitude of the oscillating solution to grow in time, reaching the stable branch with larger oscillation amplitude near point D again.

Bifurcations may be encountered also when a parameter is changed in a periodic large-signal steady state. A method for the determination of the stability of a periodic steady state has been described in the previous paragraph, and a procedure for the design of an instability has also been introduced; both are based on the conversion matrix. When a parameter of the circuit changes in such a way that the circuit starts oscillating at a frequency different from that of the large signal, a secondary Hopf bifurcation takes place, either supercritical or subcritical. If the frequency of the instability is one half that of the periodic steady state, the bifurcation is called a flip or indirect or period-doubling bifurcation; this is normally encountered in regenerative frequency dividers (see Section 6.4). A typical bifurcation diagram is shown in Figure 8.13, where the amplitude of both the fundamental frequency and subharmonic frequency of order two are plotted versus the input power of the frequency divider at fundamental frequency ω_0. When the input power is P_1, the conditions for a subharmonic of order two with $\omega = \dfrac{\omega_0}{2}$ to exist are fulfilled, and the two signals coexist within the circuit; the amplitude of the output signal at fundamental frequency decreases because of the simultaneous presence of another signal at one-half this frequency. If the subharmonic is not detected, the unstable solution with only the fundamental-frequency signal is found; this is only a mathematical solution, as said before. Similar plots are found for secondary Hopf bifurcations, where the second branch represents a signal with a frequency different from the period-doubling subharmonic. Turning points or direct-type bifurcations are also found in periodic regimes, with similar characteristics as seen above for the stability of DC regimes. Other types of bifurcations can be encountered in a nonlinear system [7], which are not described here, as they are less common to be found in practical circuits.

It is not unusual that successive bifurcations are encountered along the branches of a bifurcation diagram. This mechanism usually leads to chaotic behaviour of the circuit for higher values of the controlling parameter. A chaotic system is not a system with random solutions, strictly speaking: it is a system where two solutions starting from two initial points lying very close to one another diverge from one another. It is possible to identify the characteristics of a nonlinear system that leads to chaotic behaviour [8, 9];

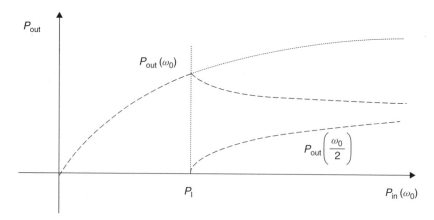

Figure 8.13 Bifurcation diagram vs input power for a regenerative frequency divider; the unstable branch is the dotted line, while the stable branches are the dashed lines

however, this is not the object of this description, as chaotic systems have not so far found any established application in electronic systems.

Other examples of bifurcations are described below, in Section 8.4.

8.3.2 Nonlinear Algorithms for Stability Analysis

In this paragraph, the nonlinear algorithms for the analysis of nonlinear circuits with autonomous oscillations and external excitation are described.

Circuits where an oscillation coexists with a forced periodic state are, in principle, quite naturally analysed by means of time-domain algorithms. No special modification is required in the formulation of the algorithm with respect to the analysis of autonomous circuits; however, the considerations made both for the case of two-tone analysis and for the case of oscillator analysis hold.

It is worth repeating, however, that the time-domain analysis always yields a stable solution, that is, the solution actually present in the circuit; moreover, the transient behaviour is correctly found. This is important, especially in the cases in which the state actually reached by the circuit is uncertain because of the presence of possible instabilities.

Harmonic or spectral balance is another viable method for this case. Formally, the electrical quantities are expressed as in the case of a two-tone analysis (see Chapter 1), with the first frequency being that of the external signal source ω_{ext} and the second being that of the oscillating signal ω_{osc}; in this case, the second basis frequency is unknown and must be added to the vector of the unknowns of the problem; however, the phase of the oscillation is undetermined and can be set to zero (see Chapter 5). Therefore, the number of unknowns is again equal to the number of equations, and the system can be solved by a numerical procedure. The problem always has a trivial solution, where the amplitudes of all phasors relative to the second basis frequency ω_{osc} and of the intermodulation frequencies $n_{ext}\omega_{ext} + n_{osc}\omega_{osc}$ are zero, corresponding to absence of oscillation. This solution can be real or can be unstable and only mathematical. With the aid of the same procedures as those described for oscillator analysis, the trivial solution can be avoided. For example, a port of the circuit where non-zero amplitude of the signal is expected to appear at the oscillation frequency is selected. Then, Kirchhoff's equation at that port and frequency can be replaced by the Kurokawa condition:

$$Y(\omega_{osc}) = \frac{I(\omega_{osc})}{V(\omega_{osc})} = 0 \tag{8.28}$$

Alternatively, Kirchhoff's equations at all nodes and frequencies can be rewritten as

$$\frac{\overline{I}_L(\overline{V}_{n_{osc}}) + \overline{I}_{NL}(\overline{V}_{n_{osc}})}{\sqrt{\sum_n |V_{n_{osc}}|^2}} = \overline{0} \tag{8.29}$$

Both the approaches remove the trivial solution, being in fact extensions of the previously described methods.

   An alternative point of view that is an extension of that illustrated in Figure 5.23 in Section 5.5 is now described [10]. A probing voltage or current at frequency ω is introduced at a single port of the nonlinear circuit driven by the large signal at frequency ω_0 (Figure 8.14). The nonlinear circuit is analysed by means of a non-autonomous, two-tone harmonic or spectral balance algorithm. Frequency and amplitude of the probing signal are swept within a suitable range; an oscillation is detected when the control quantity (the probing current or voltage respectively) is zero, indicating that an autonomous oscillating signal is present in the circuit and that the removal of the probing signal does not perturb the circuit.

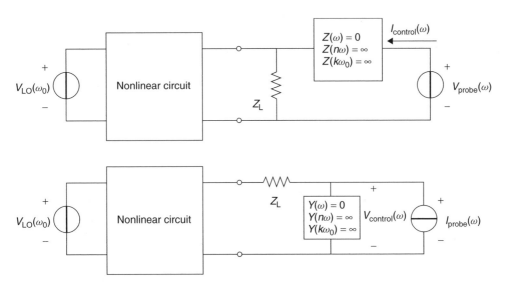

Figure 8.14   Voltage and current probes for instability detection

A filter 'masks' the presence of the probe at all other frequency components; we remark that harmonics of the probing frequency ω can be present, since a two-tone harmonic or spectral balance analysis is performed, and arbitrary amplitude of the probing tone is accounted for. As seen in the case of oscillators, both probing amplitude and frequency are *a priori* unknown; they are found from the real and imaginary parts of either of the following complex equations:

$$I_{\text{probe}}(V_{\text{probe}}, \omega) = 0 \quad \text{or} \quad V_{\text{probe}}(I_{\text{probe}}, \omega) = 0 \tag{8.30}$$

Quite naturally, the same problems in identifying a suitable starting point for easing the convergence of the analysis are present in this case also.

Volterra analysis could be used for this type of analysis; however, the authors are not aware of such an algorithm being proposed so far.

The methods described have been extensively used for the analysis of two-tone mixed autonomous/non-autonomous circuits, and in particular, they have been used for the determination of the bifurcation diagram of the circuits. To achieve this goal, continuation methods are applied, in order to 'follow' the solution of the circuit as a parameter is varied and to detect the qualitative changes in its behaviour at bifurcation points [11–13]. As stated above in Section 8.3.2, the harmonic balance method can also find unstable solutions and is therefore ideal for a complete study of the behaviour of a circuit; however, the stability of branches or solutions must be verified. In general, this is straightforwardly done by application of Nykvist's stability criterion as described above.

The analysis along a branch of the bifurcation diagram requires, in principle, simply the repeated application of the methods described above, as the value of the parameter is varied. However, problems arise both at turning points and at Hopf bifurcations. Referring to Figure 8.12, the diagram is, in principle, computed by selecting the gate bias voltage V_{gs} as a parameter, and the input power to the FET as one of the problem unknowns that identifies the branch. However, near the turning point C, the curve becomes multi-valued, and numerical problems arise. When the turning point is approached, therefore, it is advantageous to switch the role of the two axes in the plot, use the amplitude of a given frequency component at a given port (e.g. the fundamental-frequency component at the gate port, related to the input power) as a parameter and set the gate bias voltage as a problem unknown. This requires a modification of the analysis algorithm, as sketched in Section 5.5. The analysis then gets through the turning point, and the whole branch can be followed. The approaching of the turning-point bifurcation can be detected by inspection or automatically by monitoring the quantity $\dfrac{\text{d}P_{\text{out}}}{\text{d}V_{\text{gs}}}$, and setting a maximum value for it. At the turning point, the derivative becomes infinite. The two quantities can be switched back to the original role when the derivative becomes reasonably small again.

Another problem arises when a Hopf of a flip bifurcation is encountered along a branch. Referring to Figure 8.13 for a frequency divider-by-two, the diagram is plotted by starting from a low input power, where the solution is quasi-linear and no subharmonic is present; the solution is a Fourier expansion on the basis frequency ω_0. When the input power P_1 is reached by stepping the input power, a second branch appears on the diagram, representing a frequency-dividing solution, with power at frequency ω_0 and also at frequency $\dfrac{\omega_0}{2}$; however, the previous type of solution is also present as a continuation of the branch, but becomes unstable. Therefore, if the bifurcation is not detected on the way, the stable branch is overlooked and a non-physical result is found. A natural approach consists of monitoring the stability of the solution at the stepping values of the parameter by checking Nykvist's plot. After P_1, the solution becomes unstable, indicating that a bifurcation is present at a lower power (Figure 8.15).

The bifurcation can be accurately located, and the frequency of the new frequency component approximately determined, by repeating Nykvist's analysis in smaller steps around P_1, until a sufficient approximation is obtained. Then, the stable branch is followed by an analysis with a basis frequency $\omega = \dfrac{\omega_0}{2}$, which includes the frequency-divided component.

The bifurcation can be directly located in a way similar to what has been described above, once the frequency of the new branch is known from a Nykvist's plot. For a flip-type bifurcation, an analysis based on the frequency $\omega = \dfrac{\omega_0}{2}$ is performed, where the amplitude of the frequency-divided component is set to a very small value and therefore is no more an unknown, while the amplitude of the fundamental-frequency component at the bifurcation, which is related to the input power P_1 at ω_0, is unknown. In this way, the bifurcation is located with a single analysis. If the bifurcation is a Hopf-type one, where the frequency of the autonomous oscillation is not exactly known but only approximately determined from Nykvist's plot (Figure 8.15), the autonomous frequency is included in the vector of the unknowns, while the phase of the relevant phasor is arbitrarily set to a fixed value, for example, zero. The input power at the bifurcation and the autonomous frequency are therefore simultaneously determined. Obviously, a good starting point must be used for all these analyses, given the critical behaviour of the circuits.

8.4 INJECTION LOCKING

In this paragraph, some circuits using injection locking of self-oscillating signals for oscillation synchronisation or for frequency multiplication or division are described.

Injection locking is a nonlinear mechanism that synchronises a free oscillation in a circuit to an externally injected signal. In a linear circuit, signals at different frequencies are independent and coexist in the same circuit without interaction. In a nonlinear circuit, that is, in all practical active circuits, an injected signal interacts with the nonlinear active device, locking the free oscillation frequency to that of the external signal, provided that some conditions are fulfilled. In case the conditions are not fulfilled, either the two signals coexist in the circuit or the injected one suppresses the free oscillations that disappear altogether. In the following, the basic conditions for injection locking are given, and an overview of the typical behaviour of an injected oscillator will be given, with some applications.

Let us consider a generic oscillating circuit similar to that in Figure 5.13 of Chapter 5, with an added injected signal (Figure 8.16) [14–17]. Let us assume that

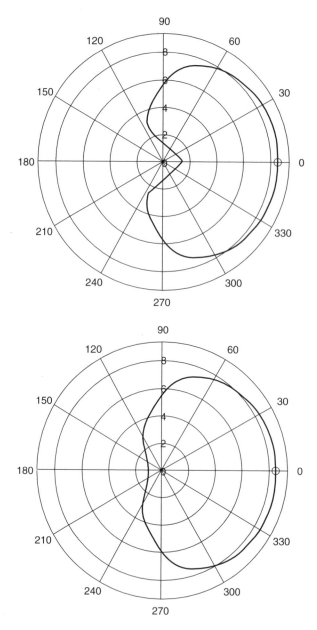

Figure 8.15 Nykvist's plots before (stable solution) and after (unstable solution) the bifurcation point

eq. (5.60a) holds.

$$Y_s + Y_d = Y_{tot}(A_0, \omega_0) = 0 \tag{8.31}$$

The network oscillates at frequency ω_0 with an amplitude A_0 when no signal is injected, that is, it is a free-running oscillator. If the frequency of the injected signal is close to that

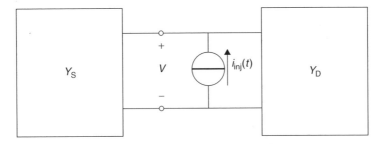

Figure 8.16 An oscillating circuit with an injected signal

of the free-running oscillator, and therefore we assume that the amplitude and frequency of the oscillation will be perturbed to a small extent:

$$A \cong A_0 + \Delta A \qquad s \cong \Delta\alpha + j(\omega_0 + \Delta\omega) \tag{8.32}$$

we can expand the total admittance in Taylor series as in eq. (5.62):

$$Y_{\text{tot}}(A, s) \cong Y_{\text{tot}}(A_0, j\omega_0) + \left. \frac{\partial Y_{\text{tot}}(A, s)}{\partial A} \right|_{\substack{A=A_0 \\ s=j\omega_0}} \cdot \Delta A + \left. \frac{\partial Y_{\text{tot}}(A, s)}{\partial s} \right|_{\substack{A=A_0 \\ s=j\omega_0}} \cdot \Delta s + \ldots \tag{8.33}$$

Remembering eq. (8.31), Kirchhoff's equation now reads as

$$I_{\text{inj}} + \left(\frac{\partial Y_{\text{tot}}}{\partial A} \cdot \Delta A + \frac{\partial Y_{\text{tot}}}{\partial \omega} \cdot \Delta\omega \right) \cdot (A_0 + \Delta A) = 0 \tag{8.34}$$

where we have assumed steady state, where I_{inj} is a complex phasor and where the derivatives are computed at the free-running amplitude and frequency. Neglecting higher-order terms,

$$I_{\text{inj}} + \left(\frac{\partial Y_{\text{tot}}}{\partial A} \cdot \Delta A + \frac{\partial Y_{\text{tot}}}{\partial \omega} \cdot \Delta\omega \right) \cdot A_0 = 0 \tag{8.35}$$

from which (see Appendix A.11) we can compute the maximum locking range:

$$|\Delta\omega|_{\text{max}} = \frac{|I_{\text{inj}}|}{A_0} \cdot \left| \frac{\dfrac{\partial Y}{\partial A}}{\dfrac{\partial Y_r}{\partial \omega} \cdot \dfrac{\partial Y_i}{\partial A} - \dfrac{\partial Y_i}{\partial \omega} \cdot \dfrac{\partial Y_r}{\partial A}} \right| \tag{8.36}$$

This formula tells us some interesting information on the attitude of a circuit to be locked and on the locking range. First of all, the locking range is proportional to the amplitude of the injected signal, as intuition suggests. Then, it is apparent that the denominator of eq. (8.36) is the same as in eq. (5.69). This term is related to the stability of the free-running oscillator: the larger the amplitude of this term the more stable the free-running oscillator; also, the narrower is its locking range, as intuition suggests as well.

Similarly (see Appendix A.11), we can compute the maximum variation of the amplitude of the oscillation in injection-locked operations:

$$|\Delta A|_{max} = \frac{|I_{inj}|}{A_0} \cdot \left| \frac{\frac{\partial Y}{\partial \omega}}{\frac{\partial Y_r}{\partial \omega} \cdot \frac{\partial Y_i}{\partial A} - \frac{\partial Y_i}{\partial \omega} \cdot \frac{\partial Y_r}{\partial A}} \right| \tag{8.37}$$

From this expression, we see that the amplitude of the locked oscillation is proportional to the relative amplitude of the locking signal; the sensitivity is inversely proportional to the stability of the free-running oscillation, again as intuition suggests. Therefore, with a suitable choice of the parameters, the injection-locked oscillator can behave as an amplifier.

As an example, let us consider a simple parallel resonant circuit with an injected signal (Figure 8.17).

The locking range is (eq. (5.96))

$$|\Delta \omega|_{max} \cong \frac{|I_{inj}|}{A_0} \cdot \left| \frac{\partial G_s}{\partial A} \right| \cdot \frac{1}{C \cdot \left| \frac{\partial G_s}{\partial A} \right|} = \frac{|I_{inj}|}{A_0} \cdot \frac{1}{C} \tag{8.38}$$

If we normalise the locking range to the free-running oscillation frequency, we get

$$\frac{|\Delta \omega|_{max}}{\omega_0} \cong \frac{|I_{inj}|}{A_0} \cdot \frac{1}{\omega_0 C} \tag{8.39}$$

The larger the tank the narrower the locking range. If we plot the relative amplitude of the locking signal as a function of the normalised locking range, we get a plot as in Figure 8.18 (in logarithmic scale), which is typical of the locking phenomenon.

From eq. (8.37), we compute the dependence of the amplitude on the input signal:

$$|\Delta A|_{max} \cong \frac{|I_{inj}|}{A_0} \cdot C \cdot \frac{1}{C \cdot \left| \frac{\partial G_s}{\partial A} \right|} = \frac{|I_{inj}|}{A_0} \cdot \frac{1}{\left| \frac{\partial G_s}{\partial A} \right|} \tag{8.40}$$

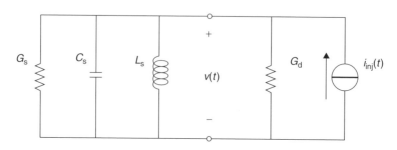

Figure 8.17 A parallel resonant circuit with an injected signal

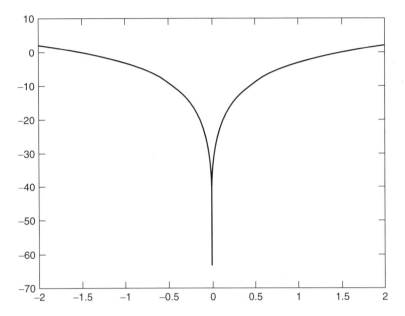

Figure 8.18 Relative amplitude of the locking signal as a function of the normalised locking range (in logarithmic scale)

where the derivative of the conductance of the active device with respect to the amplitude of the oscillation can be evaluated from the plot in Figure 5.16. This equation tells us that the smaller the sensitivity of the negative conductance to the amplitude of the applied signal the larger the sensitivity of the oscillation to the amplitude of the locking signal.

In fact, the plot in Figure 8.18 deserves a more detailed description. First of all, it must be remarked that eq. (8.37) is valid for small variations of the amplitude and frequency of oscillation with respect to the free-running values. Therefore, for large values of the input signal, the calculations do not hold any more, and different phenomena arise. Then, we consider the behaviour of the circuit in different regions of the plot in Figure 8.19.

As stated above, for small amplitudes of the input signal the curve as in Figure 8.18 determines the boundary between locked operations and free-running oscillations super-imposed to the input signal. This last operating mode has, in fact, a spectrum similar to that of a mixer, since both the free-running frequency and the input signal frequency coexist within the (nonlinear) circuit; however, the local oscillator frequency is generated within the circuit itself, as a free-running oscillation, and no external local oscillator is required. This circuit is usually called self-oscillating mixer. Within the locking range, the oscillator behaves, in fact, as an amplifier if the parameters are suitably chosen; as such, it is used for amplification or signal generation, especially at high frequency where a powerful but noisy oscillating device (e.g. an IMPATT diode) is frequency locked by a cleaner but smaller signal.

If the input signal is very large, however, it saturates the nonlinear device, and no free-running or injection-locked oscillation takes place. The instability is suppressed by

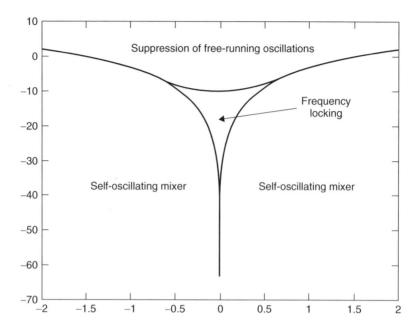

Figure 8.19 Operating regions of an injection-locked oscillator/self-oscillating mixer

the large applied signal, and the circuit behaves as a single-tone nonlinear system. This is true at all frequencies, provided that the input signal is large enough.

It must be said that the picture described so far is somewhat oversimplified. In fact, transitions between the regions can show complicated behaviours, involving bifurcations of different types [7], as those described above in Section 8.3.1; however, the general behaviour is as described.

It is clear from what has been said that self-oscillating mixers and injection-locked oscillators must be designed in different ways. The design of self-oscillating mixers must be such as to minimise the influence of the input signal on the frequency of the free-running oscillation, in order not to affect the converted frequency. This is obtained by increasing the quality factor of the resonator; as an illustration, the locking range of a simple parallel resonant circuit is shown in Figure 8.20. The free-running frequency is 1 GHz in both cases, but the capacitance is 10 pF in the case of the larger locking range and $C = 100$ pF for the narrower locking range. In a practical application, the oscillation frequency is stabilised by means of a DRO; a possible scheme including a series feedback (see Chapter 5) is shown in Figure 8.21.

The transistor should be biased near Class-B in order that the input signal drives the transistor itself in nonlinear behaviour, for efficient mixing. If suitably designed, such a circuit can exhibit conversion gain, at the expenses of the bias supply, and minimise the circuitry with respect to traditional mixers with external local oscillator. However, frequency stability is somehow more problematic, as it is affected by pulling from the input signal, even if stabilisation by means of a dielectric resonator does a lot to reduce the problem.

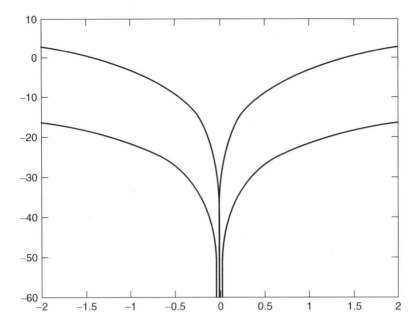

Figure 8.20 Relative amplitude of the locking signal as a function of the normalised locking range (in logarithmic scale) for two different quality factors of the resonator

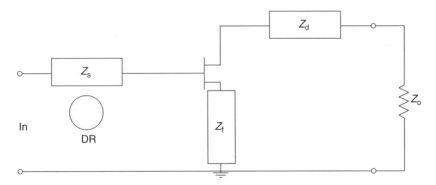

Figure 8.21 A series-feedback scheme for a self-oscillating mixer based on a DRO

An interesting extension of the concepts described allows the design of a special type of frequency multipliers and dividers that is based on the injection locking by means of harmonics of the input signal and free-running oscillation. In particular, a non-regenerative frequency divider is obtained by locking the second harmonic of the free-running oscillator by means of the fundamental frequency of the input signal [18]. The general scheme looks somewhat similar to that of regenerative frequency dividers (Figure 8.22).

However, the circuit oscillates at, or more precisely near, $\dfrac{\omega_0}{2}$ when the input signal is not present or is very small. When the amplitude and frequency of the input signal are

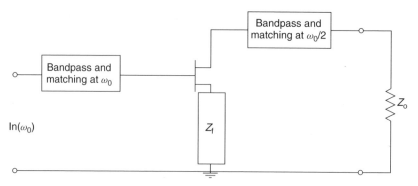

Figure 8.22   A qualitative scheme of a frequency divider-by-two based on a free-running oscillator at $\dfrac{\omega}{2}$

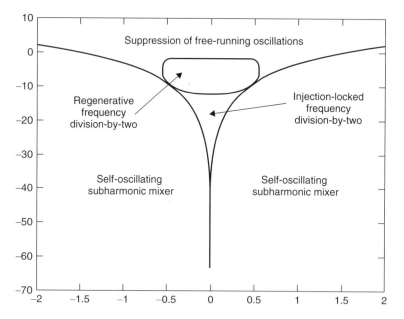

Figure 8.23   Operating regions of a non-regenerative frequency divider-by-two/self-oscillating subharmonic mixer

such that the second harmonic of the free-running oscillator is locked by the input signal, perfect division-by-two is performed by the circuit. The relevant operating regions are shown in Figure 8.23.

If the frequency of the input signal is close to the second harmonic of the free-running oscillation and the amplitude is rather large, the circuit behaves as a regenerative frequency divider rather than as an injection-locked oscillator. The nonlinearity of the active device acts as in regenerative frequency dividers (see Section 6.4). For even higher amplitudes of the input signal and/or greater frequency difference, the free-running oscillation is suppressed, and the circuit has a single-tone behaviour at ω_0. If the amplitude of the input signal is small or moderate but its frequency lies outside of the locking range, the circuit behaves as a self-oscillating subharmonic mixer (see Chapter 7). In this case, the free-running oscillation frequency must be stabilised, for example, by means of a dielectric resonator, in order to avoid frequency pulling by the injected signal. A qualitative scheme is shown in Figure 8.24 for a series-feedback free-running oscillator.

Similar considerations are made for non-regenerative injection-locked frequency multipliers: in this case, the second harmonic of the input signal locks the free-running oscillation. A qualitative picture of the operating regions is given in Figure 8.25.

An interesting feature of the injection-locking frequency multiplier is the combination of output power supplied by the (presumably) noisy oscillator and frequency control

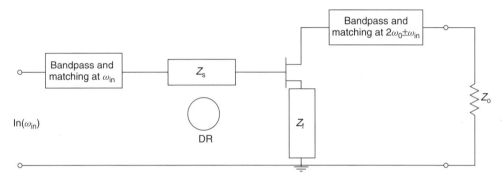

Figure 8.24   A qualitative scheme of a self-oscillating subharmonic mixer

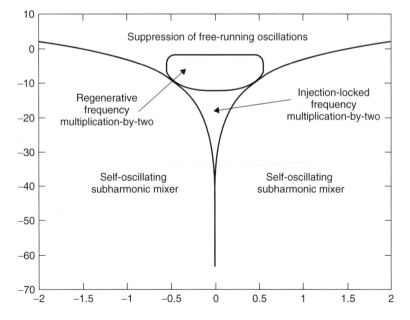

Figure 8.25   Operating regions for a non-regenerative frequency multiplier-by-two/self-oscillating harmonic mixer

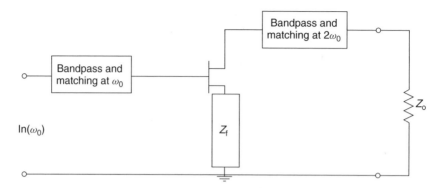

Figure 8.26 A frequency multiplier-by-two based on a free-running oscillator at 2 ω

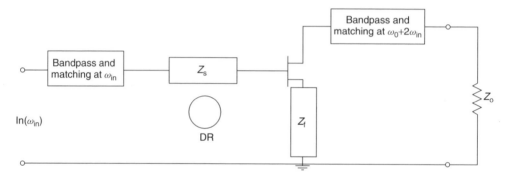

Figure 8.27 A qualitative scheme of a self-oscillating harmonic mixer

by means of a cleaner locking signal at lower frequency. A qualitative scheme is shown in Figure 8.26.

Similarly, a self-oscillating harmonic mixer has a qualitative scheme as in Figure 8.27.

8.5 BIBLIOGRAPHY

[1] R. Grimshaw, *Nonlinear Ordinary Differential Equations*, CRC Press, Boca Raton (FL), USA, 1993.

[2] T.S. Parker, L.O. Chua, *Practical Numerical Algorithms for Chaotic Systems*, Springer-Verlag, New York (NY), 1989.

[3] V. Rizzoli, A. Lipparini, 'General stability analysis of periodic steady-state regimes in nonlinear microwave circuits', *IEEE Trans. Microwave Theory Tech.*, **MTT-33**(1), 30–37, 1985.

[4] F. Di Paolo, G. Leuzzi, 'A design approach for sub-harmonic generation or suppression in nonlinear circuits', *Proc. GAAS 2002*, Milan (Italy), 2002.

[5] S.A. Maas, *Microwave Mixers*, Artech House, Norwood (MA), 1993.

[6] F. Di Paolo, G. Leuzzi, 'Bifurcation synthesis by means of harmonic balance and conversion matrix', *Proc. GAAS 2003*, München (Germany), 2003.

[7] A. Suárez, R. Quéré, *Stability Analysis of Nonlinear Circuits*, Artech House, Norwood (MA), 2003.

[8] G. Chen, T. Ueta (Ed.), *Chaos in Circuits and Systems*, World Scientific, Singapore, 2002.

[9] L.O. Chua, 'Dynamic nonlinear networks: state-of-the-art', *IEEE Trans. Circuits Syst.*, **CAS-27**, pp. 1059–1087, 1980.

[10] A. Suárez, J. Morales, R. Quéré, 'Synchronisation analysis of autonomous microwave circuits using new global stability analysis tools', *IEEE Trans. Microwave Theory Tech.*, **46**, 494–504, 1998.

[11] V. Rizzoli, A. Neri, 'Automatic detection of Hopf bifurcations on the solution path of a parameterised nonlinear circuit', *IEEE Microwave Guided Wave Lett.*, **3**(7), 219–221, 1993.

[12] V. Iglesias, A. Suárez, J.L. Garcia, 'New technique for the determination through commercial software of the stable-operation parameter ranges in nonlinear microwave circuits', *IEEE Microwave Guided Wave Lett.*, **8**(12), 424–426, 1998.

[13] S. Mons, J.-Ch. Nallatamby, R. Quéré, P. Savary, J. Obregon, 'A unified approach for the linear and nonlinear stability analysis of microwave circuits using commercially available tools', *IEEE Trans. Microwave Theory Tech.*, **MTT-47**(12), 2403–2409, 1999.

[14] K. Kurokawa, 'Some basic characteristics of broadband negative resistance oscillator circuits', *Bell Syst. Tech. J.*, **48**, 1937–1955, 1969.

[15] K. Kurokawa, 'Injection locking of microwave solid-state oscillators', *Proc. IEEE*, **61**, 1386–1410, 1973.

[16] H.C. Chang, A.P. Yeh, R.A. York, 'Analysis of oscillators with external feedback loop for improved locking range and noise reduction', *IEEE Trans. Microwave Theory Tech.*, **MTT-47**, 1535–1543, 1999.

[17] F. Ramírez, E. de Cos, A. Suárez, 'Nonlinear analysis tools for the optimised design of harmonic-injection dividers', *IEEE Trans. Microwave Theory Tech.*, **51**(6), 1752–1762, 2003.

[18] F. Ramírez, E. de Cos, A. Suárez, 'Analog frequency divider by variable order 6 to 9', *IEEE MTT-S Int. Symp. Dig.*, 2002, pp. 127–130.

Appendix

A.1 TRANSFORMATION IN THE FOURIER DOMAIN OF THE LINEAR DIFFERENTIAL EQUATION

$$C \cdot \frac{\mathrm{d}v(t)}{\mathrm{d}t} + g \cdot v(t) + i_\mathrm{s}(t) = 0 \tag{A.1}$$

By Fourier transforming the equation, we get

$$C \cdot \int_{-\infty}^{\infty} \frac{\mathrm{d}v(t)}{\mathrm{d}t} \cdot \mathrm{e}^{-j\omega t} \cdot \mathrm{d}t + g \cdot \int_{-\infty}^{\infty} v(t) \cdot \mathrm{e}^{-j\omega t} \cdot \mathrm{d}t + \int_{-\infty}^{\infty} i_\mathrm{s}(t) \cdot \mathrm{e}^{-j\omega t} \cdot \mathrm{d}t =$$

$$= C \cdot j\omega \cdot \int_{-\infty}^{\infty} v(t) \cdot \mathrm{e}^{-j\omega t} \cdot \mathrm{d}t + g \cdot \int_{-\infty}^{\infty} v(t) \cdot \mathrm{e}^{-j\omega t} \cdot \mathrm{d}t + \int_{-\infty}^{\infty} i_\mathrm{s}(t) \cdot \mathrm{e}^{-j\omega t} \cdot \mathrm{d}t =$$

$$= j\omega C \cdot V(\omega) + g \cdot V(\omega) + I_\mathrm{s}(\omega) = Y(\omega) \cdot V(\omega) + I_s(\omega) = 0 \tag{A.2}$$

Nonlinear Microwave Circuit Design   F. Giannini and G. Leuzzi
© 2004 John Wiley & Sons, Ltd   ISBN: 0-470-84701-8

A.2 TIME-FREQUENCY TRANSFORMATIONS

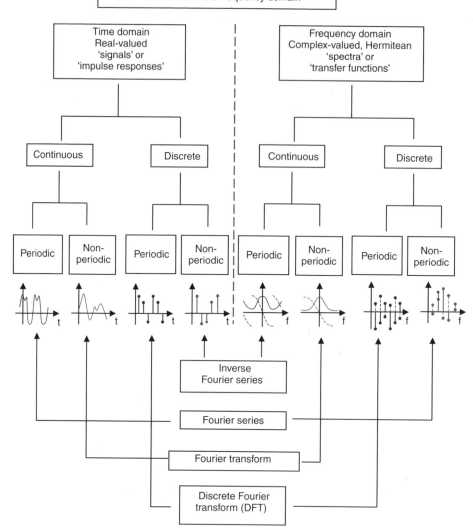

A.3 GENERALISED FOURIER TRANSFORMATION FOR THE VOLTERRA SERIES EXPANSION

$$\int_{-\infty}^{\infty} y(t) \cdot e^{-j\omega t} \cdot dt = \cdots + \int_{-\infty}^{\infty} \int_{-\infty}^{t} \int_{-\infty}^{t} h_2(t - \tau_1, t - \tau_2) \cdot x(\tau_1)$$

$$\cdot x(\tau_2) \cdot d\tau_2 \cdot d\tau_1 \cdot e^{-j\omega t} \cdot dt + \cdots$$

$$
= \cdots + \int_{-\infty}^{\infty} \int_{-\infty}^{t} \int_{-\infty}^{t} h_2(t - \tau_1, t - \tau_2) \cdot x(\tau_1) \cdot x(\tau_2) \cdot e^{-j\omega_1(t-\tau_1)}
$$

$$
\cdot e^{-j\omega_2(t-\tau_2)} \cdot e^{-j\omega_1\tau_1} \cdot e^{-j\omega_2\tau_2} \cdot e^{j\omega_1 t} \cdot e^{j\omega_2 t} \cdot d\tau_2 \cdot d\tau_1 \cdot e^{-j\omega t} \cdot dt + \cdots
$$

$$
= \cdots + \int_{-\infty}^{t} \int_{-\infty}^{t} h_2(t - \tau_1, t - \tau_2) \cdot e^{-j(\omega_1(t-\tau_1)+\omega_2(t-\tau_2))} \cdot x(\tau_1)
$$

$$
\cdot e^{-j\omega_1\tau_1} \cdot x(\tau_2) \cdot e^{-j\omega_2\tau_2} \cdot d\tau_2 \cdot d\tau_1 \cdot \int_{-\infty}^{\infty} e^{-j(\omega-\omega_1-\omega_2)t} \cdot dt + \cdots
$$

$$
= \int \cdots \int H_n(\omega_1, \ldots, \omega_n) \cdot X(\omega_1) \ldots, X(\omega_n)
$$

$$
\cdot \delta(\omega - \omega_1 - \ldots \omega_n) \cdot d\omega_n \ldots d\omega_1 = Y(\omega) \tag{A.3}
$$

A.4 DISCRETE FOURIER TRANSFORM AND INVERSE DISCRETE FOURIER TRANSFORM FOR PERIODIC SIGNALS

A periodic signal with period T can be expanded in Fourier series:

$$
x(t) = \sum_{n=-\infty}^{\infty} X_n \cdot e^{jn\omega_0 t} \tag{A.4}
$$

where the angular frequency ω_0 is

$$
\omega_0 = \frac{2\pi}{T} = 2\pi \cdot f_0 \tag{A.5}
$$

The coefficients of the Fourier series expansion are expressed as

$$
X_n = \frac{1}{T} \int_0^T x(t) \cdot e^{jn\omega_0 t} dt \tag{A.6}
$$

If the signal has a bandwidth limited to a maximum frequency $f_{max} = N \cdot f_0 = \dfrac{N}{2\pi} \cdot \omega_0$, the series can be truncated as

$$
x(t) = \sum_{n=-N}^{N} X_n \cdot e^{jn\omega_0 t} \tag{A.7}
$$

and by using Nykvist's sampling theorem, the coefficients are computed as

$$
X_n = \frac{1}{T} \sum_{k=-N}^{N} x(t_k) \cdot e^{jn\omega_0 t_k} \quad \text{with } t_k = \frac{T}{2N+1} \cdot k, \quad k = -N, \ldots, N \tag{A.8}
$$

By replacing we get

$$X_n = \frac{1}{T} \sum_{k=-N}^{N} x(t_k) \cdot e^{j2\pi \frac{n \cdot k}{2N+1}} \tag{A.9}$$

In matrix form,

$$\vec{x} = \overset{\leftrightarrow}{F} \cdot \vec{X} \qquad \vec{X} = \frac{1}{T} \overset{\leftrightarrow}{F}^{-1} \cdot \vec{x} \tag{A.10}$$

where

$$\vec{X} = \begin{bmatrix} X_N \\ \cdot \\ X_0 \\ \cdot \\ X_{-N} \end{bmatrix} \overset{\leftrightarrow}{F} = \begin{bmatrix} e^{jN\omega_0 t_{-N}} & \cdot & e^{jN\omega_0 t_0} & \cdot & e^{jN\omega_0 t_N} \\ \cdot & \cdot & \cdot & \cdot & \cdot \\ 1 & \cdot & 1 & \cdot & 1 \\ \cdot & \cdot & \cdot & \cdot & \cdot \\ e^{-jN\omega_0 t_{-N}} & \cdot & e^{-jN\omega_0 t_0} & \cdot & e^{-jN\omega_0 t_N} \end{bmatrix}$$

$$= \begin{bmatrix} e^{-j2\pi \frac{N \cdot N}{2N+1}} & \cdot & 1 & \cdot & e^{j2\pi \frac{N \cdot N}{2N+1}} \\ \cdot & \cdot & \cdot & \cdot & \cdot \\ 1 & \cdot & 1 & \cdot & 1 \\ \cdot & \cdot & \cdot & \cdot & \cdot \\ e^{j2\pi \frac{N \cdot N}{2N+1}} & \cdot & 1 & \cdot & e^{-j2\pi \frac{N \cdot N}{2N+1}} \end{bmatrix} \vec{x} = \begin{bmatrix} x(t_{-N}) \\ \cdot \\ x(t_0) \\ x(t_N) \end{bmatrix} \tag{A.11}$$

If the signal is real, the coefficients at negative frequencies are the complex conjugate of the coefficients at positive frequencies:

$$X_{-n} = X_n^* \tag{A.12}$$

Only positive-frequency or negative-frequency coefficient plus the coefficient at zero frequency are needed to completely describe the signal. The series can therefore be written as

$$x(t) = X_0 + \sum_{n=1}^{\infty} X_n \cdot e^{jn\omega_0 t} + \sum_{n=-\infty}^{-1} X_n \cdot e^{-jn\omega_0 t} = X_0 + \sum_{n=1}^{\infty} (X_n \cdot e^{jn\omega_0 t} + X_n^* \cdot e^{-jn\omega_0 t})$$

$$= X_0 + \mathrm{Re}\left\{ \sum_{n=1}^{\infty} X_n \cdot e^{jn\omega_0 t} \right\} \tag{A.13}$$

Using only real numbers, the series becomes

$$x(t) = \sum_{n=0}^{N} (X_n^r \cdot \cos(n\omega_0 t) - X_n^i \cdot \sin(n\omega_0 t))$$

$$= X_0 + \sum_{n=1}^{N} (X_n^r \cdot \cos(n\omega_0 t) - X_n^i \cdot \sin(n\omega_0 t)) \tag{A.14}$$

since the zero-frequency coefficient X_0 is real. The values of the signal $x(t)$ at the Nykvist's time instants $t_m = \dfrac{T}{2N+1} \cdot m$ with $m = 0, \ldots, 2N$ are written as

$$\vec{x} = \overset{\leftrightarrow}{R} \cdot \vec{X} \qquad \vec{X} = \frac{1}{T} \overset{\leftrightarrow}{R}^{-1} \cdot \vec{x} \qquad (A.15)$$

where

$$\vec{x} = \begin{bmatrix} t_0 \\ \cdot \\ t_N \\ \cdot \\ t_{2N} \end{bmatrix} \quad \overset{\leftrightarrow}{R} = \begin{bmatrix} 1 & \cos(\omega_0 t_0) & -\sin(\omega_0 t_0) & \cdot & \cdot & \cos(N\omega_0 t_0) & -\sin(N\omega_0 t_0) \\ \cdot & \cdot & \cdot & & & \cdot & \cdot \\ 1 & \cos(\omega_0 t_N) & -\sin(\omega_0 t_N) & \cdot & \cdot & \cos(N\omega_0 t_N) & -\sin(N\omega_0 t_N) \\ \cdot & \cdot & \cdot & & & \cdot & \cdot \\ 1 & \cos(\omega_0 t_{2N}) & -\sin(\omega_0 t_{2N}) & \cdot & \cdot & \cos(N\omega_0 t_{2N}) & -\sin(N\omega_0 t_{2N}) \end{bmatrix}$$

$$\vec{X} = \begin{bmatrix} X_0 \\ X_1^r \\ X_1^i \\ \cdot \\ \cdot \\ X_N^r \\ X_N^i \end{bmatrix} \qquad (A.16)$$

When replacing the expression for the time instants t_m, the matrix $\overset{\leftrightarrow}{R}$ becomes

$$\overset{\leftrightarrow}{R} = \begin{bmatrix} 1 & 1 & 0 & \cdot & \cdot & 1 & 0 \\ \cdot & \cdot & \cdot & \cdot & \cdot & \cdot & \cdot \\ 1 & \cos\left(\dfrac{2\pi}{2N+1} \cdot N\right) & -\sin\left(\dfrac{2\pi}{2N+1} \cdot N\right) & \cdot & \cdot & \cos\left(\dfrac{2N\pi}{2N+1} \cdot N\right) & -\sin\left(\dfrac{2N\pi}{2N+1} \cdot N\right) \\ \cdot & \cdot & \cdot & \cdot & \cdot & \cdot & \cdot \\ 1 & \cos\left(\dfrac{2\pi}{2N+1} \cdot 2N\right) & -\sin\left(\dfrac{2\pi}{2N+1} \cdot 2N\right) & \cdot & \cdot & \cos\left(\dfrac{2N\pi}{2N+1} \cdot 2N\right) & -\sin\left(\dfrac{2N\pi}{2N+1} \cdot 2N\right) \end{bmatrix}$$
$$(A.17)$$

It must be remarked that the time samples are a linear but not an analytic function of the Fourier coefficients X_n, given the presence of conjugated terms. This requires a real number representation and treatment of the harmonic balance system of equations. This loss of analyticity corresponds to special features of the electrical behaviour of the circuit, as shown in Chapters 1 and 3.

A.5 THE HARMONIC BALANCE SYSTEM OF EQUATIONS FOR THE EXAMPLE CIRCUIT WITH $N = 3$

For our example circuit, the harmonic balance (Kirchhoff's) system of equations in the nodal formulation is described in detail for the sake of clarity. First, it can be seen that the nonlinear conductance $i_g(v)$ has a linear behaviour for small voltages with a nonlinear distortion at high voltages. The two parts can be separated, and eq. (1.12) in Chapter 1 is rewritten as

$$i_g(v) = i_{max} \cdot tgh\left(\frac{g \cdot v}{i_{max}}\right) = g \cdot v + \{i_{max} \cdot tgh\left(\frac{g \cdot v}{i_{max}}\right) - g \cdot v\} = g \cdot v + i_{g,NL}(v)$$

(A.18)

The system of equations (1.80) therefore reads, for $N = 3$ as

$$g \cdot V_0 + I_{g,0} = 0$$
$$g \cdot V_1^r - \omega_0 C \cdot V_1^i + I_{g,NL,1}^r + I_s = 0$$
$$g \cdot V_1^i - \omega_0 C \cdot V_1^r + I_{g,NL,1}^i = 0$$
$$g \cdot V_2^r - 2\omega_0 C \cdot V_2^i + I_{g,2}^r = 0 \qquad (A.19)$$
$$g \cdot V_2^i - 2\omega_0 C \cdot V_2^r + I_{g,2}^i = 0$$
$$g \cdot V_3^r - 3\omega_0 C \cdot V_3^i + I_{g,3}^r = 0$$
$$g \cdot V_3^i - 3\omega_0 C \cdot V_3^r + I_{g,3}^r = 0$$

or, in matrix form as

$$\ddot{Y} \cdot \vec{V} + \vec{I}_{g,NL} + \vec{I}_s = 0$$

(A.20)

where

$$\vec{V} = \begin{bmatrix} V_0 \\ V_1^r \\ V_1^i \\ V_2^r \\ V_2^i \\ V_3^r \\ V_3^i \end{bmatrix} \quad \ddot{Y} = \begin{bmatrix} g & 0 & 0 & 0 & 0 & 0 & 0 \\ 0 & g & -\omega_0 C & 0 & 0 & 0 & 0 \\ 0 & -\omega_0 C & g & 0 & 0 & 0 & 0 \\ 0 & 0 & 0 & g & -2\omega_0 C & 0 & 0 \\ 0 & 0 & 0 & -2\omega_0 C & g & 0 & 0 \\ 0 & 0 & 0 & 0 & 0 & g & -3\omega_0 c \\ 0 & 0 & 0 & 0 & 0 & -3\omega_0 C & g \end{bmatrix}$$

$$\vec{I}_{g,NL} = \begin{bmatrix} I_{g,NL,0} \\ I_{g,NL,1}^r \\ I_{g,NL,1}^i \\ I_{g,NL,2}^r \\ I_{g,NL,2}^i \\ I_{g,NL,3}^r \\ I_{g,NL,3}^i \end{bmatrix} \quad \vec{I}_s = \begin{bmatrix} 0 \\ I_s \\ 0 \\ 0 \\ 0 \\ 0 \\ 0 \end{bmatrix} \qquad (A.21)$$

It is apparent that the admittance matrix is block-diagonal, since the linear part does not convert the frequency. The nonlinear current $\vec{I}_{g,NL}$ is obtained by means of a discrete Fourier transform from

$$\vec{I}_{g,NL} = \frac{1}{T}\ddot{R}^{-1} \cdot \vec{i} \tag{A.22a}$$

with

$$\ddot{R} = \begin{bmatrix} 1 & 1 & 0 & 1 & 0 & 1 & 0 \\ 1 & 0.6235 & -0.7818 & -0.2225 & -0.9749 & -0.9010 & -0.4339 \\ 1 & -0.2225 & -0.9749 & -0.9010 & 0.4339 & 0.6235 & 0.7818 \\ 1 & -0.9010 & -0.4339 & 0.6235 & 0.7818 & -0.2225 & -0.9749 \\ 1 & -0.9010 & 0.4339 & 0.6235 & -0.7818 & -0.2225 & 0.9749 \\ 1 & -0.2225 & 0.9749 & -0.9010 & -0.4339 & 0.6235 & -0.7818 \\ 1 & 0.6235 & 0.7818 & -0.2225 & 0.9749 & -0.9101 & 0.4339 \end{bmatrix}$$

$$\vec{i} = \begin{bmatrix} i_{max} \cdot tgh\left(\dfrac{g \cdot v(t_0)}{i_{max}}\right) - g \cdot v(t_0) \\[2mm] i_{max} \cdot tgh\left(\dfrac{g \cdot v(t_1)}{i_{max}}\right) - g \cdot v(t_1) \\[2mm] i_{max} \cdot tgh\left(\dfrac{g \cdot v(t_2)}{i_{max}}\right) - g \cdot v(t_2) \\[2mm] i_{max} \cdot tgh\left(\dfrac{g \cdot v(t_3)}{i_{max}}\right) - g \cdot v(t_3) \\[2mm] i_{max} \cdot tgh\left(\dfrac{g \cdot v(t_4)}{i_{max}}\right) - g \cdot v(t_4) \\[2mm] i_{max} \cdot tgh\left(\dfrac{g \cdot v(t_5)}{i_{max}}\right) - g \cdot v(t_5) \\[2mm] i_{max} \cdot tgh\left(\dfrac{g \cdot v(t_6)}{i_{max}}\right) - g \cdot v(t_6) \end{bmatrix} \tag{A.22b}$$

The time samples are computed from the phasors by means of an inverse discrete Fourier transform from

$$\vec{v} = \ddot{R} \cdot \vec{V} \tag{A.23}$$

where

$$\vec{v} = \begin{bmatrix} v(t_0) \\ v(t_1) \\ v(t_2) \\ v(t_3) \\ v(t_4) \\ v(t_5) \\ v(t_6) \end{bmatrix} \tag{A.24}$$

Therefore, the harmonic balance system of equations (A.20) can be written in the following form:

$$\overset{\leftrightarrow}{Y} \cdot \vec{V} + \vec{I}_{g,\text{NL}} + \vec{I}_s = \vec{F}(\vec{V}) = 0 \qquad \text{(A.25)}$$

This is a nonlinear system of equations that is solved iteratively, usually by means of a Newton–Raphson algorithm, starting from a suitable initial guess $\vec{V}^{(0)}$. The analytic Jacobian is obtained as described in next paragraph (eq. (A.30)). In particular,

$$G(v) = i_{\max} \cdot tgh\left(\frac{g \cdot v}{i_{\max}}\right) - g \cdot v$$

$$G'(v) = \frac{\partial G(v)}{\partial v} = g \cdot \left(\frac{1}{\cosh^2\left(\dfrac{g \cdot v}{i_{\max}}\right)} - 1\right) \qquad \text{(A.26)}$$

whence

$$\overset{\leftrightarrow}{G}(\vec{v}) = g$$

$$
\begin{bmatrix}
\dfrac{1}{\cosh^2\left(\frac{g\cdot v(t_0)}{i_{\max}}\right)} - 1 & 0 & 0 & 0 & 0 & 0 & 0 \\[2em]
0 & \dfrac{1}{\cosh^2\left(\frac{g\cdot v(t_1)}{i_{\max}}\right)} - 1 & 0 & 0 & 0 & 0 & 0 \\[2em]
0 & 0 & \dfrac{1}{\cosh^2\left(\frac{g\cdot v(t_2)}{i_{\max}}\right)} - 1 & 0 & 0 & 0 & 0 \\[2em]
0 & 0 & 0 & \dfrac{1}{\cosh^2\left(\frac{g\cdot v(t_3)}{i_{\max}}\right)} - 1 & 0 & 0 & 0 \\[2em]
0 & 0 & 0 & 0 & \dfrac{1}{\cosh^2\left(\frac{g\cdot v(t_5)}{i_{\max}}\right)} - 1 & 0 & 0 \\[2em]
0 & 0 & 0 & 0 & 0 & \dfrac{1}{\cosh^2\left(\frac{g\cdot v(t_6)}{i_{\max}}\right)} - 1 & 0 \\[2em]
0 & 0 & 0 & 0 & 0 & 0 & \dfrac{1}{\cosh^2\left(\frac{g\cdot v(t_7)}{i_{\max}}\right)} - 1
\end{bmatrix}
$$

$$\text{(A.27)}$$

A.6 THE JACOBIAN MATRIX

The Jacobian matrix of the harmonic balance problem is the derivative of the equation system with respect to the unknowns:

$$\overset{\leftrightarrow}{J}(\vec{V}) = \frac{\partial \vec{F}(\vec{V})}{\partial \vec{V}} \qquad \text{(A.28)}$$

The function has the form

$$\vec{F}(\vec{V}) = \vec{I}_L + \vec{I}_{NL} = \ddot{Y} \cdot \vec{V} + \vec{I}_{L,0} + \vec{I}_{NL} \tag{A.29}$$

Its derivative, therefore, is

$$\ddot{J}(\vec{V}) = \frac{\partial \vec{F}(\vec{V})}{\partial \vec{V}} = \ddot{Y} + \frac{\partial \vec{I}_{NL}(\vec{V})}{\partial \vec{V}} \tag{A.30}$$

The derivative of the vector of the phasors of the nonlinear currents is best computed analytically if the derivative of the currents as functions of the voltages is explicitly available:

$$\vec{i}_{NL}(t) = \vec{G}(\vec{v}(t)) \qquad \frac{\partial \vec{i}_{NL}(t)}{\partial \vec{v}(t)} = \frac{\partial \vec{G}(\vec{v}(t))}{\partial \vec{v}(t)} = \ddot{G}'(\vec{v}(t)) \tag{A.31}$$

In a practical harmonic balance algorithm, time-domain voltages and current are sampled at a suitable set of time instants, satisfying Nykvist's sampling theorem, and conversion from time domain to frequency domain is performed by means of a real-valued discrete Fourier transform and its inverse (see above). Therefore, the phasors of the nonlinear currents are computed as shown below:

$$\vec{v}(t) = \ddot{R} \cdot \vec{V} \qquad \frac{\partial \vec{i}_{NL}(t)}{\partial \vec{v}(t)} = \frac{\partial \vec{G}(\vec{v}(t))}{\partial \vec{v}(t)} = \ddot{G}'(\vec{v}(t)) \frac{\partial \vec{I}_{NL}(\vec{V})}{\partial \vec{V}} = \frac{1}{T} \cdot \ddot{R}^{-1} \cdot \frac{\partial \vec{i}_{NL}(t)}{\partial \vec{v}(t)} \tag{A.32}$$

from which

$$\frac{\partial \vec{I}_{NL}(\vec{V})}{\partial \vec{V}} = \frac{1}{T} \cdot \ddot{R}^{-1} \cdot \ddot{G}'(\ddot{R} \cdot \vec{V}) \tag{A.33}$$

The matrix $\ddot{G}'(\vec{v}(t)) = \dfrac{\partial \vec{G}(\vec{v}(t))}{\partial \vec{v}(t)}$ is actually a block-diagonal matrix, where the diagonals are the derivative of the function computed at the sampling time instants.

A.7 MULTI-DIMENSIONAL DISCRETE FOURIER TRANSFORM AND INVERSE DISCRETE FOURIER TRANSFORM FOR QUASI-PERIODIC SIGNALS

A quasi-periodic signal with two basic frequencies can be expressed as

$$v(t) = \sum_{n_1=-\infty}^{\infty} \sum_{n_2=-\infty}^{\infty} V_{n_1,n_2} \cdot e^{j(n_1\omega_1 + n_2\omega_2)t} = \sum_{n_1=-\infty}^{\infty} \sum_{n_2=-\infty}^{\infty} V_{n_1,n_2} \cdot e^{j\omega_{n_1,n_2}t} \tag{A.34}$$

For transformation purposes, it can be seen as a particular case of the function of two time variables:

$$v(t_1, t_2) = \sum_{n_1=-\infty}^{\infty} \sum_{n_2=-\infty}^{\infty} V_{n_1,n_2} \cdot e^{j(n_1\omega_1 t_1 + n_2\omega_2 t_2)} \tag{A.35}$$

when $t_1 = t_2 = t$. By analogy with the case of the mono-dimensional Fourier series, the coefficients are expressed as

$$V_{n_1,n_2} = \frac{1}{T_1 T_2} \sum_{k_1=-N_1}^{N_1} \sum_{k_2=-N_2}^{N_2} v(t_{k_1}, t_{k_2}) \cdot e^{j(n_1 \omega_1 t_{k_1} + n_2 \omega_2 t_{k_2})} \tag{A.36}$$

with

$$t_{k_1} = \frac{T_1}{2N_1 + 1} \cdot k_1, \quad k_1 = -N_1, \ldots, N_1, \quad t_{k_2} = \frac{T_2}{2N_2 + 1} \cdot k_2, \quad k_2 = -N_2, \ldots, N_2 \tag{A.37}$$

By replacing, we get

$$V_{n_1,n_2} = \frac{1}{T_1 T_2} \sum_{k_1=-N_1}^{N_1} \sum_{k_2=-N_2}^{N_2} v(t_{k_1}, t_{k_2}) \cdot e^{j2\pi \left(\frac{n_1 \cdot k_1}{2N_1+1} + \frac{n_2 \cdot k_2}{2N_2+1} \right)} \tag{A.38}$$

As in the case of the mono-dimensional Fourier transform, a real signal is more effectively represented by a real-number series; the detailed expressions are easily derived in analogy with the single-tone case.

A.8 OVERSAMPLED DISCRETE FOURIER TRANSFORM AND INVERSE DISCRETE FOURIER TRANSFORM FOR QUASI-PERIODIC SIGNALS

Let us come back to the real-numbered expression of a periodic signal as stated above:

$$x(t) = \sum_{n=0}^{N} (X_n^r \cdot \cos(n\omega_0 t) - X_n^i \cdot \sin(n\omega_0 t))$$

$$= X_0 + \sum_{n=1}^{N} (X_n^r \cdot \cos(n\omega_0 t) - X_n^i \cdot \sin(n\omega_0 t)) \tag{A.39}$$

In the case of a multi-tone signal (only two tones in this example), the expression becomes

$$x(t) = \sum_{n_1=0}^{N_1} \sum_{n_2=0}^{N_2} (X_{n_1,n_2}^r \cdot \cos((n_1 \omega_1 + n_2 \omega_2)t) - X_n^i \cdot \sin((n_1 \omega_1 + n_2 \omega_2)t)) \tag{A.40}$$

We can again write the transformation formula from coefficients of the two-tone Fourier series expansion and time samples (discrete Fourier transform) as

$$\vec{x} = \overset{\leftrightarrow}{R} \cdot \vec{X} \tag{A.41}$$

where

$$
\vec{x} =
\begin{bmatrix}
t_0 \\
\cdot \\
\cdot \\
\cdot \\
\cdot \\
t_{N_{\max}}
\end{bmatrix}
$$

$$
\overset{\leftrightarrow}{R} =
\begin{bmatrix}
1 & \cos(\omega_1 t_0) & -\sin(\omega_1 t_0) & \cdot & \cdot & \cos((N_1\omega_1 + N_2\omega_2)t_0) & -\sin((N_1\omega_1 + N_2\omega_2)t_0) \\
\cdot & \cdot & \cdot & \cdot & \cdot & \cdot & \cdot \\
\cdot & \cdot & \cdot & \cdot & \cdot & \cdot & \cdot \\
\cdot & \cdot & \cdot & \cdot & \cdot & \cdot & \cdot \\
1 & \cos(\omega_1 t_{N_{\max}}) & -\sin(\omega_1 t_{N_{\max}}) & \cdot & \cdot & \cos((N_1\omega_1 + N_2\omega_2)t_{N_{\max}}) & -\sin((N_1\omega_1 + N_2\omega_2)t_{N_{\max}})
\end{bmatrix}
$$

$$
\vec{X} =
\begin{bmatrix}
X_{0,0} \\
X^r_{1,0} \\
X^i_{1,0} \\
\cdot \\
\cdot \\
X^r_{N_1,N_2} \\
X^i_{N_1,N_2}
\end{bmatrix}
\tag{A.42}
$$

If $N_{\max} = N^0_{\max} = N_1 \cdot N_2 + N_1 + N_2 + 1$, the equation system is square and it can be inverted (discrete Fourier transform):

$$
\vec{X} = \frac{1}{T} \overset{\leftrightarrow}{R}{}^{-1} \cdot \vec{x}
\tag{A.43}
$$

However, problems usually arise because of the ill-conditioning of the matrix $\overset{\leftrightarrow}{R}$. A good choice of the sampling time instants $t_0 \ldots t_{N_{\max}}$ ensures the well-conditioning of the matrix. A simple choice is done by the random selection of a number of time instants in excess of the minimum number:

$$
N_{\max} = m \cdot N^0_{\max} \quad \text{with } m = 2 \div 3
\tag{A.44}
$$

An overdetermined system results, with a rectangular matrix $\overset{\leftrightarrow}{R}$. The rows of the matrix are first orthogonalised by means of a standard Grahm–Schmidt algorithm; then, the largest $N^0_{\max}$ rows are retained, while the others are discarded. The matrix $\overset{\leftrightarrow}{R}$ is now square and can be inverted for discrete Fourier transformation. If the starting number of samples is high enough and randomly selected, the matrix is usually well conditioned.

A.9 DERIVATION OF SIMPLIFIED TRANSPORT EQUATIONS

An analytical solution of Poisson's and transport equations is available if the latter are simplified. In particular, eq. (3.9) is deduced from eq. (3.5) by suitable assumptions. We rewrite eq. (3.5) as

$$\frac{d(n \cdot v)}{dt} = \frac{n \cdot q E}{m_{\text{eff}}} - \frac{\partial(n \cdot v^2)}{\partial x} - \frac{2}{3} \cdot \frac{\partial \left(\dfrac{n \cdot w}{m_{\text{eff}}} - \dfrac{n \cdot v^2}{2} \right)}{\partial x} - \left(\frac{\partial(n \cdot v)}{\partial t} \right)_{\text{coll.}} \tag{A.45}$$

In steady state, the derivative of the velocity $n \cdot v$ with respect to time is zero. Then, we assume that the derivatives with respect to space of both potential and kinetic energy can be neglected, and express the collision term by means of a relaxation time, in analogy with the charge recombination time. Eq.(A.45) becomes

$$0 = \frac{n \cdot q E}{m_{\text{eff}}} - \frac{2}{3} \cdot \frac{w}{m_{\text{eff}}} \frac{\partial n}{\partial x} - \frac{n \cdot v}{\tau_v} \tag{A.46}$$

Introducing the current density $j = n \cdot v$, the mobility $\mu = \dfrac{\tau_v}{m_{\text{eff}}}$ and the diffusion constant $D = \dfrac{3}{2} \cdot \dfrac{w \cdot \tau_v}{m_{\text{eff}}}$, and rearranging eq. (A.46) we have

$$j = n \cdot q \cdot \mu \cdot E - D \cdot \frac{\partial n}{\partial x} \tag{A.47}$$

Equation (A.47) is eq. (3.9). If for the energy we assume the simple expression

$$w = \tfrac{3}{2} \cdot k_B T \tag{A.48}$$

then we have Einstein's relation for the diffusion constant:

$$D = \frac{k_B T}{\mu} \tag{A.49}$$

A.10 DETERMINATION OF THE STABILITY OF A LINEAR NETWORK

Let a linear network with an admittance representation $Y(s)$ have a zero of its determinant at the complex Laplace parameter s_0:

$$\det(\overset{\leftrightarrow}{Y}(s_0)) = F(s_0) = F(\alpha_0 + j\omega_0) = 0 \tag{A.50}$$

The network is unstable if

$$\alpha_0 > 0 \tag{A.51}$$

Let us now assume that s_0 lies in the vicinity of the frequency ω_1 where the phase of the function $F(j\omega)$ becomes zero, that is, where the circuit reactances computed in periodic regime $s = j\omega$ resonate (see Chapter 5). We can write

$$s_0 = \alpha_0 + j\omega_0 = \alpha_0 + j(\omega_1 + \delta\omega) \tag{A.52}$$

where α_0 and $\partial\omega$ are small. We can expand the determinant in Taylor series truncated at the first order in the vicinity of $s = j\omega_1$:

$$F(s) = F(j\omega_1) + \left.\frac{\partial F}{\partial s}\right|_{s=j\omega_1} \cdot ds + \cdots$$

$$= F(j\omega_1) + \left.\frac{\partial F}{\partial \alpha}\right|_{s=j\omega_1} \cdot d\alpha + \left.\frac{\partial F}{\partial \omega}\right|_{s=j\omega_1} \cdot d\omega + \cdots \tag{A.53}$$

If the complex function $F(s) = F_r(\alpha + j\omega) + jF_i(\alpha + j\omega)$ is analytical, the Cauchy–Riemann condition states that

$$\frac{\partial F_r}{\partial \alpha} = \frac{\partial F_i}{\partial \omega} \quad \frac{\partial F_r}{\partial \omega} = -\frac{\partial F_i}{\partial \alpha} \tag{A.54}$$

Therefore, we can write

$$F(s) = F(j\omega_1) + \left(\frac{\partial F_r}{\partial \alpha} + j\frac{\partial F_i}{\partial \alpha}\right) \cdot d\alpha + \left(\frac{\partial F_r}{\partial \omega} + j\frac{\partial F_i}{\partial \omega}\right) \cdot d\omega + \cdots \tag{A.55}$$

Since we want to compute the determinant only for $s = j\omega$, we use the Cauchy–Riemann condition to get

$$F(s) = F(j\omega_1) + \left(\frac{\partial F_i}{\partial \omega} - j\frac{\partial F_r}{\partial \omega}\right) \cdot d\alpha + \left(\frac{\partial F_r}{\partial \omega} + j\frac{\partial F_i}{\partial \omega}\right) \cdot d\omega + \cdots =$$

$$= F(j\omega_1) - j\left(\frac{\partial F_r}{\partial \omega} + j\frac{\partial F_i}{\partial \omega}\right) \cdot d\alpha + \left(\frac{\partial F_r}{\partial \omega} + j\frac{\partial F_i}{\partial \omega}\right) \cdot d\omega + \cdots = \tag{A.56}$$

$$= F(j\omega_1) + \frac{\partial F}{\partial \omega} \cdot (d\omega - jd\alpha) + \cdots$$

where all derivatives are computed at $s = j\omega_1$. For $s = s_0 = \alpha_0 + j\omega_0 = \alpha_0 + j(\omega_1 + \delta\omega)$, the determinant $F(s)$ is zero; therefore,

$$F(j\omega_1) + \frac{\partial F}{\partial \omega} \cdot (\delta\omega - j\alpha_0) \cong 0 \tag{A.57}$$

We now write this equation separately for the real and imaginary part, in order to solve for α_0. Remembering that the phase, and therefore the imaginary part, of the determinant is zero for $s = j\omega_1$, we get

$$F(j\omega_1) + \frac{\partial F_r}{\partial \omega} \cdot \delta\omega + \frac{\partial F_i}{\partial \omega} \cdot \alpha_0 = 0$$

$$\frac{\partial F_i}{\partial \omega} \cdot \delta\omega - \frac{\partial F_r}{\partial \omega} \cdot \alpha_0 = 0 \tag{A.58}$$

From the second equation,

$$\delta\omega = \frac{\dfrac{\partial F_r}{\partial\omega}}{\dfrac{\partial F_i}{\partial\omega}} \cdot \alpha_0 \tag{A.59}$$

Replacing this in the first equation, we get

$$F(j\omega_1) + \left(\frac{\left(\dfrac{\partial F_r}{\partial\omega}\right)^2}{\dfrac{\partial F_i}{\partial\omega}} + \frac{\partial F_i}{\partial\omega}\right) \cdot \alpha_0 = F(j\omega_1) + \frac{\left(\left(\dfrac{\partial F_r}{\partial\omega}\right)^2 + \left(\dfrac{\partial F_i}{\partial\omega}\right)^2\right)}{\dfrac{\partial F_i}{\partial\omega}} \cdot \alpha_0 = 0 \tag{A.60}$$

whence, expliciting the real part of the Laplace parameter we have

$$\alpha_0 = -F(j\omega_1) \cdot \frac{\Im\left[\left.\dfrac{\partial F(j\omega)}{\partial\omega}\right|_{\omega=\omega_1}\right]}{\left|\left.\dfrac{\partial F(j\omega)}{\partial\omega}\right|_{\omega=\omega_1}\right|^2} \tag{A.61}$$

A.11 DETERMINATION OF THE LOCKING RANGE OF AN INJECTION-LOCKED OSCILLATOR

The equation determining the locking range is (eq. (8.35))

$$I_{inj} + \left(\frac{\partial Y_{tot}}{\partial A} \cdot \Delta A + \frac{\partial Y_{tot}}{\partial\omega} \cdot \Delta\omega\right) \cdot A_0 = 0 \tag{A.62}$$

Separating the real and imaginary parts, we get

$$I_{inj,r} + \left(\frac{\partial Y_r}{\partial A} \cdot \Delta A + \frac{\partial Y_r}{\partial\omega} \cdot \Delta\omega\right) \cdot A_0 = 0 \tag{A.63a}$$

$$I_{inj,i} + \left(\frac{\partial Y_i}{\partial A} \cdot \Delta A + \frac{\partial Y_i}{\partial\omega} \cdot \Delta\omega\right) \cdot A_0 = 0 \tag{A.63b}$$

Multiplying eq. (A.63a) by $\dfrac{\partial Y_i}{\partial A}$ and eq. (A.63b) by $\dfrac{\partial Y_r}{\partial A}$ we get

$$I_{inj,r} \cdot \frac{\partial Y_i}{\partial A} + \left(\frac{\partial Y_r}{\partial A}\frac{\partial Y_i}{\partial A} \cdot \Delta A + \frac{\partial Y_r}{\partial\omega}\frac{\partial Y_i}{\partial A} \cdot \Delta\omega\right) \cdot A_0 = 0 \tag{A.64a}$$

$$I_{inj,i} \cdot \frac{\partial Y_r}{\partial A} + \left(\frac{\partial Y_r}{\partial A}\frac{\partial Y_i}{\partial A} \cdot \Delta A + \frac{\partial Y_i}{\partial\omega}\frac{\partial Y_r}{\partial A} \cdot \Delta\omega\right) \cdot A_0 = 0 \tag{A.64b}$$

Subtracting the two equations we get

$$I_{\text{inj,r}} \cdot \frac{\partial Y_{\text{i}}}{\partial A} - I_{\text{inj,i}} \cdot \frac{\partial Y_{\text{r}}}{\partial A} + \left(\left(\frac{\partial Y_{\text{r}}}{\partial \omega} \cdot \frac{\partial Y_{\text{i}}}{\partial A} - \frac{\partial Y_{\text{i}}}{\partial \omega} \cdot \frac{\partial Y_{\text{r}}}{\partial A} \right) \cdot \Delta \omega \right) \cdot A_0 = 0 \qquad (A.65)$$

If we set

$$I_{\text{inj}} = I_{\text{inj,r}} + j I_{\text{inj,i}} = |I_{\text{inj}}| \cdot (\cos \varphi + j \sin \varphi) \qquad (A.66a)$$

$$\frac{\partial Y}{\partial A} = \frac{\partial Y_{\text{r}}}{\partial A} + j \frac{\partial Y_{\text{i}}}{\partial A} = \left| \frac{\partial Y}{\partial A} \right| \cdot (\cos \vartheta + j \sin \vartheta) \qquad (A.66b)$$

then eq. (A.65) is rewritten as

$$\frac{|I_{\text{inj}}|}{A_0} \left| \frac{\partial Y}{\partial A} \right| (\cos \varphi \cdot \sin \vartheta - \sin \varphi \cdot \cos \vartheta) = \left(\left(\frac{\partial Y_{\text{r}}}{\partial \omega} \cdot \frac{\partial Y_{\text{i}}}{\partial A} - \frac{\partial Y_{\text{i}}}{\partial \omega} \cdot \frac{\partial Y_{\text{r}}}{\partial A} \right) \cdot \Delta \omega \right) \qquad (A.67)$$

or

$$\frac{|I_{\text{inj}}|}{A_0} \cdot \left| \frac{\partial Y}{\partial A} \right| \cdot \cos(\varphi - \vartheta) = \left(\left(\frac{\partial Y_{\text{r}}}{\partial \omega} \cdot \frac{\partial Y_{\text{i}}}{\partial A} - \frac{\partial Y_{\text{i}}}{\partial \omega} \cdot \frac{\partial Y_{\text{r}}}{\partial A} \right) \cdot \Delta \omega \right) \qquad (A.68)$$

The maximum locking range then is

$$|\Delta \omega|_{\max} = \frac{|I_{\text{inj}}|}{A_0} \cdot \left| \frac{\dfrac{\partial Y}{\partial A}}{\dfrac{\partial Y_{\text{r}}}{\partial \omega} \cdot \dfrac{\partial Y_{\text{i}}}{\partial A} - \dfrac{\partial Y_{\text{i}}}{\partial \omega} \cdot \dfrac{\partial Y_{\text{r}}}{\partial A}} \right| \qquad (A.69)$$

Similarly, from eq. (A.64) we get

$$|\Delta A|_{\max} = \frac{|I_{\text{inj}}|}{A_0} \cdot \left| \frac{\dfrac{\partial Y}{\partial \omega}}{\dfrac{\partial Y_{\text{r}}}{\partial \omega} \cdot \dfrac{\partial Y_{\text{i}}}{\partial A} - \dfrac{\partial Y_{\text{i}}}{\partial \omega} \cdot \dfrac{\partial Y_{\text{r}}}{\partial A}} \right| \qquad (A.70)$$

Index

Alenia Marconi Systems, 214
AM/AM, *see* Conversion
AM/PM, *see* Conversion
Analysis
   large-signal, 143, 329, 331–332
   multi-tone, 33, 34
   single-tone, 33, 39
   two-tone, 34, 356
Angelov, *see* Model
Antiparallel diodes, 324, 327–328
Autonomous circuit, 8, 58, 230–233, 251,
   259–267, 277, 341–342, 345, 356,
   358, 369

Balance
   harmonic, 9, 22–34, 37–39, 43–49,
      57–58, 88, 142–143, 147, 151, 157,
      266, 274–277, 288, 298, 303, 313,
      339–340, 348–349, 358, 368
   spectral, 47, 49, 59, 142, 145, 259,
      262, 264–268, 332, 348–349,
      356–358
   waveform, 33, 88
Balanced
   configuration, 281, 308, 311, 318, 327
   mixer, *see* Mixer
Balun, 311
Bandwidth, 6, 11, 24, 32, 224, 254, 279–281,
   298, 306–307, 311–313, 338
Barkhausen oscillation condition, 234, 258,
   348–349
Baseband, 270, 273
Behavioural model, 96, 142, 146, 157

Bias
   condition, 125–126, 129, 192, 208, 212,
      214
   forward, 99, 128, 130, 197, 279, 319
   network, 7, 9, 63, 271
   reverse, 265, 279–280, 308, 319–320
   supply, 77, 164, 188, 262, 364
   T, 49, 63–64, 66
Bifurcation, 349, 351, 355–356, 358–359,
   364, 368
   diagram, 353–356, 358
   direct-type, 355
   flip, 355, 358–359
   Hopf, 351, 355, 358–359, 369
   turning-point, 355, 358
BJT, 92, 121, 259, 273, 319
Black-box, *see* Model
Boltzmann's equation, 84–88, 90, 92–93
Boundary, 56, 61, 363
   conditions, 85, 88, 92–93, 343–344
Breakdown, 90, 149, 170, 187, 197, 201, 205,
   218, 282–284, 287–288, 292, 294, 353

Carrier, 8, 43–45, 58, 146–147, 165–166,
   270, 273–275
Carriers, 74, 89, 94, 105, 110, 152
Chaos, 57, 229
Chaotic behaviour, 22, 229, 355–356,
   358–359
Characterisation, 22, 47, 61–62, 64, 66–67,
   74, 77, 138, 214
Circulation angle, 151,192, 208, 212,
   288–289, 295

Nonlinear Microwave Circuit Design   F. Giannini and G. Leuzzi
© 2004 John Wiley & Sons, Ltd   ISBN: 0-470-84701-8

Class
  A, 136, 164, 169–170, 180, 190–194,
    208–209, 212, 259, 282, 285, 290,
    294, 296
  AB, 169, 191–192, 194, 198, 214, 226,
    259, 296, 299
  B, 77, 169, 172, 198, 208, 212, 226, 259,
    282, 353, 364
  C, 57, 169, 194, 208, 227, 282, 285,
    287–288, 294, 296, 299
  E, 157, 169, 184, 188, 195, 208, 226–227,
    258, 277
  F, 157, 169–170, 184–188, 194–200, 208,
    211–212, 226–227, 258, 277
  FG, 198, 209, 211–223
  G, 198–201, 208, 212, 214, 222, 227
  inverse F, 187, 194–196, 226
  of operation, 169
Clipping, 184, 207, 209, 282, 285–286,
  290–296, 305,
Cold bias, 125–126, 129, 131, 155, 322
Compression, see Gain
Conduction angle, 193–194, 206–207, 209,
  212, 285, 287, 291, 293–294
Continuation method, 32, 58, 264–265, 358
Contours, see also Power contours
Convergence, 6, 31–32, 88–89, 264–269, 358
Conversion
  AM/AM, 47, 146
  AM/PM, 47, 146
  efficiency, 62, 279
  gain, 62, 279–282, 287, 294, 297–298,
    300–307, 320, 322, 331–332, 364
  loss, 281, 318, 327, 337
  matrix, 49–51, 55–56, 274, 308, 310, 329,
    333, 336–339, 342–350, 355
  noise, 273–274
Convolution, 2, 9–11, 13, 143, 262
Coupler
  delta/sigma, 308, 324
  directional, 63–64, 66–69, 73
  hybrid, 323, 327

De-embedding, 65, 132, 144
Describing function, 46–47, 142, 146–147
Desensitivisation, 18
Determinant, 230, 239, 242, 260, 344,
  346–347
Diffusion, 87, 92, 99, 280
Discretisation, 4–7, 10, 13, 93

Displacement current, 93, 98–99, 102, 113
Distortion, 11, 16, 33–34, 62, 65, 159, 161,
  165–170, 173, 209, 282, 306, 318,
  332–333
  intermodulation, 56, 95, 136, 138–139, 168,
    195
Distributed, see Element
Downconverter, 318
Duty cycle, 46, 212, 295–296
Dynamic range, 56, 160, 318, 333, 337

Ebers–Moll, see Model
Efficiency, 10, 31, 62–63, 72, 148, 163–165,
  169–170, 182–188, 191–195, 198, 201,
  224, 252, 258–259, 270, 280
  drain, 163–164, 191, 195–196, 212
  power-added, 66, 71, 164, 187, 218, 220,
    222
Element
  active, 342
  distributed, 9, 95–96, 141, 262
  extrinsic, 96
  frequency-converting, 300, 302, 308, 333,
    337, 346
  intrinsic, 96, 122, 125, 133, 230
  lumped, 95–96, 181, 262
  parasitic, 86, 123, 125, 132, 144–145, 170,
    230, 281
  passive, 83, 230, 268
  reactive, 256
  resistive, 237–238, 269
Embedding
  admittance, 329, 331
  impedance, 258, 348
  network, 246–248, 254, 258, 269
Envelope, 8, 43–45, 47, 146, 166, 276
Equilibrium point, 251, 259, 354–355
Equiripple condition, 201, 203
Error, 5–6, 11–12, 25, 31, 72, 82–83, 125,
  168, 207, 264
Euler, 5–6, 44
Extraction, 62, 80, 92, 120–121, 128–130,
  138–139, 141, 214
  procedure, 84, 95–96, 104, 108, 114, 125,
    133, 142–143, 149

Feedback, 121, 123, 170, 244, 252–259, 269,
  282, 297, 306, 364, 367
Filter, 11, 14, 47, 69, 119, 190, 267, 281, 296,
  308, 311, 315, 318, 320, 323, 325, 329,
  358

Fixed-point method, 6
Floquet, 342–343
  coefficients, 343–345
Forward-bias, *see* Bias
Fourier
  series, 13, 22–39, 48–49, 51, 54, 88, 188, 264, 310, 344, 349
  transform, 2, 15, 24, 26, 30, 37, 39, 47, 49, 52, 91, 104, 143, 208, 262
    discrete, 25–26
    inverse, 2, 11, 30
    multidimensional, 38
Frequency
  conversion, 270, 279, 281, 300, 306, 308, 333, 335, 349
  converting circuit, 300
  divider, 229, 311, 358
    non-regenerative, 341, 365
    regenerative, 308, 355, 365–366
  doubler, 296, 299–300, 306, 341
  image, 331
  locking, 229, 364
  multiplier, 70, 146, 150, 279–283, 288, 296–300, 306–307, 365, 367
  offset, 269, 273
  remapping, 38
  shifting, 315
  spurious, 162–163, 231
  subharmonic, *see* Subharmonic frequency
Fujitsu, 224
Function
  delta, 13, 15
  describing, *see* Describing function
  distribution, 86–87
  energy, 117
  fitting, 112–113, 122, 133, 136, 146
  goal, 123
  Hermitean, 25
  network, 245
  nonlinear scattering, *see* Nonlinear scattering functions
  objective, 258
  piecewise-linear, 149
  polynomial, 136, 145
  single-valued, 117–117
  wave, *see* Wave function

Gain
  compression, 65–66, 161, 164, 168, 170, 216, 352–353
  conversion, *see* Conversion
  expansion, 251, 259, 353
  maximum available, 331
  maximum stable, 331
  power, *see* Power gain
  saturation, 255
  variable, 43
Giacoletto, *see* Model

Harmonic
  balance, *see* Balance
  content, 14, 165, 184, 188, 197, 209, 215, 284, 292, 296, 302, 305
  current, 287, 292, 294–295
  generation, 207, 209–210, 229, 280, 282–283, 305
  loading, 258
  manipulation, 169, 196, 198–199, 211–212, 216, 220, 224
  reaction amplifier, 184
  terminating control, 187
  tuning, 194, 207, 220
HBT, 89, 121, 279, 319
HEMT, 89–90, 91, 94–95, 121–122, 136–137, 224, 279, 289, 319, 332
Heterojunction, 85, 89–90
High-Q resonator, 229, 245, 252, 262, 264, 266
Hilbert transform, 12
HP-MDS, 180, 216

Idler, 191, 195, 281–282, 296–297, 299, 302–304
IF, 34, 318
IMD, *see* Distortion
Impulse response, 2, 9–11, 13, 142–143
Incremental ratio, 4–6, 31
Initial
  guess, 6, 9, 265–267
  state, 7–8, 261
  value, 5, 9, 118
Injection-locked oscillator, 341, 362, 354, 366
Injection locking, 229, 359, 365, 367
Instability, 229–230, 235–239, 243, 248–250, 258, 282, 298, 302, 320, 346–349, 355, 363
  nonlinear, 71
  numerical, 264
  potential, 252, 349
  region, 304–305

Integration, 4–9, 15, 27, 85–87, 92, 104, 108, 118–120, 142, 145, 259–262, 276, 279, 332

Intercept point, 166,168

Interference, 297–298, 318

Intermodulation, *see* Distortion

Interpolation, 46, 65

Isolation, 297, 307–308, 311, 318–322, 325, 327, 332

Isothermal, 74, 77, 79, 107, 109, 111, 119

Jacobian, 31, 58, 147

Knee voltage, 74, 197, 273, 283–284

Kurokawa condition, 231, 263–264, 268, 357

Laplace
    Domain, 2, 4, 235, 239,
    Parameter, 235–236, 238, 244–347, 355

Large-signal
    measurements, 74, 80–82, 143, 157, 300, 329
    model, 80, 109, 145, 151, 153

Layout, 86, 91, 139, 218–219, 223, 296, 319

Liouville, 87

LO, *see* Local oscillator

Load
    curve, 119, 170, 172, 175–177, 179, 192–193, 216, 219, 288, 319, 342,
    line 49, 143, 172–175, 230, 282–283, 289, 319–322, 331
    pull, 62, 179–181, 207

Load/source-pull, 61–63, 67–73, 146

Local oscillator 33–34, 62, 146, 315, 322, 325–327, 329–339, 343, 349, 363–364

Locking
    Range, 361–365, 367, 369
    Signal, 362–363, 365, 368

Look-up table, 31, 46

Low-frequency dispersion, 105, 113–114, 119, 138, 145, 271

Lumped, *see* Element

Manley-Rowe relation, 281

Matching
    Network, 63, 171, 180–181, 189, 214, 216, 231, 281, 296, 299, 318
    Conjugate, 300, 302–303
    power, 180–181

Materka, *see* Model

Maximally flat condition,196, 204–205

Maxwell's equations, 85, 88

Measurement
    fast, 108, 110, 113
    large-signal,74, 80–82, 143, 157, 300, 329
    pulsed, 61, 74–75, 77, 79, 82, 107, 110, 119, 154
    small-signal, 74, 77, 108, 110, 118–119, 121

Memory, 13–14, 45, 57, 142–143, 262

MESFET, 89–90, 93–95, 105, 121–122, 180, 214, 279, 332

Mixer
    balanced, 59, 316, 318, 324–329, 339–340
    diode, 57, 318, 331, 340
    doubly balanced, 281, 329
    self-oscillating, 266, 341, 359, 363–368
    single-diode, 325, 327–329
    singly balanced, 281, 324–328
    subharmonically pumped, 59, 327–328, 339

Model
    analytical, 90, 92–93, 95, 153
    Angelov 289, 292, 313
    black-box, 74, 84, 96, 142–145, 147, 230
    drift-diffusion, 88–91
    Ebers–Moll, 92–93
    equivalent-circuit, 84, 92, 95–97, 99, 102, 122, 133
    empirical 83–84, 92, 95, 154–156
    energy-balance, 88, 91
    Giacoletto, 100–101, 230
    Hydrodynamic, 88–90, 151
    Materka, 155,180
    physical, 83–87, 89–93, 152
    quasi-static, 100–101, 116–117, 143, 156
    quasi-2D, 85, 88, 90–91, 152
    stochastic, 90

Modelling, 11, 61, 74, 82, 108–109, 137, 141

Monte Carlo, 88, 90–91, 152

Multi-harmonic manipulation, 182, 187, 197, 205–209, 212, 215, 218, 222–223

Negative
    Conductance, 233, 236, 248–249, 251, 310–311, 363
    Resistance, 244, 250–252, 268, 276–277, 311, 369

Neural network, 74, 122, 139, 146, 154, 157

Newton-Raphson, 6, 31, 58, 253, 255–267

Noise
  Phase, 108, 229, 252, 269, 271–273,
    277–278
  source, 269–270, 275, 276, 337, 339
Non-autonomous circuit, 8, 260, 263,
    267–268, 275, 358
Nonlinear Scattering Functions, 73, 147, 157,
    300, 306
Norton, 28–30, 62, 262, 275
Nuclei, 14–15, 18, 21–22, 168, 269, 335
Numerical
  model, 11, 46, 88, 151
  solution, 4–10, 25–26, 31, 33, 46, 56, 93,
    265
Nykvist
  Criterium, 6, 24, 26, 30, 38–39, 346–349,
    358
  Plot, 347, 359–360

Ohmic region, 149, 169–170, 205, 283, 287,
    320, 322
Oscillation, 7, 8, 69, 77, 229–232, 235–239,
    243–273, 276–277, 305, 314, 341–342,
    346–348, 351–367
Oscillator
  free-running, 277, 360–368
  voltage-controlled, 252, 276, 280
Oversampling, 26, 39, 41

Package, 26, 122, 139–142, 156
Parasitics, 122–123, 125, 130, 170, 183, 194,
    281
Peaking effect, 199–201, 218
Phase-locked loop, 229
Phase noise, *see* Noise
Piecewise-linear
  Analysis, 149, 169, 283,
  model, 84, 149–150, 170–171, 179,182,
    212, 284–285, 290, 296, 305
Pinch-off, 136, 169–170, 184, 187, 207,
    282–285, 288, 290–292, 296, 299, 305,
    355
PLL, *see* Phase-locked loop
Poisson's equation, 84–85, 87–88, 90,
    92–94
Power
  Amplifier, 22, 32–34, 62–65, 68, 77, 81,
    84, 148–150, 155–170, 181–182,
    187, 191, 195, 198, 212, 218–223,
    226–228, 258–259, 341

available, 159, 276, 298
consumption, 159, 163, 280, 288, 294
contour, 65, 178–179, 181, 183, 185
gain, 62–63, 66, 159–160, 164–165, 194,
    259, 353–354
meter, 62–63, 68
supply, 63, 161, 163, 262
Probe, 73, 260–262, 267–268, 357–358
Probing
  generator, 268
  method, 21, 49
  signal, 20–22, 267–269, 357
Pushing factor, 271–272

Quiescent point, 49, 74, 75, 79, 102, 105, 108,
    230

Reliability, 89, 201, 209, 280, 282, 288, 294
Rectification, 77, 164, 270
Resonant circuit, 183, 235–238, 246, 250,
    343, 350–352, 362, 364
RF, 33–34, 61, 74, 119, 138, 163–164, 183,
    188, 190, 194, 216, 230, 264, 268, 318
Root locus, 350–351

Sampling time instant, 11, 26, 37–39
Saturation, 92, 136–137, 146, 155, 165,
    169–170, 187, 255–256, 268, 270, 282,
    305
Schrödinger's equation, 84–87, 90
Self-oscillation, 269, 342
Shooting method, 8–9
Sideband, 53, 271, 273–274, 325–326, 329,
    336–337, 342, 345, 347–349
Signal
  multitone, 58–59, 81, 277, 340
  quasi-periodic, 33–34, 43
  single-tone, 17, 23, 33–34, 39, 42, 45, 65,
    138, 156, 165, 168, 332, 364, 366
  two-tone, 17, 19, 22, 34–36, 39–41, 57,
    65, 81, 138, 167, 333–337, 340,
    356–358
Snider, 157, 183, 198, 226
Solution
  degenerate, 230, 260–261, 263–268, 277
  explicit, 4–6, 21, 85
  trivial, 138, 230, 260, 339, 344–345, 350,
    356–357
Spectral Balance, *see* Balance
Spectrum Analyser, 62, 65, 139

Stability
  analysis, 252, 308, 342, 348, 351, 356
  circle, 252, 306, 331, 348
  condition, 243, 245–248, 250, 252, 348
  criterion, 238, 244, 341, 346, 348–349,
    358
  factor, 331, 348–349
  margin, 298
  unconditional, 297, 331
Subharmonic
  frequency, 308, 355
  generation, 22, 229
  mixing, 325
  signal, 311
Symmetry, 137, 293–294, 316, 318, 322–323,
  325, 327
Switch, 64, 66–67, 71, 77, 169, 188, 194,
  315–316, 319–321, 331, 358
Switching-mode operation, 184, 186, 195,
  227

Taylor series, 103, 115, 138, 139, 239, 244,
  333, 335, 361
Thévenin, 62
Time step, 5–8, 10, 44, 88, 262
Transient, 7–8, 46, 56–57, 59, 92–93, 261,
  277, 354, 356

Trap, Trapping, 74–76, 79, 105–107,
  109–110, 119, 123, 153–154
Truncation, 34–39, 42, 83, 264, 332
Tuned Load approach, 187, 192
Tuner, 63–70, 73, 80–81
Tuning
  Curve, 266, 268
  Parameter, 265–266, 268

Upconverter, 318

Varactor diode, 265
VCO, *see* Oscillator
Vector Network Analyser, 49, 67, 72–73, 108
Voltage
  Gain Function, 198–201, 203–204
  Overshoot Function, 201–202
Volterra series, 13–18, 21–22, 33, 49, 56–57,
  138, 142–143, 153–156, 166, 168,
  209–210, 222, 227, 259, 268–269,
  276–277, 332, 335–337, 340, 358

Wave
  shaping, 169, 186, 195, 206
  function, 86

Zero-searching algorithm, 6, 9